Quality Assurance in Environmental Monitoring

Edited by
Philippe Quevauviller

VCH

Other Important Titels for Quality Assurance:

Subramanian, G. (ed.)
Quality Assurance in Environmental Monitoring
Instrumental Methods

1995. Ca 335 pages. Hardcover. DM 178,–.
ISBN 3-527-28682-9

W. Funk, V. Dammann, G. Donnevert
Quality Assurance in Analytical Chemistry

1995. 238 pages with 87 figures. Hardcover. DM 125,–.
ISBN 3-527-28668-3

© VCH Verlagsgesellschaft mbH, D-69451 Weinheim (Federal Republic of Germany), 1995

Distribution:

VCH, P.O. Box 10 11 61, D-69451 Weinheim, Federal Republic of Germany

Switzerland: VCH, P.O. Box, CH-4020 Basel, Switzerland

United Kingdom and Ireland: VCH, 8 Wellington Court, Cambridge CB1 1HZ, United Kingdom

USA and Canada: VCH, 220 East 23rd Street, New York, NY 10010–4606, USA

Japan: VCH, Eikow Building, 10-9 Hongo 1-chome, Bunkyo-ku, Tokyo 113, Japan

ISBN 3-527-28724-8

Quality Assurance in Environmental Monitoring

Sampling and Sample Preatreatment

Edited by Philippe Quevauviller

VCH Weinheim · New York · Basel · Cambridge · Tokyo

Dr. Philippe Quevauviller
European Commission
Measurements and Testing Programme
Rue de la Loi, 200
B-1049 Brussels

> This book was carefully produced. Nevertheless, the editor, authors and publisher do not warrant the information contained therein to be free of errors. Readers are advised to keep in mind that statements, data, illustrations, procedural details or other items may inadvertently be inaccurate.

Published jointly by
VCH Verlagsgesellschaft mbH, Weinheim (Federal Republic of Germany)
VCH Publishers, Inc., New York, NY (USA)

Editorial Director: Dr. Steffen Pauly

Publication no. EUR 16611 EN of the European Commission

Dissemination of Scientific and Technical Knowledge Unit,
Directorate-General Telecommunications, Information Market and Exploitation of Research, Luxembourg

LEGAL NOTICE
Neither the European Commission nor any person acting on behalf of the Commission is responsible for the use which might be made of the following information.

Library of Congress Card No. applied for

A catalogue record for this book is available from the British Library

Die Deutsche Bibliothek – CIP-Einheitsaufnahme
Quality assurance in environmental monitoring. – Weinheim ;
   New York ; Basel ; Cambridge ; Tokyo : VCH.
   Sampling and sample pretreatment: [publication ... of the
   European Commission Dissemination of Scientific and
   Technical Knowledge Unit, Directorate General
   Telecommunications, Information Market and Exploitation of
   Research, Luxembourg] / ed. by Philippe Quevauviller. - 1995
      ISBN 3-527-28724-8 Gb.
NE: Quevauviller, Philippe [Hrsg.]; Europäische Kommission /
   Abteilung für Verbreitung der Wissenschaftlichen und Technischen
   Kenntnisse

© VCH Verlagsgesellschaft mbH, D-69451 Weinheim, Federal Republic of Germany, 1995

Printed on acid-free and low-chlorine paper

All rights reserved (including those of translation into other languages). No part of this book may be reproduced in any form – by photoprinting, microfilm, or any other means – nor transmitted or translated into a machine language without written permission from the publishers. Registered names, trademarks, etc. used in this book, even when not specifically marked as such, are not to be considered unprotected by law.

Printing: Strauss Ofsetdruck GmbH, D-69509 Mörlenbach
Bookbinding: Großbuchbinderei J. Schäffer GmbH & Co. KG., D-67269 Grünstadt

Printed in the Federal Republic of Germany

PREFACE

This book treats different aspects of quality assurance (QA) and quality control (QC) for environmental monitoring, with particular emphasis on sampling and sample pretreatment.

QA and QC for environmental monitoring are major features of the 90s as illustrated by the multiplication of QA guidelines, norms, accreditation systems and the number of books which have recently been published on this topic. QA is a crucial aspect of achieving accurate results in chemical analysis. In the case of environmental analysis, it is increasingly recognized that the major risks of error in environmental monitoring do not occur in the laboratory but rather during operations performed in the field (sampling) or prior to the analysis (sampling, storage, and sample pre-treatment). This book focuses on aspects of QA related to the sample collection, storage and pre-treatment of various environmental matrices (water, sediment, soil and plant) for different contaminants (trace elements, nutrients, organic compounds).

The number of analyses performed every year to monitor the quality of the environment is constantly increasing. The data produced are used to evaluate contamination trends ("trend monitoring") and to verify that the quality requirements of EC directives are respected. The results may also be used for modeling and research activities (e.g. studies of biogeochemical pathways). Accurate measurements are the basis of monitoring activities. A lack of quality means that the results are not comparable from one laboratory to another and, hence, the above mentioned purposes cannot be achieved. Consequently, measures to evaluate and guarantee the quality of a laboratory's performance were established, involving quality assurance rules and guidelines, accreditation systems and the production and use of certified reference materials. This resulted in a considerable improvement in the quality of environmental analysis at the laboratory level. However, these rules are hardly applicable to all the operations carried out before the laboratory work, i.e. sampling, sample pre-treatment (e.g. drying, stabilization, grinding etc.) which, however, represent the core of the monitoring data. It is clear that a lack of QA in field work renders the analytical data meaningless. The aim of this book is to discuss the precautions necessary to ensure good quality assurance from sampling to sample pre-treatment, based on the experience. Typical sources of pitfalls and recommendation on how to avoid these are discussed with special emphasis on the monitoring of inorganic and organic contaminants in environmental matrices.

The book is organized into ten chapters covering various aspects of environmental monitoring. The first chapter gives an overview of QA and QC principles as applied to environmental monitoring, along with regulatory aspects. The second chapter deals with sampling strategies in environmental monitoring of biological specimens. QA and QC for surface water sampling are detailed in the third chapter, and are developed with respect to nutrients in sea water in the fourth chapter. QA of sampling of sediment, particularly of marine origin, is dealt with in the fifth chapter, and a focus is given in the sixth chapter on organic compounds in various marine matrices, including sea water, sediments and biota. The seventh chapter describes QA of sampling and pretreatment for the monitoring of trace metals in soils, whereas these aspects (also in relation to trace metal determinations) are focusing on aquatic biota in the eighth chapter. QA of biomonitoring (sampling and storage of plant materials) is described in the ninth chapter. Finally, the tenth chapter proposes a holistic structure for quality management, with particular emphasis on marine environmental monitoring.

This book has been written by experienced practitioners. By its nature, it aims to serve as a practical reference for environmental chemists (and postgraduate students) who need a wide overview of sampling, storage and pretreatment techniques, along with the precautions to be undertaken to ensure good QC of the data produced in the frame of environmental monitoring studies. The critical discussions of the methods, and of the risks of errors that may occur prior to the laboratory work, makes it unique in this respect.

The editor gratefully acknowledges the authors for their time and motivation in preparing their contributions, without which this volume would not have been possible.

Brussels   Ph. Quevauviller
July 1995

CONTENTS

Preface

by Ph. Quevauviller v

CHAPTER 1.   QUALITY ASSURANCE AND QUALITY CONTROL FOR ENVIRONMENTAL MONITORING

by Ph. Quevauviller and E.A. Maier

1.1	Definitions and general principles		2
1.2	Analytical procedures		2
	1.2.1	Type of methods	3
		1.2.1.1 Calculable methods	3
		1.2.1.2 Relative methods	3
		1.2.1.3 Comparative methods	3
	1.2.2	Method selection	3
	1.2.3	Method validation	4
		1.2.3.1 Literature search	5
		1.2.3.2 Validation of final detection - calibration	5
		1.2.3.3 Matrix influence	5
		1.2.3.4 Analysis of solid material	6
		1.2.3.5 Ruggedness and robustness	6
		1.2.3.6 Control points	6
		1.2.3.7 Control charts	8
		1.2.3.8 Case study	10
1.3	How to achieve accuracy		11
	1.3.1	Comparison with a different method	11
	1.3.2	Comparison with other laboratories	12
	1.3.3	Use of certified reference materials	12
		1.3.3.1 Calibration, traceability and accuracy	13
		1.3.3.2 Equivalence between methods	13
1.4	Requirements for the preparation of RMs and CRMs		14
	1.4.1	Selection and preparation	14
	1.4.2	Homogeneity and stability	14
	1.4.3	Types of environmental CRMs	15
	1.4.4	How to establish assigned or certified values	16
		1.4.4.1 Assigning values to a RM	16
		1.4.4.2 Certification by a single laboratory approach	16
		1.4.4.3 Interlaboratory certification studies	16

		1.4.5 Producers of environmental CRMs	17
1.5	Sampling and sample handling		18
	1.5.1	Sampling strategy	18
	1.5.2	Sampling techniques	18
	1.5.3	Registration	19
	1.5.4	Storage	19
	1.5.5	Subsampling	19
1.6	Reporting and archiving		19
1.7	Regulatory aspects of QA and QC in environmental monitoring		20
	1.7.1	Good Laboratory Practice (GLP)	20
	1.7.2	Accreditation	20
	1.7.3	ISO9000/EN29000 standards	21
	1.7.4	Standardization	21
1.8	Conclusions		22
1.9	References		22

CHAPTER 2. SAMPLING STRATEGY IN ENVIRONMENTAL MONITORING OF BIOLOGICAL SPECIMENS

by G. Wagner and R. Klein

2.1	Objectives of environmental sampling		25
	2.1.1	Representativeness of environmental samples	27
	2.1.2	Repeatability and precision	28
	2.1.3	Spatial comparability	29
	2.1.4	Reliability and probative force of environmental samples	30
2.2	Principles of quality assurance in environmental sampling: requirements and methods		30
2.3	Sampling strategies for plants (passive biomonitoring)		30
	2.3.1	Fortuitous sampling	30
	2.3.2	Selected sampling	32
	2.3.3	Random sampling	32
	2.3.4	Systematic sampling (transects, grids)	33
	2.3.5	Stratified random sampling	36
2.4	Active or experimental monitoring with plants		36
2.5	Sampling strategies for soils for biological specimens		37
2.6	Sampling strategies for animals in environmental monitoring		40
	2.6.1	Indicator function of animals	40
	2.6.2	Availability of animals	41
	2.6.3	Stratified random sampling of less mobile animals	42
	2.6.4	Stratified random sampling of mobile animals	44
	2.6.5	Zebra mussels as examples of sampling under semi-artificial conditions	46
2.7	References		47

CHAPTER 3. QUALITY ASSURANCE AND QUALITY CONTROL OF SURFACE WATER SAMPLING

by G.J. Stroomberg, I.L. Freriks, F. Smedes and W.P. Cofino

3.1	Quality systems for sampling	52
3.2	Development of a sampling strategy	53

	3.2.1	Analysis of information need	53
	3.2.2	Design of a sampling strategy	54
		3.2.2.1 Representativeness of samples	55
		3.2.2.2 The sampling site	55
		3.2.2.3 Determinands	56
		3.2.2.4 Methodology	56
		3.2.2.5 Statistical design	57
3.3	Documentation and planning		65
3.4	Validation of sampling		66
3.5	Execution of the sampling plan		68
	3.5.1	Processes affecting sample quality	68
	3.5.2	Equipment, chemicals and cleaning procedures	70
		3.5.2.1 Materials	70
		3.5.2.2 Chemicals	71
		3.5.2.3 Sampling devices	72
		3.5.2.4 Cleaning procedures	77
		3.5.2.5 Considerations in the field	78
	3.5.3	Filtration	79
	3.5.4	Transportation and storage	81
3.6	Quality control practices		83
	3.6.1	Quality control samples	84
	3.6.2	Reference locations	85
	3.6.3	Special sampling trips	85
	3.6.4	Plausibility checks	85
		3.6.4.1 Inspection of data with graphical means	86
		3.6.4.2 Comparison with reference data	86
3.7	References		86

CHAPTER 4. QUALITY ASSURANCE OF PRE-DETERMINATION STEPS FOR DISSOLVED NUTRIENTS IN MARINE SAMPLES

by A. Aminot

4.1	Outline of the context		92
	4.1.1	Introductory remarks	92
	4.1.2	Nautical means	92
	4.1.3	Water characteristics	93
	4.1.4	Processes altering nutrient concentrations	93
		4.1.4.1 General comments	93
		4.1.4.2 Processes	94
4.2	Sampling		95
	4.2.1	Definition	95
	4.2.2	The ship	95
	4.2.3	The water sampler	96
4.3	Subsampling		97
	4.3.1	Definition	97
	4.3.2	The sample bottles	97
		4.3.2.1 The bottle material	98
		4.3.2.2 The bottle closure	98
		4.3.2.3 Bottle cleaning	99
	4.3.3	The subsampling procedure	100

4.4	Filtration		100
	4.4.1	Definition	100
	4.4.2	Is filtration necessary?	101
	4.4.3	Vacuum or pressure filtration?	101
	4.4.4	Filtration unit	102
	4.4.5	Filters	103
	4.4.6	Alternative to filtration	104
		4.4.6.1 Centrifugation	104
		4.4.6.2 Pre-filtration	104
4.5	Storage and preservation		105
	4.5.1	Introductory remarks and definition	105
	4.5.2	Short-term preservation	106
	4.5.3	Long-term preservation	106
		4.5.3.1 General comments	106
		4.5.3.2 Poisoning	106
		4.5.3.2.1 Acidification	106
		4.5.3.2.2 Chloroform	107
		4.5.3.2.3 Mercuric chloride	107
		4.5.3.2.4 Summary of poisoning	107
		4.5.3.3 Freezing	108
		4.5.3.3.1 Usefulness	108
		4.5.3.3.2 Mechanism	108
		4.5.3.3.3 Operational	109
		4.5.3.3.4 Specific problems	110
	4.5.4	Should each analyst develop storage studies?	110
4.6	From sampling to analysis: typical schemes		110
4.7	References		111

CHAPTER 5. QUALITY ASSURANCE OF SEDIMENT SAMPLING

by M. Perttilä and B. Pedersen

5.1	Sampling objectives		114
5.2	Sampling site selection		114
	5.2.1	General considerations	114
	5.2.2	Transport processes and hydrography	115
	5.2.3	Sea bed structure (topography)	115
	5.2.4	Ship positioning	117
	5.2.5	Sampling period frequency	118
	5.2.6	Replicate samples	118
5.3	Sampling		119
	5.3.1	Sampling devices	119
	5.3.2	Sampling procedure	120
	5.3.3	Subsampling	120
	5.3.4	Sample treatment	121
	5.3.5	Drying, grinding and sieving	121
	5.3.6	Handling atmosphere/room facilities	121
5.4	Sample storage		122
5.5	Internal quality control procedures		123
	5.5.1	Split samples	123
	5.5.2	Blanks	123
	5.5.3	Documentation	124
5.6	References		125

CHAPTER 6. QUALITY ASSURANCE OF ANALYSIS OF ORGANIC COMPOUNDS IN MARINE MATRICES: APPLICATION TO ANALYSIS OF CHLOROBIPHENYLS AND POLYCYCLIC AROMATIC HYDROCARBONS

by R.J. Law and J. de Boer

6.1	Sampling		129
	6.1.1	Sea water	130
	6.1.2	Sediments	132
	6.1.3	Biota	133
6.2	Sample pre-treatment and storage		133
	6.2.1	Sea water	134
	6.2.2	Sediments	134
	6.2.3	Biota	135
6.3	Analytical aspects		136
	6.3.1	Chlorobiphenyls	137
		6.3.1.1 Extraction	137
		6.3.1.2 Clean-up and fractionation	138
		6.3.1.3 Calibration and analysis	141
		6.3.1.4 Injection	142
		6.3.1.5 Detection	143
		6.3.1.6 Calibration	144
		6.3.1.7 Long-term stability	145
		6.3.1.8 Interlaboratory studies	145
		6.3.1.9 Reference materials	146
	6.3.2	Polycyclic aromatic hydrocarbons	147
		6.3.2.1 Extraction	147
		6.3.2.2 Clean-up	147
		6.3.2.3 Analysis	148
6.4	References		152

CHAPTER 7. QUALITY ASSURANCE OF SAMPLING AND SAMPLE PRETREATMENT FOR TRACE METAL DETERMINATION IN SOILS

by R. Rubio and M. Vidal

7.1	Steps in the analytical process		159
7.2	Soil sampling in trace element analysis		159
	7.2.1	Design of a sampling plan	159
	7.2.2	Sampling strategies	160
		7.2.2.1 Non-probability approach: judgmental sampling	160
		7.2.2.2 Probability approach: random and systematic sampling	160
		7.2.2.3 Geostatistics and data analysis techniques applied to soil sampling	161
	7.2.3	Soil solution sampling	162
	7.2.4	Number of samples	162
	7.2.5	Composite samples	163
	7.2.6	Sample depth	164
	7.2.7	Materials and sampling techniques	164
	7.2.8	Sources of uncertainty	165

	7.2.9	Quality control samples	166
	7.2.10	Sampling report	168
7.3	Soil pretreatment and storage conditions		168
	7.3.1	Drying processes	168
	7.3.2	Fractionation and subsampling	170
	7.3.3	Storage conditions	172
		7.3.3.1 Storage of soil solution	173
	7.3.4	Pretreatment procedures used in the preparation of reference materials (RMs) and interlaboratory trials	173
7.4	Conclusions		174
7.5	References		175

CHAPTER 8. QUALITY ASSURANCE OF SAMPLING AND SAMPLE HANDLING FOR TRACE METAL ANALYSIS IN AQUATIC BIOTA

by K.J.M. Kramer

8.1	Biomonitoring approaches		180
	8.1.1	Selection of species	181
		8.1.1.1 Bivalves	182
		8.1.1.2 Fish	182
		8.1.1.3 Macrophytes	182
		8.1.1.4 Other groups of organisms	182
	8.1.2	Approaches in the chemical monitoring of biota	183
		8.1.2.1 Collection of wild populations	185
		8.1.2.2 Translocation of organisms (caging)	185
	8.1.3	Approaches in biological effect monitoring	186
8.2	Contamination control		187
	8.2.1	Materials	187
	8.2.2	Containers	187
	8.2.3	Tools	188
	8.2.4	Clean room - clean bench	189
	8.2.5	Cleaning of materials	189
8.3	Sampling		189
	8.3.1	Location, number of samples, frequency	190
		8.3.1.1 Bivalves	190
		8.3.1.2 Fish	191
		8.3.1.3 Macrophytes	191
	8.3.2	Period of sampling, seasonal effects	191
		8.3.2.1 Bivalves	191
		8.3.2.2 Fish	191
		8.3.2.3 Macrophytes	192
	8.3.3	Collection methodology	192
	8.3.4	Selection of species, size and age, sex	193
	8.3.5	Sample size and subsampling	195
		8.3.5.1 Bivalves	196
		8.3.5.2 Fish	196
		8.3.5.3 Macrophytes	196
	8.3.6	Total organism or selected organs and/or tissue	197
		8.3.6.1 Bivalves	197
		8.3.6.2 Fish	197

		8.3.6.3	Macrophytes	198
	8.3.7	Cleaning of organisms		198
		8.3.7.1	Bivalves	198
		8.3.7.2	Fish	199
		8.3.7.3	Macrophytes	199
	8.3.8	Physical factors affecting sampling		199
8.4	Labeling, transport, and storage			200
	8.4.1	Labeling		200
	8.4.2	Forms		200
	8.4.3	Transport conditions		201
		8.4.3.1	Temperature	201
		8.4.3.2	Transport time	202
	8.4.4	Storage		202
8.5	Sample preparation, homogenization			203
	8.5.1	Tissue preparation		203
		8.5.1.1	Bivalves	203
		8.5.1.2	Fish	204
		8.5.1.3	Macrophytes	204
	8.5.2	Preservation		204
		8.5.2.1	Addition of preservatives	204
		8.5.2.2	Irradiation sterilizing	205
		8.5.2.3	Oven drying	205
		8.5.2.4	High temperature ashing	205
		8.5.2.5	Freeze drying (lyophilization)	205
		8.5.2.6	Low temperature ashing (LTA)	206
	8.5.3	Homogenization		206
		8.5.3.1	Fresh tissue	206
			8.5.3.1.1 Blender	206
			8.5.3.1.2 Cryogenic homogenization technique	206
		8.5.3.2	Dried material	207
		8.5.3.3	Testing for homogeneity	207
8.6	Additional measurements - Units of expression			207
	8.6.1	Bivalves		207
	8.6.2	Fish		209
	8.6.3	Macrophytes		209
8.7	Validation and protocols			209
8.8	References			210

CHAPTER 9. QUALITY ASSURANCE OF PLANT SAMPLING AND STORAGE

by B. Markert

9.1	The individual analytical steps and their influence on the accuracy and precision of the analytical results		223
9.2	Identification of problems and analytical planning		225
	9.2.1	Area of investigation	225
	9.2.2	Object of investigation	225
	9.2.3	Schedule and accompanying studies	225
	9.2.4	Chances of success and financial support	225
9.3	Representative sampling and plant specimens		226

Contents

	9.3.1	Classification and examples of element concentration differences and fluctuations in plants	227
	9.3.2	Development of a general sampling strategy	235
	9.3.3	Sampling for specific disciplines of plant analysis, e.g. biomonitoring trace metal status by means of plants	238
		9.3.3.1 Use of poplar leaves (Populus nigra "italica") for biomonitoring of heavy metals	238
		9.3.3.2 Use of mosses for large-area biomonitoring of heavy metals	242
9.4	Cleaning and drying the sample material		249
9.5	Homogenization, aliquots and storage		251
9.6	References		252

CHAPTER 10. AN HOLISTIC STRUCTURE FOR QUALITY MANAGEMENT: A MODEL FOR MARINE ENVIRONMENTAL MONITORING

by D.E. Wells and W.P. Cofino

10.1	Introduction		255
	10.1.1	Continuous assessment	258
	10.1.2	International comparative assessment	258
	10.1.3	Stepwise improvement	258
	10.1.4	Field studies and sampling	259
	10.1.5	The holistic project "QUASIMEME"	259
10.2	The holistic philosophy		260
	10.2.1	Cooperative interaction	260
	10.2.2	Interlinked resources	261
	10.2.3	Advice and guidelines	261
	10.2.4	Flexible framework	261
	10.2.5	Collaborative improvement	262
10.3	The elements of an holistic project		263
	10.3.1	Focus	263
		10.3.1.1 Focus on the determinands	263
		10.3.1.2 Focus on the environment	264
		10.3.1.3 Focus on the participation	265
	10.3.2	Framework	266
		10.3.2.1 The participants	266
		10.3.2.2 The project team	266
		10.3.2.3. The customers	268
		10.3.2.4. The sponsors	269
10.4.	Functions		269
	10.4.1.	Management	269
	10.4.2.	Operation	271
		10.4.2.1. Target objectives	273
		10.4.2.2. Laboratory testing and learning schemes	273
		10.4.2.3. Reference materials	275
		10.4.2.4. Documentation - data retrieval	277
		10.4.2.5. Data assessment	278
		10.4.2.6. Performance criteria	278
		10.4.2.7. Storage	281
		10.4.2.8 Sampling and sample handling	281
10.5.	Communication		282
	10.5.1.	Within laboratory communication	284

10.6	Future developments		285
	10.6.1. Involving wider participation		286
	10.6.2. Developing new strategies		287
	10.6.3. Developing new links		287
10.7.	References		288

Subject index 291

1.

Quality assurance and quality control for environmental monitoring

Ph. Quevauviller and E.A. Maier
European Commission, Standards, Measurements and Testing Programme, rue de la Loi 200, B-1049 Brussels, Belgium

Environmental monitoring involves several types of discipline, e.g. analytical chemistry, geology, and biology, but is usually based on chemical analyses which are performed routinely by a number of organizations. Awareness of quality assurance (QA) and quality control (QC) has considerably increased in the past few years, as illustrated by the proliferation of QA guidelines, accreditation systems, proficiency testing schemes, etc. In many cases, however, accurate analytical results are far from being achieved. Many examples have been given of the lack of accuracy that can be demonstrated by interlaboratory studies dealing with inorganic or organic substances or chemical species [1], and which is a result of various sources of error that may occur in the laboratory. Environmental monitoring, however, covers a broad range of sample handling steps prior to the laboratory work, e.g. collection, field treatment, and storage, which enhances the risks of multiplication of errors. Whereas great emphasis has been put on the QA/QC within the laboratory, there have been few systematic attempts to evaluate risks of inaccuracy related to field manipulations. It is suspected that the sources of error likely to occur prior to the laboratory are much higher than the pitfalls linked to lack of accuracy within the laboratory; this obviously leads to a lack of comparability of data produced for environmental monitoring. This implicitly means that, in many cases, the results are not trustworthy, which may result in economic losses, e.g. extra analyses, and may lead to misinterpretation of environmental processes.

Valuable tools exist for evaluation of within- or between-laboratory performance, e.g. interlaboratory studies, use of certified reference materials (CRMs), etc. In addition, a good quality control system has to be introduced in each laboratory. This chapter addresses some of the main aspects of quality control and gives examples of sources of discrepancies which may occur within the laboratory. QA aspects specifically related to sample collection and storage are dealt with in the other chapters of this book.

1.1 Definitions and general principles

The growing importance of chemical measurements in modern society has rendered urgent the development of measures to improve the quality of analytical results and to guarantee their quality to the end users. All the actions carried out to plan the proper performance of the analytical task are covered by the term "quality assurance" (QA).

ISO defines quality as "the totality of features and characteristics of a product or service that bear on its ability to satisfy stated or implied needs" [2]. Quality assurance is defined as "all those planned and systematic actions necessary to provide adequate confidence that a product, process or service will satisfy quality requirements".

Quality can be assured only if extensive quality control (QC) measures are taken; these involve both process monitoring and the elimination of causes of unsatisfactory performance [3]. Quality control is defined as "the operational techniques and activities that are used to fulfil requirements for quality" [2]. QA and QC are components of the "quality system" [4] which includes "the organizational structure responsibilities, procedures and resources for implementing quality management".

In order to achieve accuracy, the analyst should be involved from the beginning of the process until the delivery of the final report, which implies that he should, in practice, be involved in the choice of the parameter to be monitored and the level of accuracy (trueness and precision) required for the purpose. This will enable him to decide upon an appropriate procedure based on scientific and economic arguments. A compromise has to be found between the level of sophistication of analytical procedures and the level of accuracy in relation to the parameter(s) to be determined and the purpose of the study (e.g. rough estimate of contamination, trend monitoring, study of environmental pathways). The quality system should then include all aspects of the approach to achieve the best quality possible within the defined question [3]. The aspect of global quality assurance including a broad spectrum of activities is increasingly accepted by analytical chemists [5]. A great deal of effort has, however, still to be made to involve the analytical chemist in decisions regarding sample collection and storage which are often not his responsibility. The analytical community has recognized this need to work in concert with analysts in order to avoid irreversible errors which may occur during field work [6].

In this chapter, only the part dealing with QA within the laboratory will be developed, in particular the means to achieve and verify accuracy of analytical procedures. Some aspects related to sampling, sample pre-treatment and reporting of results are described briefly.

1.2 Analytical procedures

The range of analytical procedures, matrices, and parameters to be determined for the purpose of environmental monitoring is very large and there is no "tailor-made" system which is adapted to each case. General rules for setting up a quality system, e.g. EN 45000 [7] or ISO 25 [8] may have to be adapted to individual situations without fundamental changes. The different aspects related to personnel, management, allocation of tasks, quality assurance manual and committee, infrastructure, and equipment, have been described in detail elsewhere [3]. We will focus here on the technical aspects related to the assessment of the performance of analytical procedures.

1.2.1 Type of methods

Chemical methods may be classified into three categories with regard to the calibration procedure [9]:

1.2.1.1 Calculable methods

"A method that produces the anticipated result by performing a calculation defined on the basis of the laws governing the physical and chemical parameters involved, using measurements taken during the analysis, such as: weight of the test sample, volume of titration reagent, weight of precipitate, volume of titration product generated".

This type of determination includes, e.g., titrimetric, coulometric, and gravimetric methods.

1.2.1.2 Relative methods

"A method which compares the sample to be analysed with a set of calibration samples of known content, using a detection system for which the response (ideally linear) is recognised in the relevant working area (without necessarily being calculable by theory). The value of the sample is determined by interpolation of the sample signal and with respect to the response curve of the calibration samples".

Spectrometric methods belong to this type of method which implies pre-treatment of the samples, matrix-matching of the calibration sets, elimination of interferences, etc.

1.2.1.3 Comparative methods

"A method where the sample to be analysed is compared to a set of calibration samples, using a detection system which has to be recognised to be sensitive not only to the content of elements or molecules to be analysed but also to differences of matrix".

Calibration of such methods requires certified reference materials with a known matrix composition similar to the matrix of the sample (e.g. X-ray fluorescence spectrometry).

1.2.2 Method selection

The selection and development of an analytical procedure adapted to the purpose of environmental monitoring of different substances will be based on the experience of the analyst and on investigation of the scientific literature. Analytical procedures are based on a succession of actions [3]: (i) sampling, storage, and preservation of representative material(s), (ii) pretreatment of a sample portion for quantitation, (iii) calibration, (iv) final determination, and (v) calculation and presentation of results.

Method selection will depend on the degree of precision and trueness required. Precision is achieved when random errors are minimized. Trueness is reached when systematic errors are eliminated. As a reminder, the following definitions are relevant:

- Random error: a component of the error of a measurement which, in the course of a number of measurements of the same measurand, varies in an unpredictable way [10]
- Systematic error: a component of the error of a measurement which, in the course of a number of measurements of the same measurand, remains constant or varies in a predictable way [10]

- Precision: the closeness of agreement between the results obtained by applying the same experimental procedure several times under prescribed conditions [10]
- Trueness: the closeness of agreement between the "true value" and the measured value [11]
- True value (of a quantity): a value which would be obtained by measurement, if the quantity could be completely defined and if all measurement imperfections could be eliminated [10]

The selection of a method may also depend on regulatory requirements, e.g. official methods issued by an official agency (e.g. USEPA), or reference or standard methods developed and validated by standardization bodies (e.g. ISO, CEN, DIN, BSI, AFNOR). A recent compilation lists official and standardized methods of analysis [12].

1.2.3 Method validation

The different categories of method differ in the link between the signal produced by the substance and the signal obtained from the calibration material. The link can be related to an amount of substance of established purity and stoichiometry (for calculable and relative methods when all the analytical steps are well established) or, as mentioned above, to a CRM in the case of comparative methods. This link, if established through an unbroken chain of procedures, is called traceability ("the ability to trace the history, application or location of an item or activity, or similar items or activities by means of recorded identification" [2]). The primary objective of the validation of an analytical method is to establish its traceability to a recognized reference (pure substance or matrix CRM). This means that all steps of the analytical procedure should be performed and recorded in such a way that all essential information is made readily available; in other words the results of the determination (not only the final detection step) should be linked through an unbroken chain of comparisons to appropriate measurement standards, generally international or national standards [13], e.g. basic SI units, CRMs, and this link should be demonstrated. Basic requirements are (i) precision (repeatability and reproducibility), (ii) trueness, and (iii) sensitivity; second level requirements include, e.g., specificity, range of linear response, robustness and ruggedness.

In chemical analysis, the element or substance to be analyzed is rarely directly accessible to the detector; in most cases sample pretreatment must be applied, but without changing the matrix, which would result in loss of traceability to the predetermined reference (e.g. fundamental units) [3]. Typical pretreatment procedures are digestion, extraction, purification or separation steps; calibration and final detection consitutes two other key steps of analytical procedures. The source(s) of error(s) possibly occurring at the different analytical steps should be identified and, where possible, minimized; as a consequence, each individual step should be well studied. One approach consists in verifying the performance of the method using simple (e.g. calibrant) solutions at a start and analyzing samples of increasing difficulty (e.g. cleaned extracts, raw extracts, real samples). In this way, the analyst may control and optimize each step of the analytical procedure. In addition to the reliability of the method, the analyst must evaluate how it will behave in the real situation of daily "routine measurements" [3].

Standardized methods are often described in such a manner that validation can be achieved rapidly, consisting mainly in the verification of accuracy, sensitivity, and ruggedness [14].

Some aspects of the stepwise approach for the validation of analytical methods are described below. Examples taken from different fields (inorganic, speciation, organic) can be found in the literature [3,15].

1.2.3.1 Literature search

Descriptions of analytical methods found in the scientific literature are rarely sufficiently detailed to be applied directly. In addition, quality control aspects are often not given. A recent compilation of official and standardized methods of analysis offers good coverage of methods, with a good description of the way they should be applied [12]. In many cases, however, these methods are applicable to a certain type of sample and cannot be transferred directly to types of matrices other than those described without full validation, e.g. a standardized method for drinking water cannot be applied without change to waste water analysis; similarly a method valid for sediment analysis is not readily applicable to sludge analysis [3]. Consequently, method validation must often be based on adaptations of existing methods, following general guidelines [3].

1.2.3.2 Validation of final detection - calibration

Many interlaboratory studies conducted by the BCR (European Commission) have demonstrated that the main source of disagreement between laboratories was a result of calibration errors [3]. The errors can be simple calculation errors, e.g. of concentrations, dilution errors, mistakes during tranfer of volatile solvents, use of impure (not verified) non-stoichiometric primary calibrants, contamination, uncontrolled interferences, inappropriate internal standards, etc. A study of the system with a solution of pure compounds is necessary to detect and remove possible source(s) of error(s) occurring at the final detection step, in order to verify the reliability of the signal produced in terms of specificity, linearity, traceability to adequate pure substances, sensitivity and accuracy of the calibration [3]. For this step, stock solutions of calibrants with known purity and stoichiometry should be prepared. The choice, handling, storage, and preparation of such solutions have been fully described for organic compound determinations [16] and inorganic calibration [17]. Precautions regarding protection from light (e.g. amber glass vials) and elevated temperatures (e.g. refrigerator or deep-freezer) should be taken. Before and after taking an aliquot of the standard the analyst should weigh the vial so that losses resulting from, e.g., evaporation can be detected [3]. Stock solutions should be replaced regularly, depending on the stability of the compounds and the frequency of use.

1.2.3.3 Matrix influence

This step concerns the analysis of an extract or a digest solution in which the matrix is unknown. All the verifications performed in the validation of the final detection have to be carried out again (calibration, linearity, chromatographic conditions, and performance etc.), because the matrix will generate new interferences [3]. This problem is particularly acute for trace organic contaminants when determined with nonspecific detectors (e.g. ECD, FID, UV), i.e. possibly affected by interfering compounds; in this case the reliability of the signal is totally dependent on the prior chromatographic separation. Examples of the importance of matrix matching for calibration have recently been demonstrated in an interlaboratory study on trace elements in soil samples; results obtained for some trace elements were significantly lower without matrix matching than results obtained with matrix matching [18].

In organic trace analysis, the extract is usually cleaned before chromatographic separation in order to remove all coextracted bulk materials such as lipids, sulfur, pigments, etc. and potentially interfering compounds [3]; it is essential to estimate losses possibly occurring during clean-up, e.g. by spiking a raw extract or by use of standard additions.

1.2.3.4 Analysis of solid material

The analysis of solid material(s) will enable extraction or digestion efficiencies to be estimated. Several examples have shown the precautions to be taken for complete destruction of the matrix for the determination of certain elements/compounds, e.g. for refractory elements or when compounds have to be decomposed prior to derivatization (e.g. decomposition of arsenobetaine prior to hydride generation for the determination of total As in fish tissue [19]). In the case of organic or organometallic compounds, compromises have to be found between the best extraction efficiency and the preservation of the compounds which might be degraded if the attack were too strong.

The validation of the extraction step is often unsatisfactory in that it is nearly impossible to prove that all the compounds have been extracted. Two approaches are usually followed [3]. The first consists in successive extractions with fresh solvent and determination of residual traces of organic compounds (e.g. PCBs) in successive extracts; when the extract is free from PCBs one can assume (but not guarantee) that the extraction is complete. The second approach consists in a standard addition procedure with increasing spikes [20] and verification that the added compounds have been recovered after extraction. This method is valid only when sufficient time is given for equilibration of the spike with the matrix (at least overnight). By repeating the spiking at each level of enrichment it is possible to estimate the reproducibility and efficiency of the extraction procedure. It should be noted that this method does not give a guarantee of full extraction.

1.2.3.5 Ruggedness and robustness

"The ability of a method to be relatively insensitive to minor changes in the procedure, to the quality of reagents or to the environment" [14]. A method can be applied in a testing laboratory only when it is sufficiently robust, i.e. when it can be certain that common variations in the method do not affect its reliability. These variations can be of several types, e.g. small variations of temperature during pre-treatment, aging of chromatographic columns, replacement of part of the equipment, small variations of sample matrix, variations in the environment (temperature, humidity, atmospheric pressure). All conditions which seem to be critical should be evaluated and controlled. Chemometric tools and expert sytems have been proposed as tools for estimation of the ruggedness of methods [14,21].

1.2.3.6 Control points

When the different steps of an analytical method have been evaluated, it is good practice to define control points related to sensitive items which should be systematically verified to detect the origin of possible errors. Examples of such control points for inorganic determinations, which were used in certification campaigns organised by the BCR, are illustrated in Table 1-1; as exemplified, each method requires specific control points. Some standardized methods recommended by the USEPA contain performance criteria for individual control points [22].

Table 1-1: Some possible sources of error in inorganic analysis and their elimination

DESTRUCTIVE METHODS

Analytical step	Systematic error by	Contribution	Elimination by
preparation	weighing	+/-	calibrated balance
	volumetric manipulation	+/-	dilution, etc, carried out with calibrated glassware, temperature control
moisture	adsorption/ desorption	+/-	correction to dry mass according to chapter 11
digestion/ oxidation	volatilisation	-	for volatile elements (e.g. As, Se...) treatment carried out in closed systems
	adsorption/ desorption	+/-	acid-washed containers of hard glass; PTFE, or HDPE prerinsed surfaces
	incomplete digestion	-	many participants used pressurized digestion with oxidising acids; in some cases the residue was checked to verify the total digestion of the matrix
	reagent contamination	+	reagents of appropriate purity were chosen; verification with blank determinations
	contamination by tools/vials	+	acid washing as appropriate: when contents below 1 μg/g were determined: steaming; verification by blank determinations
	contamination from laboratory air	+	use of clean benches or clean room; care in performing methods under cover or in closed systems; verification with blank determination
sample preparation/ clean up/ preconcentration	adsorption/ irreversible precipitation	-	pH control and/or addition of complexing agents if necessary
	contamination	+	
	incomplete conversion	-	excess of reagents; methods a priori verified

Table 1-1 (continued)

NON-DESTRUCTIVE METHODS

Analytical step	Systematic error by	Contribution	Elimination by
calibration		+/-	reagents of suitable purity and stoichiometry were chosen; where necessary precautions were taken to obtain and verify stoichiometry; many participants verified their calibrants; different calibration methods were chosen when possible; calibration graphs, where necessary matrix-matched calibrant solutions and standard additions within the limits of this method
	peak overlap intrinsic irradiation	+/-	deconvolution; selection of proper decay times; RNAA as an alternative
	high background	+/-	was not observed in the exercise
	geometry	+/-	calibrant and unknowns were both in the same form, e.g. solution or powder; they were measured in identical vials at the same distance of the detector, etc.
irradiation	self-shielding	+/-	contents are such that for the elements investigated shielding does not occur
calibration	changes of flux	+/-	flux monitors were added in the irradiation process
		+/-	the same remarks as for destructive methods are valid; additional care was given to the stability on irradiation of the calibrant

1.2.3.7 Control charts

Statistical control systems through the establishment of control charts are useful for monitoring the performance of analytical methods with time. Implementing such a system implies that representative, homogeneous, and stable reference materials are analyzed at regular intervals. The requirements for the preparation of such materials are discussed in the Section 4 of this chapter. A description of control charts is given in detail elsewhere [23] and only the basic principles will be covered here.

When a laboratory works at a constant level of high quality, few random errors persist and fluctuations in the results are small. Only at this stage can the method be considered as fully validated and a statistical control system involving control charts can be initiated, enabling detection of the introduction of new systematic error and/or monitoring of the method's precision when replicate measurements are performed [3]. Reference materials are analyzed at regular intervals and the results of the determinations of the substance of interest are reported on a graph. At the start of the control chart, several replicate determinations have to be carried out in order to determine the mean value and standard deviation which represent the repeatability of the method. This value will enable the establishment of acceptance limits, e.g. warning and alarm levels, as shown in Figures 1-1 and 1-2. When one result exceeds the warning limit no action is considered to be necessary; if however, the alarm limit is exceeded once or the warning limit is exceeded twice consecutively, this implies that the cause of the deviation should be investigated. A succession of results always on the same side of the mean value should also alert the analyst to the possible introduction of a systematic error. Beside Shewhart or X-charts, R charts may be applied where the result reported consists of the difference between two replicate measurements.

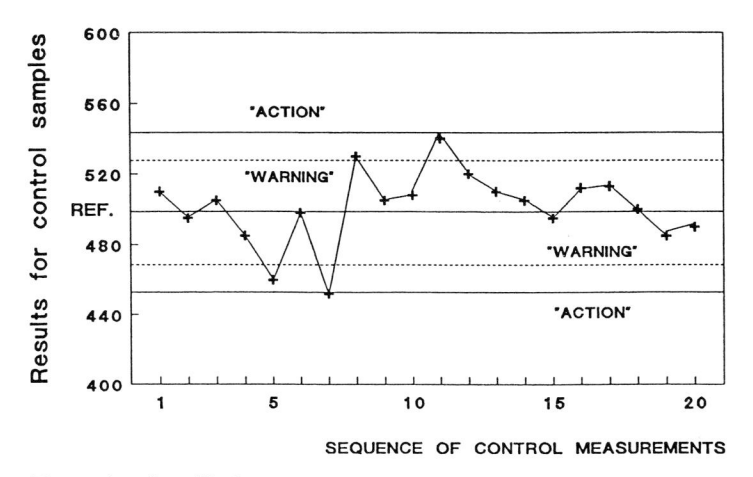

Figure 1-1: Example of an X-chart

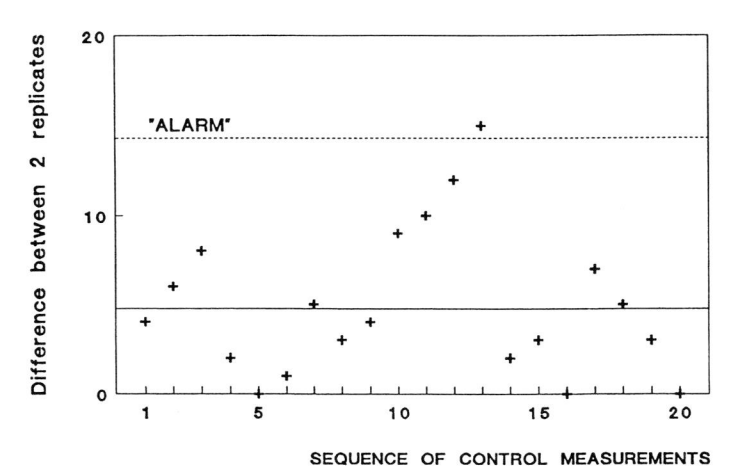

Figure 1-2: Example of an R-chart

1.2.3.8 Case study

An example has been reported of a scheme which enabled clear improvement of the determination of methyl mercury [24]. The majority of techniques used for the determination of methyl mercury (MeHg) consists in an extraction, separation, identification and quantification steps. Extraction is often performed with a lipophilic solvent, or the organic sample is destroyed in an alkaline solution followed by a selective reduction of inorganic mercury (e.g. with $SnCl_2$). The separation and identification can be carried out by gas chromatography, or by ion exchange or high-pressure liquid chromatography. Techniques such as cold-vapor atomic absorption spectrometry (after selective separation), electron capture detection, and mass spectrometry are generally used for final detection. The complexity of the methods and the multiplicity of analytical steps enhance the risks of errors.

In view of the urgent need to evaluate and possibly improve the state of the art, a step-by-step approach such as that described above was followed by BCR from 1988 to 1991; 15 laboratories from 10 EC countries participated in this program. A first intercomparison dealt with the analysis of three solutions; two contained MeHg in toluene together with possible interferents, the other was an aqueous solution; this exercise did not reveal any major discrepancies in the methods used for final determination. A second interlaboratory study involved the analysis of raw fish extract, raw extract spiked with MeHg, cleaned extract, and an aqueous solution; in this exercise difficulties were experienced mainly because many laboratories lacked of good long-term reproducibility. A Youden plot of raw and spiked extract demonstrated clearly that systematic errors had occurred (Figure 1-3).

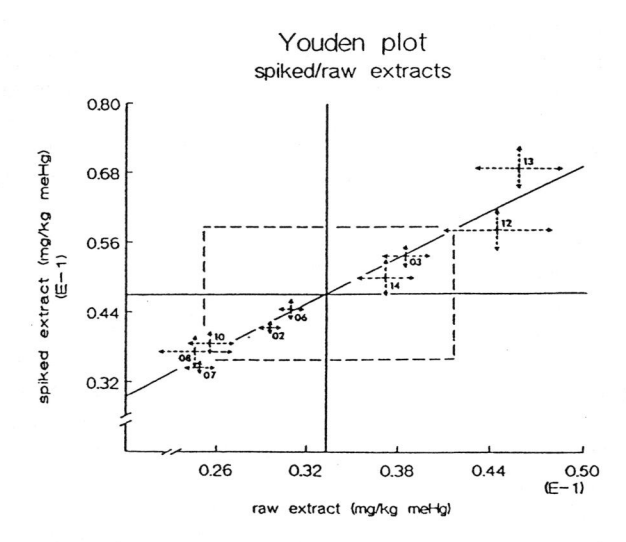

Figure 1-3: Youden plot. MeHg in spiked extract versus MeHg in raw extract. The horizontal and vertical continuous lines are the means of the laboratory means, the broken lines being the standard deviations of these means. The lengths of the bars are equal to the standard deviations (5 replicates) of the laboratories (after [24]).

A third exercise repeated the analysis of raw and raw spiked extract, and added the analysis of mussel and tuna fish samples; the main source of discrepancies was attributed to the inadequacy of the packed chromatography columns used by some laboratories. Capillary GC columns were found to be more suitable for this type of determination. The results obtained for the mussel and fish analysis stressed the need to remove impurities to limit interfering effects (e.g. using cysteine paper). Finally, the last step of the project was the certification of MeHg in two tuna fish reference materials. No major sources of error were detected in this last exercise, which enabled the production of two new CRMs (BCR CRMs 463 and 464).

1.3 How to achieve accuracy

The term accuracy covers precision and trueness in the new ISO definition. Whereas precision may be achieved within the laboratory by working under strict QA and QC rules and by properly validating the analytical method, trueness is demonstrated by external means. Trueness is necessary to produce data which are comparable in time and from one laboratory to another. Environmental monitoring is a good example for which accurate data are needed, e.g. for evaluation of contamination trends or assessment of environmental quality. Too many scientists have stated that good reproducibility was sufficient for such studies; this overlooks possible improvements in methodologies and the worldwide comparability of data. Trueness can be demonstrated by three different means, (i) comparison with a different method, (ii) comparison with other laboratories, and (iii) use of certified reference materials.

1.3.1 Comparison with a different method

Each analytical method has it own sources of errors. For spectrometric methods (e.g. AAS, ICP-AES) errors may occur at the matrix digestion step, a step which is not necessary, e.g., for neutron activation analysis; the latter may, however, suffer from other sources of errors, e.g. insufficient separation of gamma-ray peaks. Therefore, an independent method such as INAA may be used to control results obtained by spectrometric methods; if results obtained on similar samples by both methods are in good agreement, it may be considered that the values found are true. This conclusion is stronger when results from both methods differ widely; if there are similarities, e.g. the pretreatment step, the comparison may overlook a possible systematic error arising from the common step.

It should be stressed that a valid comparison relies on good quality control of both methods. A lack of QC or analyst experience may only create confusion because the risk of uncontrolled sources of errors. Therefore, the cross-comparison of two methods for the evaluation of their trueness is rarely performed within a single laboratory. As described below, the principle of interlaboratory comparison is preferred as it enables checking both of the analytical procedure and the way it is applied in the laboratory (technician's skill).

An example of sources of discrepancies which were detected in a systematic comparison of different methods within a single laboratory has been illustrated by an evaluation of the use of microwave oven systems for environmental analysis [25]. In most cases microwave digestion was found to be adequate for digestion of a wide variety of environmental matrices. In the case of mercury in fish tissue, however, significant differences were observed in relation to the digestion procedure used which raised suspicions about incomplete solubilization of the matrix (recovery of 80% of the certified value of a fish CRM); the final detection was carried out by

ICP-AES. Further investigations were carried out with different detection techniques (using different digestion techniques, including the one with low recovery) and good recoveries were systematically obtained for all digestion procedures when using CVAAS or CVAFS, whereas lower values were observed for ICP-AES and ICP-MS. Additional checks led to suspicions that interferences probably occurred during ionization, e.g. because of the high organic mercury content. This comparison thus revealed that the digestion itself was not responsible for the low results obtained but that investigations should rather focus on the detection of possible interferents in the determination of Hg by ICP-AES or ICP-MS.

1.3.2 Comparison with other laboratories

Participating in interlaboratory studies is an useful tool for improving the quality and performance of a laboratory. These intercomparisons consist in exercises in which several laboratories determine one or more substances in a similar sample provided by a central organizing body. Comparison of different methods applied in different laboratories enables detection of sources of errors as a result of a specific procedure or a given laboratory. A good practice is to discuss the results in a technical meeting so that the expertise of the participants may be shared and the exercise may be fully considered as a mutual learning process.

Such studies are often called laboratory performance studies [26]. Proficiency testing is an equivalent term which is used in regulations [7,8]. By participating in properly organized interlaboratory studies it is possible to evaluate fully the accuracy of the laboratory, depending on (i) the motivation and the degree of preparation, (ii) the motivation of the other participants, and (iii) the way the study is organized; this requires a broad knowledge in many fields. The organization of interlaboratory studies has been described in detail elsewhere [3,27,28]. It involves the preparation of clear instructions to be sent to the participants, the production and distribution of reference materials of verified homogeneity and stability, the collection, treatment, and presentation of the results, and, possibly, the organization of a technical meeting with all the participants. The benefit of each participant in the study depends strongly on the way the data received have been treated and presented, e.g. use of properly prepared reporting forms for the description of methods and reporting of results. Technical scrutiny will enable discussion of the possible sources of disagreement between results. The statistical treatment of the data will not enable outlying results to be explained nor give information about accuracy; it is only useful to analyze the data population and inform on its statistical properties.

1.3.3 Use of certified reference materials

Before considering the aspect of use of CRMs, it is important to recall the definitions of reference materials and certified reference materials as given by ISO [29]:

A Reference Material (RM) is defined as "a material or substance one or more of whose property values are sufficiently homogeneous and well established to be used for the calibration of an apparatus, the assessment of a measurement method, or for assessing values to materials".

A Certified Reference Material (CRM) is defined as "a reference material, accompanied by a certificate, one or more of whose property values are certified by a procedure which establishes its traceability to an accurate realization of the unit in which the property values are expressed, and for which each certified value is accompanied by an uncertainty at stated level of confidence".

Requirements for the preparation of RMs and CRMs are described in Section 1.4. The several categories of RM and CRM exist which are briefly described below:
- pure substances or solutions for calibration and/or identification (CRMs) or test of procedures in improvement schemes (RMs);
- materials of known matrix composition for the calibration of comparative methods (CRMs);
- matrix materials which as far as possible represent the matrix being analyzed by the user, either used in interlaboratory studies (RMs) or for the verification of a measurement process (CRMs);
- methodologically-defined reference materials for parameters such as, e.g., extractable fractions of elements (e.g. aqua-regia, EDTA etc.); the values are defined by the applied method following a strict analytical protocol.

As described above, RMs are used in interlaboratory studies or statistical control schemes (control charts). The use of CRMs is the easiest way to demonstrate accuracy and to verify the performance of the laboratory at any desired moment [30].

1.3.3.1 Calibration, traceability and accuracy

CRMs are mainly used to demonstrate the accuracy of an analytical method. In other words they enable demonstration of the traceability of the results to the units in which the certified value of the CRM is expressed. As stressed below, some CRMs may also be used to calibrate instruments directly.

For the verification of a method's performance, the CRM should resemble the unknown sample as closely as possible. For calibration purposes, the uncertainty of the CRM is of primary importance since it will affect the final uncertainty of the measured value in the unkown sample. In the case of certified pure substances or calibrant solutions, the uncertainties of the certified values are usually negligible in comparison with the method's uncertainty. The larger uncertainty of certified values in matrix CRMs may create difficulties in the case of calibration of comparative methods which may be limited to semi-quantitative measurements.

CRMs should not be used for calibration of relative methods owing to the larger uncertainties in comparison with pure calibration compounds. As said before, matrix CRMs should be reserved for verification of the accuracy of the method; this implies similarity of these materials with unknown samples. This similarity is not always achievable and it might be necessary to use different types of CRMs to estimate the degree of accuracy of the method on real samples. The use of matrix CRMs for the validation of the trueness of relative methods is fully described in ISO Guide 33 [30].

1.3.3.2 Equivalence between methods

CRMs are useful for comparing the performance of a method newly introduced into the laboratory with that of the same method applied in another laboratory, or with that of another method used in the same laboratory. If the CRM has been certified in an interlaboratory study, and if a method similar to the one tested has been used in the certification, the laboratory may compare its results with those of the certifying laboratory. This implies that the CRMs are made available with a detailed certification report, e.g. reports prepared by BCR [3].

1.4 Requirements for the preparation of RMs and CRMs

Stringent requirement have to be respected for the production of RMs and CRMs, in particular the materials should be similar to samples which are currently analyzed in the laboratories so that analytical errors can be detected. This implies in most cases similarity of (i) matrix composition, (ii) analyte contents, (iii) manner of binding of the substances of interest, (iv) fingerprint pattern of possible interferences and, (v) physical status of the material.

1.4.1 Selection and preparation

The similarity of RMs and CRMs cannot always be entirely achieved in practice. The most important aspect is to ensure that the material prepared is homogeneous and remains stable over a sufficiently long period. Homogenization and stabilization involve treatments which may slightly alter the representativeness of the materials in comparison with natural samples, and compromises have to be found; recent studies have shown, e.g., that the only way to produce CRMs of water containing unstable pesticides was to freeze-dry the solutions in order to ensure their stability and to reconstitute the sample prior to analysis [31].

The amount of raw material to be prepared should be adapted to the purpose of the study; it may vary from 100 vials in an interlaboratory study to 1000 or even 5000 vials of CRMs to ensure a sufficiently lasting stock. Needless to say, in the latter case, the producer needs to be fully equipped to treat large amounts of materials.

Stabilization is one of the most critical steps in the preparation of a RM or a CRM, and is more likely to affect the representativeness of the material. Solid materials are usually dried to achieve a moisture content below 5% in order to avoid chemical or microbiological changes; usual drying procedures are heating or freeze-drying. Some materials may be sterilized by gamma-irradiation, addition of chemical preservatives, or pasteurization; irradiation cannot, however, be used for organic or organometallic compounds since this process may degrade the substances of interest [32,33]; pasteurization, as used to stabilize TBT in a coastal sediment [33], may then be the method of choice. Stabilization by simple freeze-drying is also possible but this procedure leads to difficulties in transport and storage; the material may, in addition, loose homogeneity upon thawing. Homogenization is another important step and should be carried out in such a way that the inhomogeneity of the material should not significantly affect the total uncertainty of the measurement.

1.4.2 Homogeneity and stability

The homogeneity and stability of reference materials should be verified. Whereas this verification is not systematically carried out for reference materials, it is one of the major requirement for CRMs.

The homogeneity should be verified and the minimum sample intake should be stated. Such study implies that a highly reproducible method is used so that even a small homogeneity can be detected. The homogeneity is assessed by comparing the uncertainty of the method (several replicates of a digest or extract) with the results of replicate analyses of samples taken in one vial (within-bottle homogeneity) and results of replicate analyses taken from different vials (between-bottle homogeneity). Ideally the study should be performed on several (decreasing) intake sizes until a significant effect of inhomogeneity can be detected.

1.4 Requirements for the preparation of RMs and CRMs 15

After the homogeneity has been verified, the stability of the material should be monitored over a given period of time (usually 12 to 18 months) to verify that the sample's integrity is maintained. This verification can be carried out at various temperatures to simulate different storage conditions (e.g. accelerated aging conditions by storing the material at, e.g., 40 °C over long period of time). The results of such study will enable design of the minimum storage conditions that should be applied to the material. Figure 1-4 shows a study which has been carried out for TBT in a coastal sediment [33]; as shown, the stability could be verified at 20 °C but not at 40 °C which justified storage of the material at a minimum of 4 °C.

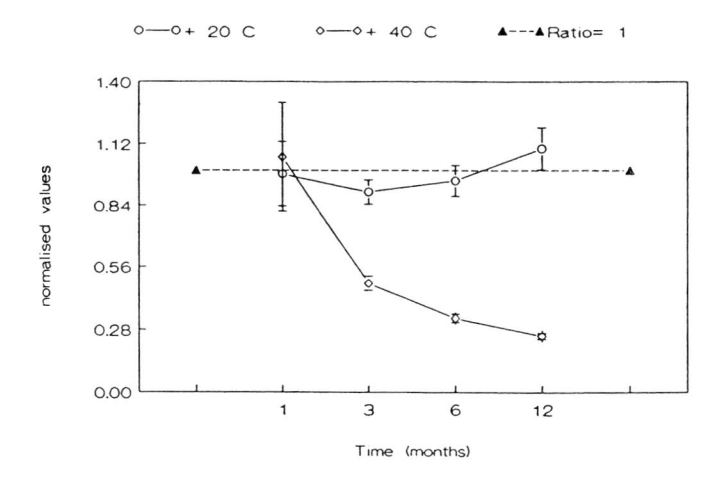

Figure 1-4: Stability tests of TBT in coastal sediment CRM 462: $R_T = X_T/X_{20°C}$ where X_T is the mean of 10 replicates at temperature T (+20 or +40°C)

1.4.3 Types of environmental CRMs

CRMs used for the quality control of chemical determinations in environmental monitoring cover very different matrices and substances of interest. The hydrosphere is controlled through the analysis of sediment and water samples. The atmosphere may be monitored using methods for gas analysis, e.g. SO_2, NO_x, etc., or fine particles, e.g. fly ashes from thermal plants such as incineration plants. For assessment of the contamination of the biosphere, several target organisms have been identified which may give an indication of contamination of the food chain; e.g. shellfish, fatty fish, algae, mosses, tree leaves or needles, lichen, etc., by toxic compounds, e.g. heavy metals, chlorinated organic compounds, organometallic substances etc. Land is monitored on two levels, firstly throught the analysis of mainly agricultural (including forest) soil for very low traces of contaminants (i.e. before amendment by e.g. sewage sludges), secondly through heavily contaminated soils, sludges or waste materials, which require a wide range of CRMs of different matrices and contents of elements/compounds. Details on CRMs available for environmental studies are given in the literature [34-36].

1.4.4 How to establish assigned or certified values

1.4.4.1 Assigning values to a RM

Procedures for assigning or certifying levels in RMs or CRMs are very similar. In the case of artificial mixtures, e.g. aqueous or organic solutions of elements or compounds (of known and verified purity and stoichiometry), the best assigned value is the result of weighing. Artificially enriched samples are difficult to reproduce accurately and are not representative of naturally-bound substances; assigning a value can only be performed by measuring the final content of the substance, which should be carried out by highly experienced laboratories using properly validated procedures. Assigned values are more reliable when they are established by a group of laboratories using different validated techniques.

The certification of reference materials follows strict rules which are described in detail in a special ISO Guide [13] which states that "the certified value should be an accurate estimate of the true value with a reliable estimate of the uncertainty compatible with the end use requirements". The approach followed for certification may differ according to the objectives of the study. Whereas pure compounds or calibrant solutions may be prepared gravimetrically, as stressed above, the so-called "matrix CRMs" pose the most difficult task to certifying bodies [3]; certification may be carried out either within a single laboratory or through an interlaboratory study.

1.4.4.2 Certification by a single laboratory approach

Two approaches may be followed for certification within a single laboratory, i.e. using a so-called definitive method or using two or more so-called reference methods (applied by two or more independent analysts). The laboratory should always be of the highest, and proven, quality. This single laboratory approach does not eliminate the risk that a systematic error has been left undetected and cross-laboratory confirmation is advisable. In addition, certification of matrix materials using one single definitive method (e.g. IDMS for pluri-isotopics elements) does not give the user who does not apply such technique in his daily practice a fair estimate of the uncertainty achievable by more classical methods. Furthermore, such definitive methods are limited in their field of application and still do not exist for a variety of organic or organometallic compounds, or elements with different oxidation states; for these analytes an interlaboratory approach is the only satisfactory solution.

1.4.4.3 Interlaboratory certification studies

Certification studies may be organized in the same way as interlaboratory studies, with the exception that they should involve only highly specialized laboratories. All participants should preferably have demonstrated their quality in prior exercises (e.g. in improvement schemes or simple interlaboratory studies as developed by BCR). The organizer should also be recognized as capable of organizing such studies and should provide the participants with detailed instructions for participation and reporting forms (for results and method description). An interlaboratory certification study should involve several methods applied by different laboratories. If the results of entirely independent methods such as e.g. for inorganic determinations, IDMS, AAS, voltammetry (between-method bias) applied by laboratories working independently (between-laboratory bias) are in good agreement, it can then be concluded that the bias of each method is negligible and the mean value obtained is the best approximation of the true value. Figure 1-5 shows bar-graphs from an interlaboratory study organised by BCR. As exemplified, several techniques have been used in this exercise, most being applied by different laboratories. This type of evaluation has proven to be very powerful for detecting systematic bias of a single technique or demonstrating that a given laboratory did not apply a method properly.

```
           BAR-GRAPHS FOR LABORATORY MEANS AND 95% CI
                1.8       2.0       2.2       2.4       2.6       2.8
                +......+......+......+......+......+......+......+......+......+......+
                                                  !
     09 INAA                     <------------------------*------------------------>
                                                  !
     08 ICPMS                                     <------------*---------->
                                                  !
     06 IDMS                                <------*------>
                                                  !
     02 ICPAES                               <----------*---------->
                                                  !
     08 ICPAES                                  <--------*------>
                                                  !
     02 DCPAES                             <------*------>
                                                  !
     05 ETAAS                           <------------*------------>
                                                  !
     02 DPASV                              <----------*-------->
                                                  !
     !MEANS!                                <---M--->
```

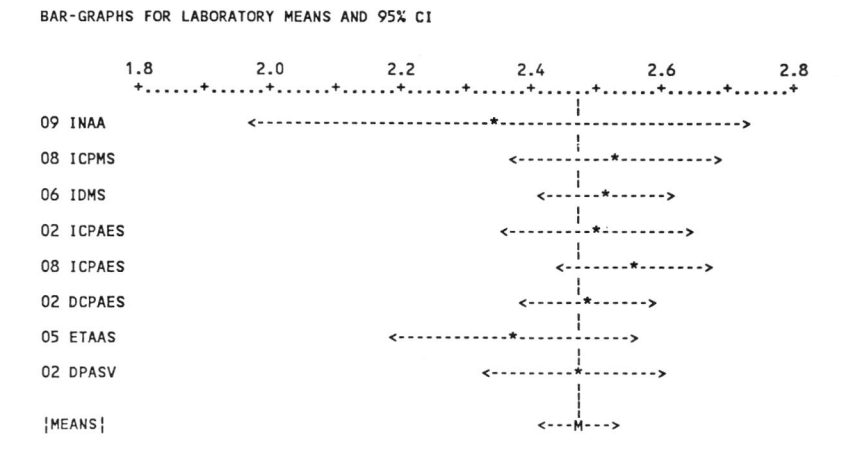

Figure 1-5:   Bar-graph of nickel in a lichen reference material (CRM 482). The laboratory codes are indicated along with the methods used. The results plotted correspond to five replicate determinations. M is the mean of laboratory means.

1.4.5   Producers of environmental CRMs

Environmental CRMs are produced by a variety of organisms, the main ones being the National Institute of Standards and Technology (NIST, USA); the Community Bureau of Reference, BCR of the European Commission (Brussels, Belgium) which works closely with two Joint Research Centres, namely the Institute for Reference Materials and Measurements, IRMM (Geel, Belgium) and the Environment Institute (Ispra, Italy); the National Research Centre of Canada, NRCC (Ottawa, Canada) and the National Institute for Environmental Studies (Japan). The International Atomic Energy Agency (IAEA) provides mainly environmental CRMs for the nuclear field.

A list of CRMs and CRM producers has recently been compiled by ISO/REMCO [37]. General information is available through the COMAR Data Bank which is a joint initiative of the Laboratoire National d'Essais (Paris, France), the Bundesanstalt für Materialforschung und Prüfung (Berlin, Germany) and the Laboratory of the Government Chemist (Teddington, United Kingdom). A procedure of accreditation for these producers is presently being examined by ISO [38], under the guidance of the ISO 9000 series of Standards and ISO Guide 25; a draft for the accreditation of RM and CRM producers has been produced, covering requirements on, e.g., management, quality system, training of staff, contracts, storage of materials and verification of long term stability, production control (preparation, traceability, measurement equipment, homogeneity/stability assessment, data treatment, and certification), and recording/reporting. CRM producers guarantee long term availability of CRMs and in most cases their replacement when the stock is exhausted; this is a basic prerequisite since these materials are used in the laboratories' QA and QC scheme.

1.5 Sampling and sample handling

The aspects of QA in sampling and sample handling are treated in detail in the other chapters of this book. Main considerations only will be covered in this section.

Data obtained by environmental monitoring are reliable if adequate samples are available. In most cases this is only possible when the analyst is fully involved in the sampling and sample handling procedures, which will depend upon the sampling strategy and techniques. Sampling remains the primary source of error in environmental monitoring. Detailed guidelines are difficult to set up, even for quality assurance and quality control, since the variety of samples and substances is very large. Some considerations on sampling techniques and strategies, e.g. on air pollutants [39], organic pollutants in waters samples [40] and sediments or biota [41], have been published in a book edited by D. Barceló.

The sampling strategy defines the location of the sampling, the number of samples to be collected, the frequency of sampling, etc., in relation to the objective of the monitoring and technical and scientific arguments, as well as economic considerations [3]. Sampling techniques will determine the sampling tools, sample transport, and sample pretreatment (e.g. homogenization before subsampling, storage).

1.5.1 Sampling strategy

To date there are no standards or prescribed procedures for sampling strategy in environmental analysis (in contrast with other fields, e.g., biomedical analysis) and the analytical team itself has to establish a sampling plan. Statistical sampling tools may be used [42,43] but are often neglected in practice. The different strategies used in various fields of environmental monitoring are examplified in this book.

1.5.2 Sampling techniques

The nature of the sample and of the substance to be monitored will dictate the choice of sampling technique. Basic principles are that the tools and containers used should not contain the substance to be measured nor major interfering compounds which could interfere with final detection. Contamination is a major risk for trace inorganic and organic analysis but losses may also occur, e.g. owing to adsorption on glass surfaces or polymers or evaporation. The risks are increased when manipulations have to performed in the field (e.g. for biological or chemical oxygen demand of water). In some cases, samples need to be stabilized, e.g. by addition of stabilizers or by deep-freezing. Detailed discussions on trends in sampling and sample treatment for soil and sediment have been published [44]. Other considerations from groups of experts also exist for, e.g., water and plant monitoring [45].

1.5.3 Registration

When received, the samples should be properly registered and stored. This includes information necessary for proper analysis and interpretation, as well as reporting. Modern management involves computerized systems, e.g. laboratory information management systems (LIMS), for the proper identification and follow-up of the sample within the laboratory.

1.5.4 Storage

The possible effects of storage on sample stability should be fully investigated prior to the establishment of storage conditions. This aspect of sample pre-treatment is also fully developed in the different chapters of this book.

1.5.5 Subsampling

Samples are usually available in a size which is sufficient to enable several replicate determinations, and storage for possible future analysis. Subsampling requires that the aliquot taken is representative of the whole sample; this may be achieved by proper homogeneization of the total sample. Difficulties are experienced for naturally inhomogeneous samples which may require a selection of a given grain size fraction, e.g. the silt-clay fraction for sediment analysis for inorganic determinations [46], or of a part of the sample, e.g. fatty tissues, liver or kidneys for organic contaminants [47].

The risk related to sample inhomogeneity has to taken into account in the analytical strategy, e.g. by optimizing the number of replicate determinations to include the inhomogeneity parameter. The preparation of composite homogeneous bulk from several individual samples is a way of achieving representative subsampling. This part of the analytical process has to be validated in the same strict manner as the other steps and primary investigations should be performed on separate samples [3].

1.6 Reporting and archiving

Reporting is a way to translate the analytical information into an understandable form which is usable by the end-user [3]. The report should, therefore, contain detailed information on:

- the identification of the end-user;
- the objective of the study;
- the sampling strategy and sampling techniques used, the staff involved and the sample location;
- registration data (dates of arrival, storage conditions);
- description of the analytical procedures, QC measures, indications on how and where this information can be retrieved, identification of the staff;
- analytical data with information of the method's performance (e.g. precision, sensitivity, limit of determination, etc.);
- evaluation/interpretation of the results in terms of their analytical significance, in reponse to the initial objective of the study;
- questions and possible unsolved problems which may have arisen from the study.

Preestablished standardized forms often exist and are given by the end-user for environmental analyses carried out for legal, health or commercial purposes (water, air, waste, etc.). The BCR has prepared sets of forms for reporting details of the analytical techniques used in certification exercises for inorganic and organic determinations.

Proper archiving of the information is necessary, enabling all the data to be traced back. Here again the use of computerized systems (LIMS) is helpful.

1.7 Regulatory aspects of QA and QC in environmental monitoring

The respect of regulations requires that the analytical data are of guaranteed quality. Several systems have therefore been established at the international level; these are briefly summarized in the following paragraphs.

1.7.1 Good Laboratory Practice (GLP)

GLP regulations were developed by the US Food and Drug Administration in 1978. Similar regulations, issued by the US Environmental Protection Agency (EPA), followed in 1983, for the production of agricultural and industrial toxic chemicals. The Organization for Economic Cooperation and Development (OECD) also considered the GLP guidelines which became mandatory in all OECD member countries for QC by producers of pharmaceuticals and toxic chemicals (e.g. pesticides) [48]. Finally, the Council of the European Community has adopted a series of Directives for the harmonization, inspection and verification of GLP regulations (87/18/EEC, 88/320/EEC and 90/18/EEC).

GLP regulations essentially focus on organizational aspects of the laboratory, the strategy and the way the studies are planned. These rules have been set up for the chemical industry producing toxic compounds and pharmaceuticals and are, therefore, not widely used in environmental monitoring. They are actually intended to guarantee that actions undertaken in the course of product development and toxicity testing are traceable.

1.7.2 Accreditation

Other control systems have been implemented by authorities to evaluate the quality of general testing laboratories, e.g. administrative inspections, proficiency testing, or both simultaneously. In several countries, the control systems for testing laboratories is based on ISO Guide 25 [8]. Within the European Union, Resolution 90/C/10/01 of the Council (21 December 1989) on "a global approach to conformity assessment" adopted guidelines developed in the European series of standards EN 45000 [7] for testing laboratories and EN 29000 series (derived from ISO 9000) [4] for good manufacturing practice. The Council resolution promotes mutual recognition of accreditation systems developed in the various member states. The EN 45001 series lists a number of QA and QC practices to be implemented by testing laboratories, e.g. recommendations on laboratory management, infrastructure, competence of personnel, equipment, and working procedures, and detailed aspects related to calibration and use of reference materials. Participation in proficiency testing is left to the judgement of the accreditation body. The EN 45002 Standards even states that the accreditation of a laboratory cannot be granted or maintained solely on the basis of results obtained in proficiency testing.

To date, EU Member States, as well as countries which ratified the European Free Trade Agreement (Iceland and Norway) and Switzerland have adopted the same system. Accreditation bodies are cooperating at an international level within the International Laboratory Accreditation Cooperation (ILAC). Other international scientific societies, e.g. IUPAC, AOAC, also cooperate to develop a common vocabulary and set up common guidelines for the improvement of the quality of testing laboratories.

1.7.3 ISO 9000/EN 29000 standards

The ISO 9000 series of standards is intended to ensure quality for commercial purposes and all types of product; these standards mainly concern industry and address items such as quality system specifications and requirements for production and installation or for final inspection and testing, and guidance on quality management. They do not directly concern analyses related to environmental monitoring.

1.7.4 Standardization

Written standards (norms) are used to establish minimum quality requirements and to improve the comparability of analytical results; they also represent the first step of introduction into regulations. Written standards have been successfully applied for, e.g., sampling strategies and techniques, measurements of operationally-defined parameters, etc.

In some fields the existence of written standards bound to regulations is criticized by end-users since standardized methods may be outdated but still mandatory when regulations have not been revised. It may happen that the state of the art of a certain field of chemical analysis has considerably improved but that a laboratory is still obliged to use old fashioned methods for legal reasons.

Using a standardized method does not guarantee that no errors will be made, owing to possible difficulties in the application of the standard. Standardization bodies have recognized this difficulty and the necessity to allow progress in analytical science to flow into standardized methods. Therefore, general method performance characteristics are described in recent written standards, stressing the need to control the performance by using CRMs.

Standardization bodies exist at regional, national, and international levels, and are founded on public authorities, or professional or commercial organizations. For European countries (EC and EFTA) CEN/CENELEC/ETSI produce European standards which are replacing national standards. CEN has passed an agreement with ISO in 1991 to avoid overlap of tasks, i.e. CEN adopts some written standards developed by ISO when they fulfil mandatory requirements.

1.8 Conclusions

Quality assurance and quality control for environmental monitoring are essential for ensuring good comparability of data. An increasing awareness of the need for quality in this field of chemical analysis will enable considerable progress to be made. This means that all organizations performing environmental monitoring should take actions to implement a quality system within their laboratory; this policy may cost a lot of effort and money and it is not uncommon for 20% of the resources for chemical analysis to be devoted to QA and QC. These efforts will be the only way to enable progress in analytical science. This should be recognized by all scientists, policy-makers, and end-users, of whom the first question should be "are these data of sufficient quality to satisfy the aim of the study?". This question should be kept in the minds of referees of scientific papers, users of data of environmental monitoring for mapping, environmental policies, data banking (e.g. the European Environmental Agency), or policy-makers (e.g. the European Commission's services).

1.9 References

[1] Ph. Quevauviller, E.A. Maier and B. Griepink (Eds.), *Quality Assurance for Environmental Analysis - Method Evaluation Within the Measurements and Testing Programme (BCR)*, Elsevier Science Publishers, Amsterdam, 1995, pp.649.

[2] ISO, Quality Vocabulary (ISO/IEC Standard 8402), International Organization for Standardization, Geneva, Switzerland (1986)

[3] E.A. Maier and B. Griepink, *Quality Assurance and Quality Control*, in: Eurocurriculum (1995)

[4] ISO, *Quality management and Quality Assurance Standards. Guidelines for Selection and Use (ISO/IEC Standard 9000)*, International Organization for Standardization, Geneva, Switzerland (1990)

[5] M. Válcarcel and A. Rios, *Anal. Chem.*, **65**(18), 781A (1993)

[6] B.E. Broderick, W.P. Cofino, R. Cornelis, K. Heydorn, W. Horwitz, D.T.E. Hunt, R.C. Hutton, H.M. Kingston, H. Muntau, R. Baudo, D. Rossi D., J.G. van Raaphorst, T.T Lub, P. Schramel, F.T. Smyth, D.E. Wells, and A.G. Kelly, *Mikrochim. Acta*, **II(1-6)**, 523 (1991)

[7] CEN, *General Criteria for the Operation of Testing Laboratories (European Standard 45001)*, CEN/CENELEC, Brussels, Belgium (1989)

[8] ISO, *Guidelines for assessing the Technical Competence of Testing Laboratories, ISO Guide 25-1978 E*, International Organization for Standardization, Geneva, Switzerland (1978)

[9] A. Marshal, *Calibration of Chemical Analyses and Use of Certified Reference Materials*, Draft ISO Guide 32, ISO/REMCO N 262, International Organization for Standardization, Geneva, Switzerland (1993)

[10] BIMP/IEC/ISO/OIML, *International Vocabulary of Basic and General Terms in Metrology*, International Organization for Standardization, Geneva, Switzerland (1984)

1.9 References

[11] ISO, *Statistics - Vocabulary and Symbols. Part 1: Probability and General Statistical Terms*, revision of ISO 3534-1977, International Organization for Standardization, Geneva, Switzerland (1977)

[12] C. Watson (Ed.), *Official and Standardized Methods of Analysis*, Royal Society of Chemistry, Cambridge, 3rd Edition, 1994, pp.778

[13] ISO, *Certification of Reference Materials - General and Statistical Principles, ISO Guide 35-1985 E*, International Organization for Standardization, Geneva, Switzerland (1985)

[14] D.L. Massart, B.G.M. Vandeginste, S.N. Deming, Y. Michotte and L. Kaufmann (Eds.), *Chemometrics: a textbook. Data Handling in Science and Technology*, Vol.2, Elsevier Science Publishers, Amsterdam, The Netherlands (1988)

[15] E.A. Maier, In: *Quality Management in Chemical Laboratories*, Cofino W. and Griepink B. (Eds), Elsevier Science Publishers, Amsterdam, The Netherlands (1995)

[16] D.E. Wells, E.A. Maier and B. Griepink, *Intern. J. Environ. Anal. Chem.*, **46**, 255 (1992)

[17] J.R. Moody, R.R. Greenberg, K.W. Pratt, and T.C. Rains, *Anal. Chem.*, **60**(21), 1203A (1988)

[18] E.A. Maier, B. Griepink, H. Muntau and K. Vercoutere, *Report EUR 15283 EN*, European Commission, Luxembourg (1993)

[19] B.E. Broderick, W.P. Cofino, R. Cornelis, K. Heydorn, W. Horwitz, D.T.E. Hunt, R.C. Hutton, H.M. Kingston, H. Muntau, R. Baudo, D. Rossi, J.G. van Raaphorst T.T Lub, P. Schramel, F.T. Smyth, D.E. Wells, and A.G. Kelly, *Mikrochim. Acta*, **II**(1-6), 523 (1991)

[20] D.E. Wells, In: *Environment Analysis Techniques, Applications and Quality Assurance*, Barceló D. Ed., Elsevier Science Publishers, Amsterdam, The Netherlands, **13**, 79-105 (1993)

[21] M. Mullholland, N. Walker, J.A. van Leuven, L. Buydens, F. Maris, H. Hindriks, and P.J. Schoenmakers, *Mikrochim. Acta*, **II**(1-6), 493 (1991)

[22] AOAC, *Official Methods of Analysis*, 15th Ed., Assoc. of Official Analytical Chemists, Arlington VA, USA (1990)

[23] T.H. Hartley, *Computerized Quality Control: Programs for the Analytical Laboratory*, Ellis Horwood, Chichester, 2nd Edition, 1990, pp.99

[24] Ph. Quevauviller, I. Drabaek, H. Muntau and B. Griepink, *Appl. Organometal. Chem.*, **7**, 413 (1993)

[25] Ph. Quevauviller, J.L. Imbert and M. Ollé, *Mikrochim. Acta*, **112**, 147 (1993)

[26] W. Horwitz, *Nomenclature for Interlaboratory Studies*, 4th Draft, IUPAC, Analytical Chemistry Division, Commission V1 (1991)

[27] Analytical Method Committee, *Analyst*, **112**, 679 (1987)

[28] E.A. Maier, Ph. Quevauviller and B. Griepink, *Anal. Chim. Acta*, **283**, 590 (1993)

[29] ISO, *Terms and definitions used in connection with reference materials, ISO Guide 30*, revised version, International Organization for Standardization, Geneva, Switzerland (1991)

[30] ISO, *Uses of Certified Reference Materials, ISO Guide 33*, International Organization for Standardization, Geneva, Switzerland (1989)

[31] D. Barceló, W.A. House, E.A. Maier and B. Griepink, *Intern. J. Environ. Anal. Chem.*, **57**, 237 (1994)

[32] E.A. Maier, In: *Quality Management in Chemical Laboratories*, Cofino W. and Griepink B. Eds, Elsevier Science Publishers, Amsterdam, The Netherlands (1995)
[33] Ph. Quevauviller, M. Astruc, L. Ebdon, V. Desauziers, P.M. Sarradin, A. Astruc, G.N. Kramer and B. Griepink, *Appl. Organometal. Chem.*, **8**, 629 (1994)
[34] A.Y. Cantillo, *Standard and reference materials for marine science*, NOAA, National Status and Trends Program for Marine Environmental Quality, U.S. Department of Commerce, Rockville MD, USA, Third Edition, 1992, pp.575
[35] Ph. Quevauviller, E.A. Maier and B. Griepink, *Fresenius' J. Anal. Chem.*, **345**, 282 (1993)
[36] Ph. Quevauviller, In: *Quality Management in Chemical Laboratories*, Cofino W. and Griepink B. Eds, Elsevier Science Publishers, Amsterdam, The Netherlands (1995)
[37] S.D. Rasberry, *Worldwide Production of Certified Reference Materials*, ISO/REMCO, Status Report (1994)
[38] CRM Producers' Registration under ISO 9002, *Draft ISO/REMCO N 288*
[39] R. Niessner, In: *Environmental Analysis Techniques, Applications and Quality Assurance*, Barceló D. (Ed.), Elsevier Publ., Amsterdam, **13**, pp.3-22 (1993)
[40] M.C. Hennion and P. Scribe, In: *Environmental Analysis Techniques, Applications and Quality Assurance*, Barceló D. (Ed.), Elsevier Publ., Amsterdam, **13**, pp.24-72 (1993)
[41] D.E. Wells, In: *Environmental Analysis Techniques, Applications and Quality Assurance*, Barceló D. (Ed.), Elsevier Publ., Amsterdam, **13**, pp.79-105 (1993)
[42] F.M. Garfield, *Quality Assurance Principles for Analytical Laboratories*, 2nd edition, AOAC International (Ed.), Arlington VA, USA (1991)
[43] P.M. Gy, *Mikrochim. Acta*, **II**, 457 (1991)
[44] Ph. Quevauviller, E.A. Maier and A. Boenke, *Quím. Anal.*, **13**, S264 (1994)
[45] Ph. Quevauviller and E.A. Maier, *EUR Report*, **16000 EN**, European Commission, Luxembourg (1994)
[46] U. Förstner and W. Wittman, *Metal Pollution in the Aquatic Environment*, Springer-Verlag, Berlin, 1983, pp.648
[47] E.A. Maier, B. Griepink and G.U. Fortunati, *Fresenius' J. Anal. Chem.*, **348**, 171 (1994)
[48] OECD, Decision of the Council: C81/30, Annex 2, *OECD guidelines for Testing Chemicals, "OECD Principles of Good Laboratory Practice"*, Paris, France (1981)

2.

Sampling strategy in environmental monitoring of biological specimens

G. Wagner and R. Klein
University of the Saarland, Centre for Environmental Research, Institute for Biogeography, D-66041 Saarbrücken, Germany

Sampling is the first and the least corrigible step in environmental monitoring and analysis. While methodology and techniques for physico-chemical measurements of emission and the impact of pollutants are well developed and standardized, the measurement of effects, including the accumulation of pollutants in the habitats, food, and bodies of living organisms is still in progress. This chapter concentrates on approved strategies for sampling environmental specimens such as animal tissues, plants, and soil as matrices for chemical analysis of nutrients, trace elements, and pollutants for purposes of environmental monitoring. Special emphasis is placed on possibilities, approaches, and methods for quality assurance and quality control in planning and field work.

2.1 Objectives of environmental sampling

The general objectives of environmental monitoring are to recognize environmental pollution problem as early as possible, to analyze the type, source, and path of contamination and effects, and to develop recommendations for solving the problem before serious effects become manifest. Another important task is to control the effectiveness of remedial measures. These tasks require sophisticated strategies and methods for selecting and methods for selecting environmental samples and collecting them [1-9]. Table 2-1 gives an overview of different sampling strategies, classified according to their spatial design and preferred utilization, with examples and references as explained by example way in the following sections.

2 *Sampling strategy of biological specimens*

Table 2-1: Sampling strategies in environmental monitoring

Spatial design	Use, employment	Measures or indicators	Examples	References
Passive monitoring: Sampling by chance	monitoring of accidental effects of unknown origin	carcasses, unfertile eggs, fish, plants, soils	acute impacts on plants, lethal effects on raptors, fish, seals....	Ellenberg 1981, Elliott 1984 Müller 1988
selected sampling	research, long-term observation, background monitoring, reference stations, microdistribution, organotropy	physico-chemical measurements passive and active biomonitors	BDF (soil long-term observation) ecosystem research UBA monitoring network	SAG 1991 Ellenberg et al. 1985 UBA 1988, Zimmer 1993
transect/gradient sampling	reach of source-related pollution	physico-chemical measurements passive and active biomonitors	environmental impact assessment	Ernst & Leloup 1987
grid sampling	large-scale surveys/inventories/ screenings	passive biomonitors soils	ECE moss monitoring soil inventories, ESB SOPsoils	Bau 1992, Rühling & Tyler 1969 Hodgeson 1985, Scholz et a. 1993 Wagner & Sprengart 1994
random sampling	small-scale surveys of randomly distributed phenomena	passive biomonitors	population biology	Green 1979, Wagner 1994b
stratified random sampling	spatial distribution of pollutants in populations of bioindicators or monitor organisms	passive biomonitoring and environmental specimen banking (ESB)	ESB SOP's	Fischer 1991 Paulus et al. 1994 UBA 1989, BMU/UBA 1994
Active or experimental monitoring: exposition of sampling devices, bioindicators or artificial habitats	source-related pollutant impact assessment	physico-chemical measurements, active biomonitoring and ESB stanardized cultures of ryegrass, green cabbage, moss, lichens etc.	VDI-guidelines 3792-1-5, ESB SOP's	Scholl 1974, VDI 1978, Winter et al. 1992, Altmeyer 1993 Bromenshenk 1978

Wagner, UoS, 7/1994

The task of a monitoring project may be, for example, to characterize and evaluate the sulfur dioxide (SO_2) or cadmium burden of a certain area. The concentration of SO_2 in the ambient air or the content of cadmium in the dustfall measured over a certain period of time at a single spot selected in some way will give only poor information about the real objective, because the real SO_2 burden depends not only on this concentration in the air, but mainly on its impact on the organisms and materials present in the area, which is influenced by the specific sensitivity of the different species and their age, and stage development, their exposure, the wind velocity, temporal changes and spatial differentiation of SO_2 concentrations, and the concentrations of interfering compounds such as, e.g., oxides of nitrogen (NO_x) and ozone (O_3). In the case of cadmium, the spectrum of possible synergistic influences is probably even greater. The current deposition may have a certain influence, but the main effect is more likely to result from cadmium deposits in the soil modified by the numerous factors regulating the solubility and bioavailability of the metal. There is no fairly simple model available which would enable a realistic calculation of the real effects resulting from the total concentration of a pollutant analyzed in a single environmental medium. This means that a spectrum of samples has to be taken which is representative of all possible or intended targets present in the area to be studied. One possible way to reduce the effort of such an inventory to reasonable proportions is to use suitable (biological) indicators [5,8,33-35].

Moreover, the objectives of environmental monitoring studies seldom relate to the properties or pollution impact of a specific point or spot in the landscape. Rather, a more or less extended and carefully delimited area is to be characterized. Nevertheless, most technical measurements or sampling procedures for air, water, or other environmental media are carried out at discrete spots, which are expected to be more or less "representative" of the area. The data obtained are then taken for the whole study area or extrapolated to "the surroundings" of this point, but it is seldom determined whether the data really represent the quality of the whole area to be investigated. Thus the internal spatial heterogeneity of the area and the variability of the data both remain unknown.

2.1.1 Representativeness of environmental samples

The term *representative* is a keyword in environmental monitoring. However, this frequently used term is seldom exactly defined or specified. Violation of representativeness may cause systematic errors of high magnitude (see [23] for more details). In effect it refers therefore to the accuracy of analytical results in the broadest sense. Neglecting the several different aspects of representativeness is the main reason for bad quality and costly accumulation of often rather useless data.

As we deal with areas and spatially distributed phenomena, the geographical or spatial aspect of representativeness is the first to be discussed. Geographical representativeness means that the whole area is characterized by a set of samples or data. This is often claimed by describing the quality of a more or less extended area by the central tendency and variation of the relevant parameters. However, in the case of significant heterogeneity of the area, the mean is not an adequate measure [3,4,22,36]. In this case, the distribution pattern has to be described by a spatial definition of gradients or by dividing the area into homogeneous subareas from the results of preliminary sampling and ascertaining the mean and variability for each of them by measurements or samples from an adequate number of spots randomly or systematically distributed over the area.

Ecologically representative samples should represent the total target population or ecosystem. This requirement may be relatively easy to fulfil in the case of areas dominated by one or few species, as in the case of agriculture and forestry. "There is no better indicator for the health of a spruce forest, than Norway spruce itself, nothing better than beech trees in a beech forest..." [37]. In diverse near-nature ecosystems like tropical rain forests it is extremely difficult and laborious, if not impossible, to obtain ecologically representative samples of the whole vegetation, because the different species may have very different accumulation behavior and resulting concentrations of ingredients and pollutants. In such cases only stratified random sampling strategies can lead to useful results (see Section 2.3.4).

Standardized biological indicator systems react with known specificity and sensitivity to different pollutants and have the ability to randomize or integrate pollutant inputs spatially and temporally. Such indicator systems can be used vicariously as representatives instead of the spatially, physiologically, and temporally often extremely variable total population. Of course the accumulation behavior of this indicator system has to be similar or comparable to that of the majority of the total population of the ecosystem to be characterized. In other words, it has to be biologically and physiologically representative.

Genotype representativeness may be a problem in genetically very heterogeneous populations, or when comparing areas with genetically different populations. This is particularly the case, for example, in spruce forests which have been planted without considering the origin or provenance of the trees, as observed, e.g., for many spruce or pine forests planted after the first and the second world wars. Differences in the growth rates and phenotypical properties caused by different physiological adaption of the genotypes can influence the accumulation behavior and resulting chemical composition. However, this seems to be of relatively minor importance compared with other factors of natural variability.

Another problem is linked with the adaptation of populations to specific contamination or long-term pollution of their habitats. There are different possible adaptation strategies [2,38] but as an effect the accumulation behavior and/or the distribution of the pollutant in the body is often changed, which may limit the comparability of the resulting concentrations of these contaminants in biological tissues.

2.1.2 Repeatability and precision

In order to establish temporal trends, e.g. of atmospheric pollution, strictly quantitative comparability of consecutive samples of a series is the first priority. Optimizing repeatability means reducing the influence of natural (spatial, biological, temporal) variability of the sampled population in order to yield maximum precision and repeatability of the samples. This demand can only be achieved by stratifying the population to be sampled, standardizing the sampling procedure, and defining a very homogeneous sampling area. However, most of the measures claimed here to improve repeatability and precision generally reduce the spatial representativeness of the samples (compare Sections 2.1.4 and 2.6.3, Figure 2-1). The more the size of a sampling area is reduced in order to furnish homogeneous conditions, the more effort has to be spent to protect it against accidental changes, because the continuation of the trend monitoring is very sensitive to such changes in the case of a small sampling area or population.

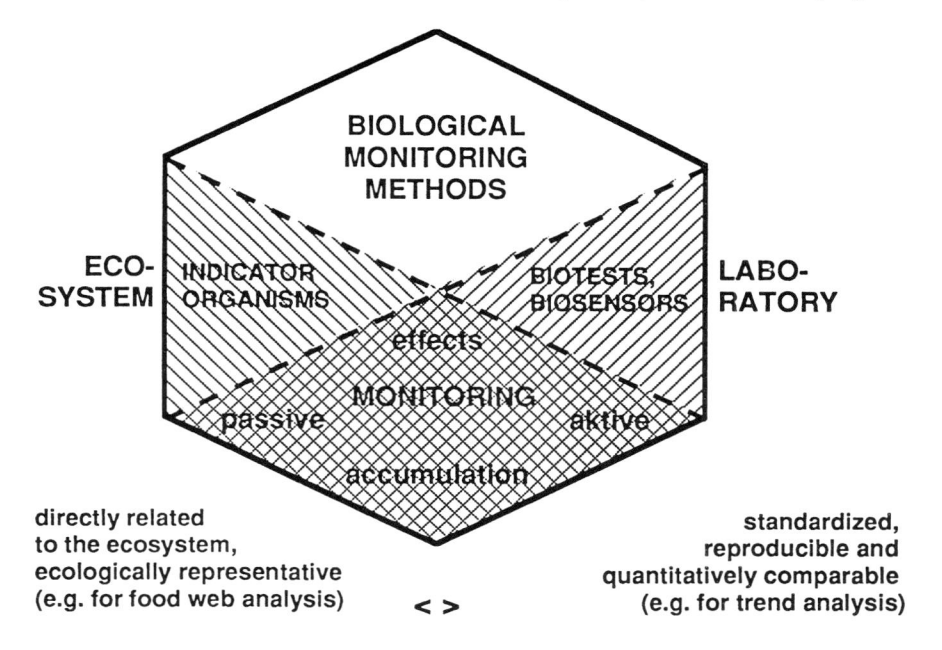

Figure 2-1: Spectrum of methods in environmental monitoring (adapted from [33,34])

2.1.3   Spatial comparability

Another demand on the results of environmental monitoring relates to the necessary comparison of contamination levels of different, sometimes far apart areas by quantitatively comparing monitoring data. This comparability may be more easily achieved with, for example, the concentrations of chemicals in water or ambient air. The problems become more acute for soil monitoring, because soils are hardly identical and additional information about the composition and chemical properties of the soil is a key aspect of understanding, e.g., the possible effects of contaminants on biota or their geochemical pathways. The amounts of chemicals in biological tissues used as biomonitors are also influenced by the local complexity of environmental factors and spatial differences in the genotypical and phenotypical composition of the populations. On the other hand, the potential and real effects of identical amounts of a substance, such as, e.g., sulfur dioxide in the ambient air in different areas, can also be very different, depending on environmental factors such as moisture, insulation, and the simultaneous presence of other additives like ozone. In the light of these relations the "biologically standardized" reactions and accumulation behavior of well known and genetically comparable monitor organisms have great potential for improving the spatial comparability of pollution monitoring data related to their effects. Requirements for use of this potential are careful selection and stratification and standardized sampling methods.

For supra-regional or large-scale comparisons across different climatic zones, the basis should, in principle, not be restricted to the assessment of the levels of desired and undesired substances in environmental media and commercial goods like food and feed products, but should also be extended to their specific importance, hazards, and effects on human health as well as on the functioning and structure of prevailing ecosystems. The geographical basis for monitoring and comparison on an ecosystem level are preferably small, well-defined ecosystems with a minimum of diffuse input-output relations like small watersheds [39-41]. This approach is thoroughly described in the guidelines for the "Integrated monitoring of the Nordic countries" [42].

2.1.4 Reliability and probative force of environmental samples

There is a broad spectrum of biological methods available for monitoring and evaluating bioaccumulation as well as toxicological and ecotoxicological effects of pollutants. Figure 2-1 gives a survey of principles of the methodological spectrum as well as implications of the different strategies. The borders of biological monitoring methodology are marked on the passive (left) side by indicator organisms showing directly observable effects of a certain influence on their habitat, such as specific growth form and colouring of plants growing on soils with high contents of heavy metals. Passive biological monitoring is carried out by analyzing organisms which are sampled on the site where they developed (natural habitat or normal cultivation). These organisms, if adequately selected and sampled, give reliable information directly related to their respective ecosystem. As far as direct influence or contamination can be excluded and quality assurance criteria are observed (see Section 2.2), such samples possess high probative force, but often precision may be rather poor because of effects of various sources of natural variability.

The active or experimental (right) side of this spectrum is characterized by standardized cultivation of well selected and genetically homogeneous organisms which are exposed to environmental influences of the location to be monitored under standardized conditions for a specific period of time before they are analyzed. However, these organisms are taken out of the context of their ecosystem and many of these efforts made for improving the precision and reproducibility of the results by standardizing the organism itself and its conditions of growth and exposure to the environment result inevitably in an increasing danger of external contamination and a decrease in ecological representativeness [23,34].

2.2 Principles of quality assurance in environmental sampling: requirements and methods

Quality assurance in environmental monitoring for biological specimens, as in other fields, is based on four principal steps:
1) planning, including preliminary investigations;
2) standardization, stratification, task-specific sampling protocols;
3) documentation; and
4) control.

Planning includes several phases, from collecting and evaluating all the data available for the area to be studied, delimitation and if necessary subdivision of the area, selection of specimen types and methods, and preliminary sampling (screening) as a basis for developing the statistical sampling design.

The *use of standardized methods* is a mean of achieving comparability of results from different studies, threshold values, or other standards and is an absolute requirement for all studies carried out in the context of official or legal procedures. As far as possible already available guidelines or standards should be used. In any case, all the methods used have to be laid down in the form of written *standard operating procedures (SOP)*. These SOPs, together with a clear definition of the task and its objectives and the statistical sampling design, form the basis for a *project-related sampling plan*. This document specifies all measures to be taken in a specific case and, in the case of commercial investigations, should be part of the final contract between the orderer and the investigator.

Complete documentation of the whole procedure of sampling and sample treatment in the field as well as transport and storage up to the final analysis in the laboratory is an absolutely necessary requirement for any quality control. The record has to contain all possibly relevant data about the area and the population to be sampled, the tools and materials used, the time, weather conditions, and, should the occasion arise, any disturbances or trouble during sampling as well as deviations from the sampling plan; last but not least, a comprehensive description of the sample itself is required. As far as automated records are possible, e.g. time, weather conditions, temperatures of different media, geographical position, and so on, such methods should be preferred for documentation. In all other cases, formal data sheets should be prepared as part of the SOP or the project-related sampling plan, and used in the field.

The documentation of the records and all data and, as far as necessary, storage of replicates or subsamples have to be achieved in a manner which enables effective subsequent control for a certain period of time defined in the project-related sampling plan.

2.3 Sampling strategies for plants (passive biomonitoring)

The examples of sampling strategies explained in this chapter are intended as a collection of approved methods and special recommendations rather than a systematic overview of existing procedures.

2.3.1 Fortuitous sampling

The roots of bioindication and effect monitoring are undoubtedly located in the observation of pollution impacts and fortuitous sampling of affected tissues. In the meantime, information, knowledge, awareness, and methodology have improved, but fortuitous sampling is still an important source of information about manifest, hitherto unknown or unexplained effects like, for example, suddenly appearing, unexplored diseases in human populations, unaccountable damage to vegetation, fish mortality in various water bodies without obvious reason, seabird mortality, or the seal mortality some years ago in the European North Sea (e.g. [10,11]).

Beyond the general problem of subsequent, investigative research on already manifest damage, the samples received by this kind of sampling are often already severely changed by autolysis, decay, and other influences [43,44] and so provide only coarse or incomplete information about the possible cause of the event.

2.3.2 Selected sampling

Preliminary investigations or initial research into a new problem often require the analysis of samples selected for special purposes. In the case of the effects of air pollution on plants, selected sampling is often used for comparing pairs of affected and unaffected plants or plant organs, e.g., ill and healthy-looking trees, twigs or leaves with and without chlorosis, necrosis, aphid infestation, and so on [15].

For screening purposes, samples are often selected according to gradients or factors present in the area or population, which are potentially correlated with the concentration of a sought-after substance, e.g., health status of the organisms, elevation, exposition, edaphic factors, human influences etc., in order to obtain information about sources of variation or the potential spans of concentrations of substances in the population.

The distribution of a substance in different organs or tissues of an animal or plant (organotropy) is also often analyzed using accidentally available or especially selected individuals. This knowledge is a precondition for any attempt to reduce variability or to increase the predicative power of the samples by stratification and/or standardization of the sampling (e.g. [6,36,45-48]).

2.3.3 Random sampling

The main criterion for differentiating random from selected sampling designs is the requirement that in the first case regions or individuals to be sampled are not selected by subjective criteria in the field. "In completely random sampling designs, each potential sampling unit must have an equal chance of being collected" [4]. Random sampling strategies offer the advantage that subjective selection criteria are eliminated and most of the common statistical methods are suitable for analyzing and describing the results [22]. From the aspect of quality assurance, true random sampling designs have some disadvantages in practical work:

(1) the designated sampling points in the plan are often difficult to locate exactly in the field or the coordinates of the randomly chosen points are difficult to determine exactly;

(2) spatial patterns or gradients may not be recognized because of the uneven density of sampling points;

(3) it is often not possible to control whether the realization in the field is in full harmony with the principles of the random sampling design. In other words, random sampling is often postulated or pretended, but not always executed in the strict sense of the word in practical field work.

With respect to the difficulties of (1) and (3) improvement is promising thanks to the increasing availability and accuracy of Geographical Positioning Systems (GPS) combined with the use of Geographical Information Systems (GIS). These tools provide new possibilities and chances of increased use of random sampling designs as well as quality assurance and quality control in environmental monitoring in the future [23].

2.3.4 Systematic sampling (transects, grids)

The most essential aspect common to both random and systematic sampling designs is that they exclude subjective decisions and selection criteria in the individual case during sampling. Systematic sampling is generally preferred when spatial structures or patterns of a parameter have to be analyzed, e.g., for large-scale screening, surveys, or inventories (e.g., [17-20,49]).

The simplest form of systematic sampling is to place a transect following a known or supposed gradient, for example from a point source of pollution in the main wind direction, and to take samples either equidistant on this line or at distances from the source preselected in some other ways (e.g., [16]). Other common systematic sampling designs are, for example, more or less sophisticated polar grids for source-related monitoring and regular, rectangular, triangular, or hexagonal grids to cover a given area. The optimal density of the grids has to be determined by preliminary sampling if no specific experience is available. For geostatistical analysis of the results the grid should be at least so dense that, generally speaking, the concentrations of neighbouring points lie nearer together than those of distant points [50-52]. The sampling points have to be previously determined in the sampling protocol and then identified in the field as accurately as possible without looking for any additional criteria. This of course is plain theory, because nobody, for example, can collect a meaningful soil sample from a preselected spot if a house, a street, or the trunk of a tree covers the surface. If such places are excluded and, for example, a minimum distance of 10 m from buildings and roads is set up, we come automatically to a stratified sampling design. This means that the results of our investigation will not be valid for the total area but only for that part of the area which is more than 10 m from buildings and roads. Most back gardens, parks, etc., as well as the edges of the streets, are excluded.

This example indicates that in practice true unrestricted, systematic, or random sampling is the exception rather than the rule. Decisive requirements have therefore to define clearly:
(1) the area to be investigated with all the exceptions and exclusion criteria (and to consider these exceptions in the interpretation and evaluation of the data); and
(2) whether such points are totally excluded or how they may be replaced (in order to prevent any subjective decisions on the spot).

The critical point is not *how* these determinations are made, but that they are made *previously* in the sampling plan and things are not left to chance, which can mean subjective decisions in the particular case. Those subjective decisions can severely bias the results; as Green says: "The assumption of independence of errors is the only one in most statistical methods for which violation is both serious and impossible to cure after the data have been collected" [22].

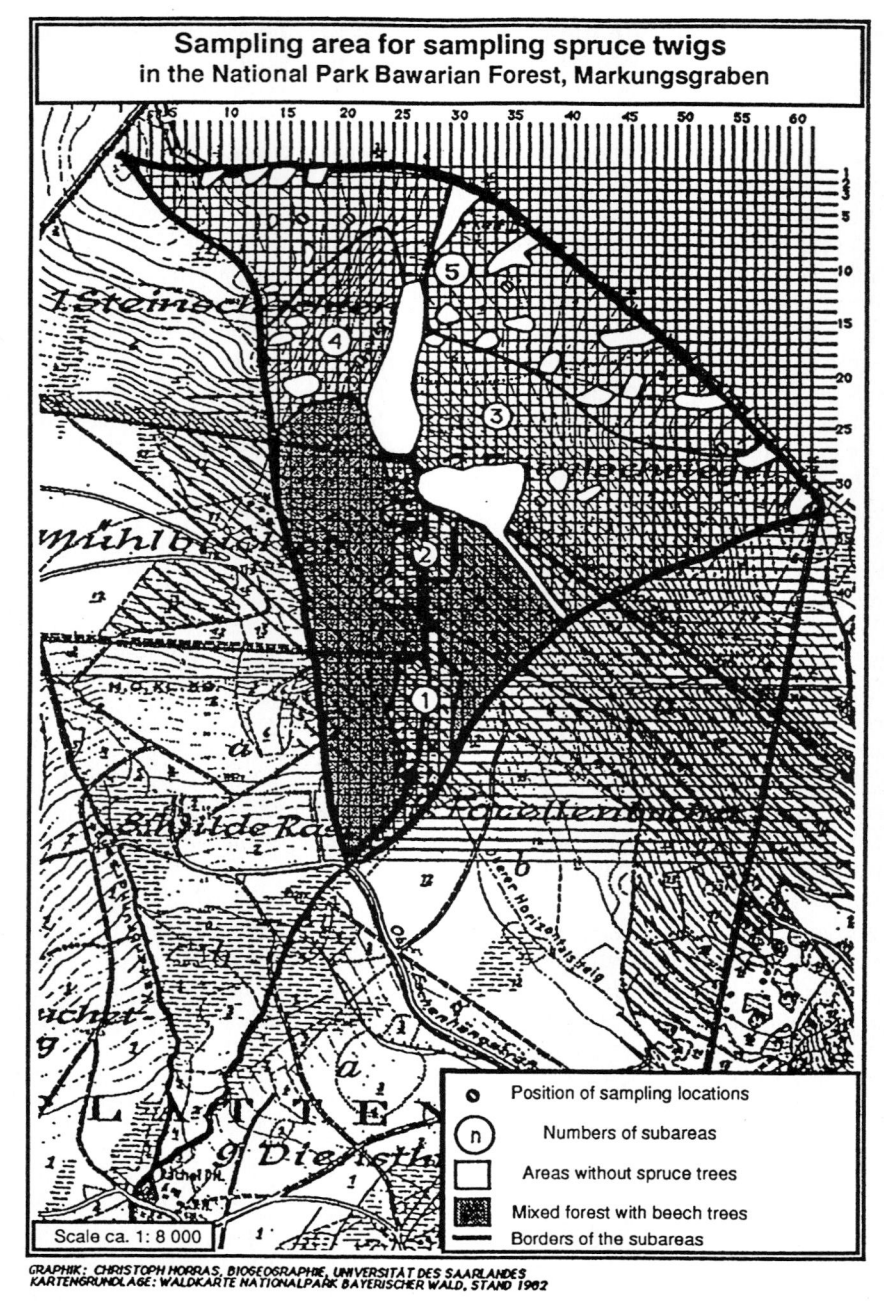

34   2 *Sampling strategy of biological specimens*

Figure 2-2: Determination of individuals for sampling by area-related random selection (from [41])

2.3 *Sampling strategies for plants* 35

GUIDELINE FOR SAMPLING SPRUCE TWIGS (*PICEA ABIES*)

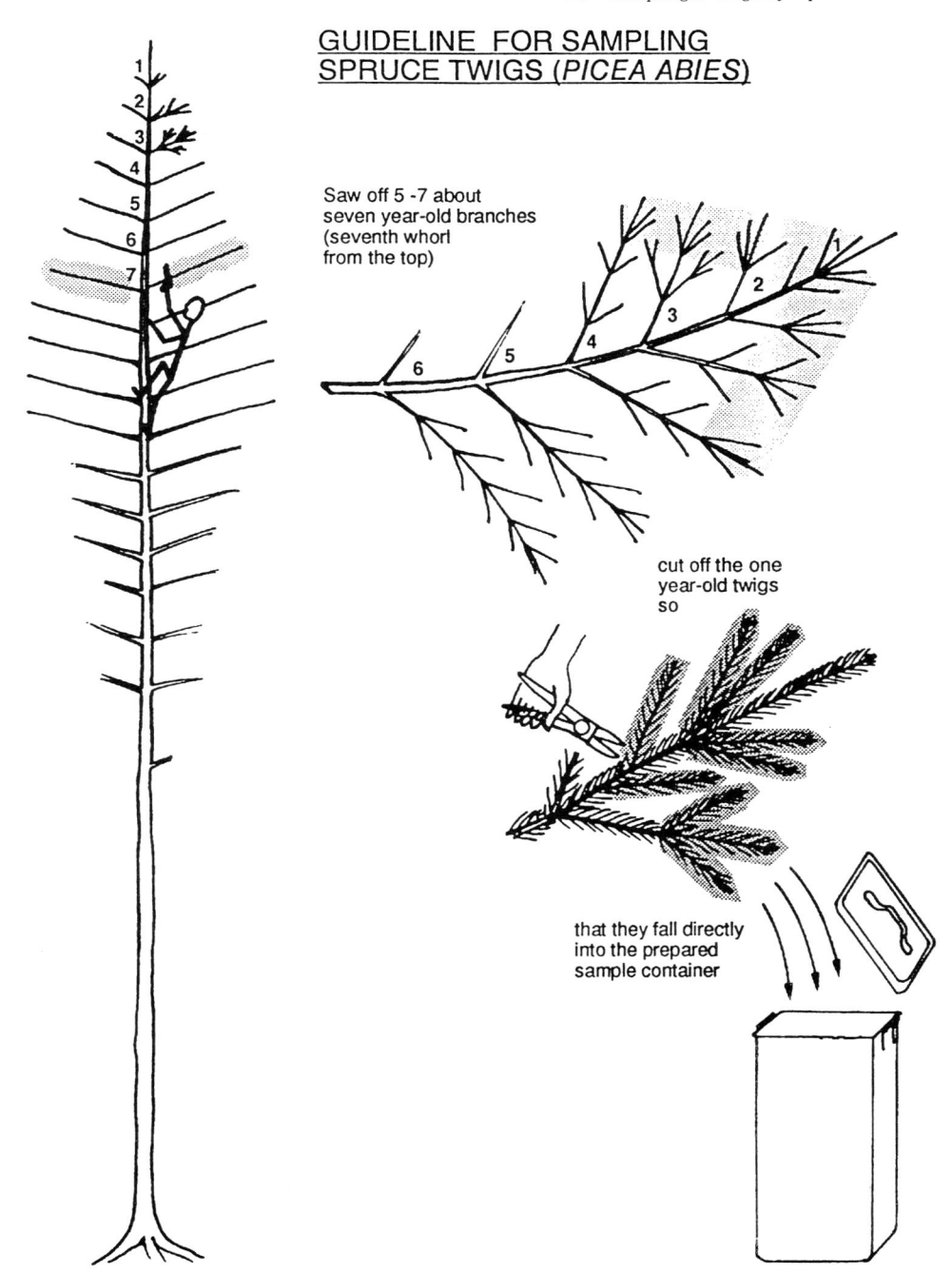

Saw off 5 -7 about seven year-old branches (seventh whorl from the top)

cut off the one year-old twigs so

that they fall directly into the prepared sample container

Figure 2-3: Sampling procedure for *Picea abies* (from [36])

2.3.5 Stratified random sampling

Stratification can be used as a strategy to improve reproducibility or to reduce the necessary number of samples in the case of diverse systems on several different levels. The principles of stratification can be explained with an example of a guideline for sampling spruce (*Picea abies*) shoots [53]. The first level of stratification carried out by mapping the area to be analyzed and dividing it into homogeneous subareas (see Figure 2-2). For area-related stratified random sampling, the subareas covered with spruce trees overlaid with a regular rectangular grid. The number of crossings in each subarea is counted and the total number of samples divided in relation to the sizes of the subareas. For each subarea the sampling locations are chosen by selecting the appropriate number of line crossings by use of random numbers. These points are transferred to an aerial photo map and the spruce tree next to such a point in each case is identified and chosen for sampling.

The next level of stratification depends on the spruce tree itself. The main objective in this case was not to characterize the whole tree but to obtain a reproducible and meaningful sample, in which the concentrations of several substances could be related to the pollution and health status of the tree. According to previous monitoring programs (e.g., [54,55]) and based on comprehensive knowledge about the distribution of different substances in the spruce tree (e.g. [47,56]) the one-year-old shoots of the seven year-old branches near the top of the tree were chosen for sampling (see Figure 2-3). Further standardization of the sampling protocol in order to improve reproducibility and comparability of the samples from year to year and between different areas requires, for example, the definition of exclusion criteria for individual trees (e.g., age >40 years, average health status), the sampling period (between snow thaw and beginning of sprouting in early spring), and so on. These measures bring about stratification effects on further levels which have to be considered in the evaluation of the results.

Analogous sampling designs have been developed for the German environmental specimen banking program for pine shoots *Pinus sylvestrius* and for leaves of beech trees (*Fagus sylvatica*) and Lombardy poplars (*Populus nigra 'italica'*) [26]. The actual sampling guidelines will be published soon [27].

2.4 Active or experimental monitoring with plants

Further standardization as well as the use of biological indicators or monitors for assessing places where they cannot be sampled by chance are possible by growing such plants and exposing them in the field under standardized conditions (active or experimental monitoring, see Fig. 2-1, e.g. [33,57,58]). The most frequently used plant species for active biomonitoring are grass (*Lolium multiflorum*) cultures for the assessment of environmental pollution by fluorine and lead, the Standardized Grass Culture Method [28,29]; the exposure of lichens (*Hypogymnie physodes*) [59,60]; and the use of kale for passive accumulation of polycyclic aromatic hydrocarbons (PAHs), dioxins, and furans [61,62].

2.5 Sampling strategies for soils for biological specimens

Soils are one of the most important media for environmental monitoring of biological specimens and risk assessment. Status and changes of pollutant concentrations and other soil parameters in the different soil horizons of a site are valuable indicators of environmental quality, change, and ecosystem integrity. The greatest problem connected with the use of soils is their high structural, morphological, and functional variability in both space and time [63-65]. It is, therefore, difficult and time-consuming to obtain useful answers from soil analysis for purposes of environmental monitoring of biological specimens. If soil samples are collected to obtain reasonable information about a certain area the following general requirements have to be observed:
- the samples must represent the soils of the area or ecosystem to be characterized and give information about the central tendency and variation of the desired parameters for the whole area or homogeneous subareas;
- the samples must not be contaminated or changed in any way during sampling, sample treatment, transport, and storage;
- the samples must refer to the different soil horizons rather than to sampling depth;
- the sampling method and sample characterization should enable the analytical results to refer not only to sample mass but also to a specific fraction such as fine soil (<2 mm), clay, or organic matter as well as to surface units (e.g./m²) or to volume units (e.g. mg/l).

There are numerous more or less sophisticated guidelines and standard procedures for collecting soil samples for various purposes; most meet these requirements only partly. The International Standards Organization is now working on a series of standards for soil quality, which will give guidance on sampling strategy, sampling techniques, preservation of samples, safety precautions, and procedures for the investigation of natural, near-natural and cultivated sites [66,67].

Even more than for plants and animal tissues, soil sampling needs in principle to prove the homogeneity of the sampling area prior to pooling individual samples. In the organic layers and the uppermost horizons, soil properties are usually spatially heterogeneous to such a degree that analytical results of a sample composed or pooled of several individual samples taken selectively, randomly, or systematically from an extended area do not make much sense if their homogeneity is not proven [64].

Of course the selection of a sampling strategy and sampling design has to follow the objectives of the investigation. For example, for testing the heavy metal contamination of agricultural areas according to the German sewage sludge decree [20], pooling of a fixed number of systematically taken samples is specified and the procedure has to follow exactly the legal instructions. For scientific or monitoring purposes it will generally make more sense to cover an area with a grid of adequate density (based on preliminary sampling or specific experience) and to analyze the samples separately. In many cases it is technically much easier to find and locate a potentially present spatial pattern using a systematic grid rather than random or selected sampling.

If an area has to be analyzed by very expensive methods, for example for the assessment of dioxin contamination levels, the number of samples is often restricted to a small quantity. In such cases attempts are often made to collect samples from small subareas or spots assumed to be "typical" or "representative". In any case it would be much better to sample the whole area systematically and to use less expensive detectable parameters like soil morphology, humus content, and quality, and the concentrations of related or correlated substances (in the case of dioxins, for example, hexachlorobenzene, HCB, which is a good indicator for the probable presence of dioxins [68,69]). It may sometimes even make scientific and economic sense to use suitable plants or animals as accumulating indicators to screen an area prior to soil sampling. The results of such screening can then be used to establish a spatial pattern or gradients of the contamination of the sampling area. This can be used as a basis to decide on selecting or selected pooling of soil samples, or to limit the area in which expensive analysis is necessary [34,52]. In any case, comparing the results of different areas will only make sense if, in addition to the mean or median values, the standard deviations or the limits of confidence for the individual areas are also known [22].

The method and equipment used for sampling often exert a profound influence on the quality of the samples and their possible use. As mentioned above, the sampling method and sample characterization should enable us to refer the analytical results to sample mass, to surface units, or to volume units. This can be done in the traditional manner by determining the soil density using separately taken, volume-related samples with soil sample rings of well defined volume. This is a useful method for deeper soil layers but hardly or sometimes not at all applicable for humus layers and topsoil. An alternative method for strictly area-related sampling has been developed for the German environmental specimen banking program. The same procedure is also used for the soil sample bank of the Saarland within the framework of the nation-wide long-term observation program for soils (BDF) in the Federal Republic of Germany [12]. The operational steps and tools are briefly explained below. The full version of the SOP [21,70] will soon be published [27].

(1) The sampling area is selected and delimited by making full use of existing maps of all kinds, aerial photographs, etc. Additional mapping in the field is performed by soil core boring, mapping of visible landmarks and vegetation, elimination of disturbed places. The soil type representative of the area is defined as a basis for delimitating an homogeneous sampling area of about 2500 m², if possible with uniform soil type, land use, vegetation, and history.

(2) Preliminary examinations to develop a site-specific sampling plan and protocol:
- pedological and analytical characterization of sites and samples;
- calculation of the number of individual samples necessary to reach the required sample mass;
- measures are necessary to guarantee long-term repeatability:
  -- installation of a soil profile pit as well as interpretation of soil cores and cleaning of borers has to be done outside the sampling area;
  -- avoiding any disturbing material deposition as well as land use for other purposes;
  -- long-term demarcation of the borders;
- establishing a regular sampling grid, which is successively systematically displaced for subsequent sampling.

2.5 *Sampling strategies for soils*   39

Figure 2-4:   Tools and procedure for sampling the organic layers of forest soils

Figure 2-5:   Tools and procedure for sampling topsoil in relation to area and volume

(3) Sampling on the basis of a regular sampling grid to guarantee spatially representative random sampling. Special tools and methods are used for surface and volume-related sampling:

Sampling of the organic layers using a steel frame (20x20 cm) by carefully removing the organic layers inside the frame with tweezers and spatula (see Figure 2-4). In the case of more than 3 cm of humic layer (O_h) in the mean, the different organic layers should be separated by subsequent removal of the litter and fermentation layer (L + O_f) and separate sampling of the humic layer in the same manner.

Sampling of top soil with a split-tube-sampler (diameter 5 cm, see humus layer; Figure 2-5). If the humus topsoil (A_h layer) is less than 10 cm thick, the different soil horizons in the top 10 cm of mineral soil should be sampled separately. The thickness of each horizon has to be measured and recorded at each sampling spot. The individual samples of organic layers and soil horizons are quantitatively collected and manually mixed in steel tubs.

For the special purpose of specimen banking, the process of drying and sieving the material is not adequate because of possible losses or changes of soil constituents. In this case the coarse soil fractions and plants are manually separated for separate analysis using steel tweezers. All the samples taken for specimen banking and analytical characterization are then deep-frozen on the spot, using liquid nitrogen; transportation, homogenization, subsampling, and long-term storage are carried out in the gas phase above liquid nitrogen. The samples are stored in 10 g portions, at temperatures below -140 °C, ready for analysis [71]. Soil texture and moisture content are subsequently detected using representative subsamples and taken into account in the calculation of substance concentrations. Trace and bulk elements, toxic metals, polycyclic aromatic hydrocarbons (PAH), and chlorinated hydrocarbons are analyzed routinely for sample characterization in the sense of analytical fingerprints. The banked samples are held ready for deferred examination for any chemical substance or property. For data processing and evaluation, documentation and quality assurance, an information system has been especially developed for environmental specimen banking.

2.6 Sampling strategies for animals in environmental monitoring

There are some important differences between plants and animals in relation to indicator function and availability which influence the sampling strategy in an important way.

2.6.1 Indicator function of animals

The mobility of many animal species is the most important fact in this context because mobile animals may be regarded as integrators of the spatial burdening of xenobiotic chemicals. This means that spatial representativeness has a different quality and a different importance compared with plants. Therefore, the sampling strategy for mobile animals should not necessarily be based upon spatial representativeness. It has to be emphasized that the efficiency of spatial integration depends not only on the surface of home range but also on the number and density of relations to the site over a period of time. In fact, there is a wide range of efficiency related to the integration of the spatial burdening.

For example, a mature earthworm with a small home range represents one limit of this range, which may under certain conditions be comparable with that of plants, whereas a predator like the red fox, older than two years, shows the other limit by integrating over a large area and over at least one year [72].

Furthermore, the ecological function of animals are a special feature. Whereas many (higher) plant species have more or less the same function of primary production, the diversity of functions of animal species is much higher. This means that animals are exposed to xenobiotic chemicals in a different way to plants and the differences of exposure within animals are normally higher than those within plant species [73].

A third aspect influencing the indicator function refers to toxicokinetics, which vary considerably between animals and plants. Differences in accumulation are a consequence of variations in uptake and clearance, and concern the parts or organs of the organisms where the pollutants are accumulated.

Overall, the type of information about environmental pollution stored in plants and animals is often very different. Animals indicate pollution of the environment "closer" to man, i.e., results from these are more relevant to human beings. First, suitable animal indicators show the hazards of xenobiotics especially for the final consumers as a result of the phenomena of bioaccumulation and, secondly, their reaction to the pollutants is more similarl to man's. In this context it is important to emphasize that the main goal of environmental monitoring is to show the hazards of xenobiotics to man. It should be stressed that plants and animals can never represent each other but they should always be used together in an environmental monitoring program. Therefore a set of suitable specimens should include plants and animals of different trophic levels in the context of their use as accumulation indicators [9,34,74,75].

2.6.2   Availability of animals

The availability (second level) refers to several aspects. First it concerns the number of individuals which may be collected at one site on one sampling date [4]. A large and stable population forms the basis of this aspect of availability. The final consumers cause the most problems in this context because the population density decreases naturally with the increase in the trophic level. The number of species that are suitable representatives of this ecological function is limited. Many populations of such species are reduced in number or destroyed by man or have to live in unsuitable habitats [76].

Secondly, availability means a sufficient quantity of sample for the chemical analysis that has to be carried out. This meaning of availability expresses itself in two ways. On the one hand, organs like liver and kidneys as target tissues are, e.g. for small fish species or small mammals, sometimes too small for analyzing a high number of different chemical compounds. On the other hand, individuals themselves may be too small for determination of any pollutant of interest. In addition, they are also too small to be determined with certainty in the field. A lot of reducers like bacteria, fungi or many "worms" (not in a taxonomic sense) are examples of this second meaning. This is the reason why only some species of the genus *Lumbricus* or *Aporrectodea* are suitable indicators at the trophic level of reducers.

Thirdly, availability refers to the sampling methods. Each method is selectively applicable so that only certain parts of an animal population may be sampled. This is a very important problem in the context of sampling strategy for animals in environmental monitoring because it concerns the representativeness of samples in relation to a population. How is it possible to obtain representative samples if it is not possible to evaluate the structure of a population statistically? This question will be discussed in Section 2.6.4. Apart from representativeness, the sampling methods should fulfil a further important requirement. They should guarantee that the specimen type must be readily and reliably available. This requires standardized sampling methods that influence the sampled population and the sampled individuals at a very low level. Every capturing method puts animals under stress, which is why a certain unwanted influence on the samples always exists. To ensure the quality of animal samples this kind of influence has to be minimized.

Fourthly, availability has a legal aspect. Collections must conform to applicable laws and ethical considerations. The final consumers are the most endangered species and very often protected by law, so that again they are hardly available.

2.6.3 Stratified random sampling of less mobile animals

Mature individuals of deep-burrowing anéctique earthworm species like *Lumbricus terrestria* or *Aporrectodea longa* (*Lumbricidae, Oligochaeta*), which have a tendency to K-selection in relation to their population growth, are very loyal to their habitat but do not really integrate the burden in a spatial dimension any differently from plant species. The sampling strategy has to take this fact into account if spatially representative specimens are to be sampled.

Furthermore, mature individuals of both species could be accepted as being almost randomly distributed in suitable habitats outside the breeding season [77]. This is important because it makes it possible to select the collection sites randomly but it should not be mistaken for existing habitat preferences. Mature individuals are not distributed randomly in all habitats in a given area. These facts should be the starting point for developing a suitable sampling strategy.

The first step is to look for a large-scale environmental pattern in the area to be sampled (Figure 2-6). An environmental pattern results from the combination of all parameters that are able to characterize an area, such as vegetation types, soil types, land use, or meso-climatical types. Afterwards, the area has to be broken up into relatively homogeneous subareas [22].

The next step is screening to study the distribution and abundance of the indicator species in the area [25]. For this purpose each subdivision is overlaid with a grid (Figure 2-6). Each unit of the grid has the same surface (size) depending on the capturing method [74], which has to be the same for the actual sampling of specimens. The units to be sampled must be selected randomly. The size of sample in each subdivision has to be proportional to its part of the surface of the area and must be high. The result will in most cases be that the species is not distributed and not abundant in all subareas of the environmental pattern (Figure 2-6). If the indicator species occurs abundantly in one dominant subdivision it would be able to characterize the burdening of this part of the area. The use of the species in subdivisions where it is not well distributed and abundant is not relevant because there will be no basis for representative and standardized sampling. The right method is to use other indicator species for characterizing the whole area. This is an additional argument for a set of species.

2.6 *Sampling strategies for animals*   43

First step:

Breaking down the area to be sampled in relatively homogenous subdivisons at the base of the environmental pattern

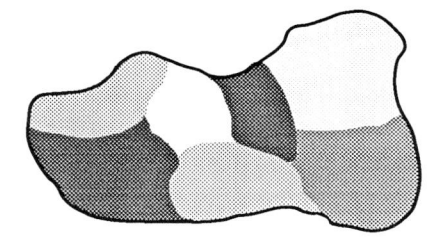

Second step:

Screening on the base of stratified random sampling to study the availibility of the bioindicator in the area

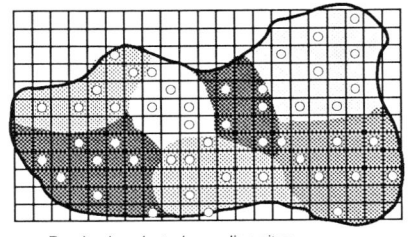

○ = Randomly selected sampling sites

Third step:

Selection of suitable stratum according to the distribution and abundance of the bioindicator

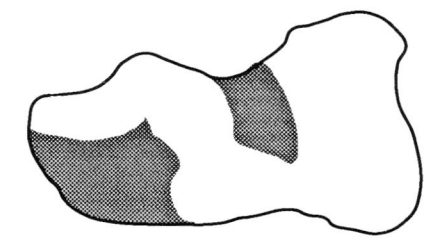

Fourth step:

Random sampling in the selected stratum

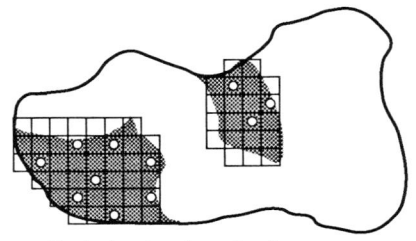

○ = Randomly selected sampling sites

Figure 2-6: Scheme of the basic sampling strategy for earthworms (see text)

Screening is the basis for obtaining representative samples with the actual sampling. The procedure of the actual sampling corresponds in principle to the screening but the spatial reference basis is different. The suitable subdivision(s) of the environmental pattern is (are) overlaid by a grid and units to be sampled are selected randomly as mentioned above (Figure 2-6). The number of selected units depends on the surface and homogeneity of the subdivision(s), on the one hand, and on the size of each sample and the time spent on sampling, on the other hand.

This procedure is not proper random sampling nor proper stratified random sampling related to the whole area, but it may be regarded as a combination of both. The area is divided into strata, at least one stratum is selected according to the distribution and abundance of the target species and random sampling is carried out within the selected stratum. Therefore, this selection could be regarded as subjective but it is the only possibility of obtaining samples which provide correct information on the burden of xenobiotic chemicals (see above). In relation to the subdivision, this sampling strategy is real random sampling and it is, therefore, suitable for pooling all samples from one subdivision if the quantity and quality of samples is equal. In the case of *Lumbricus terrestris* or *Aporrectodea longa*, the limitation to matured individuals and to one sampling date in autumn ensure the comparable quality of samples.

Potential modification of sampling strategy depends on the criteria for defining subdivisions and/or on the number of suitable subdivisions according to the distribution and abundance of the indicator species. Only two of the possibilities are to be mentioned here.

If the spatial distribution of pollution within an area is interesting, the definition of the environmental pattern should include the prior knowledge or assumptions concerning this spatial burdening. For example, if a plant emitting pollutants is dominant in or nearby the area to be sampled, the predominant wind direction is one important criterion for the definition of subdivisions of the environmental pattern. This point is of minor interest or rather should not be a criterion if the burdening of the area is studied in comparison to others.

If there are a lot of suitable subdivisions in an area for one species, the sampling strategy can be modified in such a way that a random sampling is realized within each spatial stratum. If this is done, the differences between the subdivisions become clearer and the samples may be pooled under the above-mentioned requirements with regard to quantity and quality of samples. The advantage of this elevated sampling in relation to sample size is the reduction in the number of chemical analyses and, ultimately in the cost.

2.6.4 Stratified random sampling of mobile animals

Roe deer (*Capreolus capreolus*) and fish species like bream (*Abramis brama*) and roach (*Rutilus tutilus*) are suitable accumulation indicators [78-81]. Because of their mobility they are able to integrate the pollutants over a certain area. This is why spatial stratification does not have to be carried out and the sampling itself may be regarded as random in relation to space and individuals. Nevertheless, this sampling strategy must be based upon a necessary standardization at the population level to ensure the quality of samples.

The big problem with the representation of population is that the statistical reference basis is not really known. The reasons for this are mentioned in Section 2.6.1 (see also [23]). The conclusion is very simple: representativeness of samples referring to population could not be realized from a statistical point of view. The consequence could be either to realize it as well as possible according to the sampling techniques used or to perform sampling without it. The decision is based on the question of whether it is necessary to represent a population in the context of biological indication.

What we want to know concerns the information about the environmental pollution stored in the organisms, i.e., the ability to indicate. The variation of the contamination concentration in a population is not the purpose here. Therefore, the right question is: how is it possible to obtain representative samples in relation to their ability to indicate the environmental pollution correctly? To avoid misunderstanding, the knowledge of accumulation behavior within a population is very important for studying the suitability of an animal species as indicator and constitutes the basis for its use in monitoring programmes.

So far, the decision should be to perform sampling without the need to represent a population. If we do not have to represent a population, and are not able to, the possibility exists to standardize the samples. Reproducibility and representativeness join at this point. This means that standardization does not reduce the representativeness of samples related to the information content of organisms. The principle of stratification is suitable for obtaining the "best" stratum of the population. There are several criteria for stratifying a population, such as age, sex, weight, length, and genotype [4]. Age is the most important criterion for selecting individuals of freshwater fish species and roe deer because accumulation is very often age-dependent [81-84].

A suitable stratum in roe deer population is relatively quickly found using age as a stratifying criterion. The age can only be determined with certainty if the individuals are not older than one year. Reliable age determination is very important in the context of quality assurance. The selection of suitable specimens within a group of individuals not older than one year is based upon consideration of the advantage and disadvantage of fawns and exactly one-year-old individuals. Fawns are only suitable up to an age of about seven to eight months because younger fawns are suckled. The consequence is that lipophilic chemicals are overrepresented, whereas metals like cadmium are underrepresented in suckled fawns because they do not pass the placenta [81] and are normally accumulated during a long period without reaching a saturation concentration. The problem with older fawns is that their availability is not equally satisfactory in all hunting areas because in some areas fawns are already hunted in September and October. One-year-old individuals are mostly hunted between mid-May and the end of June, directly after the opening of the hunting season. Sampling one or two months earlier would be desirable because until that time the young males (not only the females) are also loyal to their habitat, where they have lived since their birth. Later in the year, the probability is higher that they are driven away by older males. The above mentioned fish species cause more problems when a suitable age class has to be selected. This has several reasons. First, age cannot be determined with certainty in the field during sampling. It causes problems even in the laboratory [85]. Weight and length are the indicators of age during sampling, but they are specific for each water body. Therefore, screening to study the relation between age and weight and age and length has to be carried out if prior information is not available. Generally, eight to twelve year-old individuals of the bream and four- to seven-year-old individuals of the roach seem to be suitable because of their availability in different water bodies and the fact that age-dependent accumulation is not recorded within these groups.

Random sampling may be carried out within the selected age classes and is generally ensured by the sampling techniques used [86,87]. In conclusion, the sampling strategy for mobile animals like freshwater fish and roe deer may, because one stratum is selected subjectively, be regarded as a modification of stratified random sampling. It has to be emphasized that the selected stratum is assumed to represent the information content of the indicator species correctly.

2.6.5 Zebra mussels as examples of sampling under semi-artificial conditions

In contrast to the previous sections in which sampling strategy is strictly based upon sampling of specimens under natural conditions, this strategy is founded on the manipulation of specimens in such a way that the samples are more readily available with the disadvantage that the specimens are living on an artificial substratum. The sessile zebra mussel (*Dreissena polymorpha*), which has been well studied as an accumulation indicator [76,88-91], is an example of sampling under semi-artificial conditions.

The manipulation is realized by using an artificial substrate for colonization with mussels. Uncolonized plates are exposed in order to sample specimens originating from the water body to be studied. The technique of sampling is described in Figure 2.7. The plates are made from additive-free polyethylene (PE) and have a surface size of 900 cm². Seven to twelve plates are screwed together at a distance of 7 cm each to form a so-called plate tower or rack. The spatial steel screw connections are concealed by spacers made from polytetrafluorethylene (PTFE). These plate racks can be installed in the water either individually or in a group of several racks connected to each other. It is important that all the materials used have no influence on the concentrations of pollutants in the mussels.

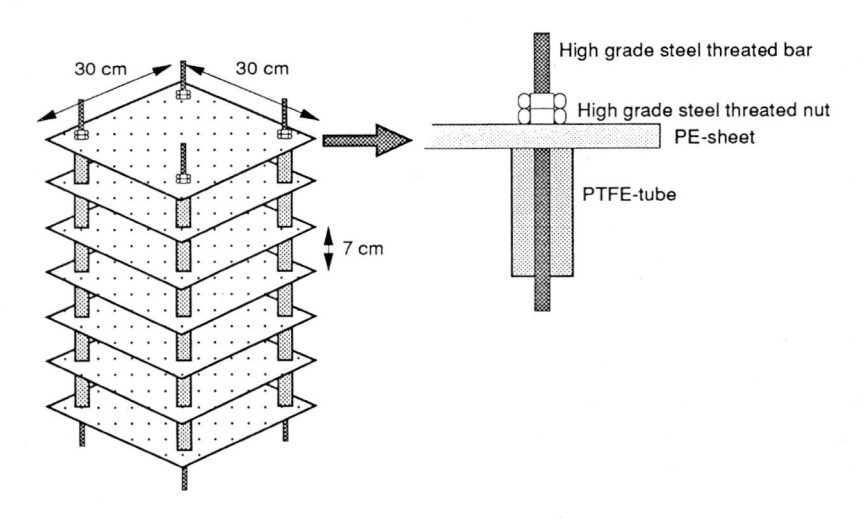

Figure 2-7: Exponates for colonization by Zebra mussels (see text)

The advantage of this method is that individuals of zebra mussels from one specific suitable age or length class are more readily available with regard to a sufficient number of individuals, easy handling during collection of the specimens, and use of zebra mussels as indicators even in water bodies where they do not occur naturally. This is realized by colonizing the plates with zebra mussels in a water body where they are abundant and transferring the colonized plates to another water body to be monitored where mussels are absent.

It has to be stressed that this method is only semi-artificial because the mussels become acclimatized to their new surroundings. This does not change even if the mussel on the plates are replaced. The difference from natural conditions is minimized by exposing the plates as near to the bed of the water bodies as possible. It is well known that the proximity to sediment is an important factor in the context of accumulation [91].

Similar techniques for sampling under semi-artificial conditions are used for other species, e.g. feral pigeons [31] and honey bees [6,32].

Acknowledgements

We want to thank our colleagues of the Environmental Specimen Banking programme (UPB) in Saarbrücken, Jülich, Münster, Neuherberg, Grosshansdorf and Berlin for valuable help and discussion, the Ministry for Environment and the Environmental Protection Agency for financial support, and Mrs. J. Wheeler for language correction.

2.7 References

[1] H. Ellenberg, *Ökol. Vögel Sonderheft*, **3**, 83 (1981)

[2] W.H.O. Ernst, *In: Sampling of environmental materials for trace analysis*, B. Markert (Ed.), VCH-Publisher, Weinheim, New York, Tokyo, 381-394 (1994)

[3] L.H. Leith (Ed.), *Principles of Environmental Sampling*, Amer. Chem. Soc., ACS Prof. Ref. Book (1987)

[4] R.A. Lewis, *Guidelines for Environmental Specimen Banking with Special Reference to the Federal Republic of Germany: Ecological and Managerial Aspects*, U.S. Department of the Interior, National Park Service, U.S. MAB Report No. 12.

[5] B. Markert (Ed.), *Plants as biomonitors - Indicators for Heavy Metals in Terrestrial Environment*, VCH-Weinheim, 285-328 (1993)

[6] P. Müller and G. Wagner, in:*BMFT, Bundesministerium für Forshung und Technologie*, Springer-Verlag, Berlin, Heidelberg 158 pp (1988)

[7] M. Stoeppler, *Probenahme und Aufschluß - Basis der Spurenanalytik*, Springer-Verlag, 1994

[8] G. Wagner, *Sci. Total Environ.*, **139/140**, 213 (1993)

[9] G. Wagner, *Ecoinforma*, **4**, 71 (1993)

[10] J.E. Elliott, in: *Environmental specimen banking and monitoring as related to banking*, R.A. Lewis, N. Stein and C.W. Lewis (Eds.), Martinus Nijhoff, The Hague (1984)

[11] P. Müller, *Zeit. Ang. Umweltforsch.*, 209 (1988)

[12] SAG, Sonderarbeitsgruppe, *Informationsgrundlagen Bodenschutz*, Boden-Dauerbeobachtungsflächen (BDF), Arbeitshefte Bodenschutz 1, Bayer. Staatsministerium für Landesentw. und Umweltfragen, München (1991)

[13] H. Ellenberg, O. Fränzle and P. Müller, *Ökosystemforschung im Hinblick auf Umweltpolitik und Entwicklungsplanung*, BMI Bonn (1978)
[14] UBA (Umweltbundesamt), *Immissionsmeβnetze in der Bundesrepublik Deutschland*, Monatsberichte aus dem Meβnetz 10/87, Berlin, 31-47 (1988)
[15] M. Zimmer, *Thesis*, Universität des Saarlandes, Saarbrücken (1993)
[16] W.H.O. Ernst and S. Leloup, *Chemosphere*, **16**, 233 (1987)
[17] H. Bau, *VDI-Berichte*, **901**, 37 (1992)
[18] A. Rühling and G. Tyler, *Bot. Notizer*, **122**, 248 (1969)
[19] J.M. Hodgeson, *Soil Survey Field Handbook: Describing and Sampling Soil Profiles*, Batholomew Press, Dorking, Harpenden (1985)
[20] R.W. Scholz, N. Nothbaum and T.W. May, *Ecoinforma*, **4**, 229 (1993)
[21] G. Wagner and J. Sprengart, in: *Soil Responses to Climate Change*, M.D.A. Rounsevell and P.J. Loveland (Eds.), NATO ASI Series I, vol.23, 247-254 (1994)
[22] R.H. Green, *Sampling design and Statistical Methods for Environmental Biologists*, John Wiley & Sons. New York, 257 (1979)
[23] G. Wagner, *Sci. Total Environ.*, submitted
[24] P. Fischer, *Unpublished Thesis*, University of Saarbrücken, Dept. of Biogeography (1991)
[25] M. Paulus, M. Altmeyer, R. Klein, A. Hildebrandt, K. Oxynos and M. Rossbach, *UWSF-Z. Umweltchem. Ökotox*, **6**(5) (1995)
[26] UBA (Umweltbundesamt/Federal Environmental Agency Ed., *Guide for Environmental Specimen Banking of the Federal Republic of Germany*, US/German Seminar on "Environmental Specimen Banking", Ottawa (1989)
[27] BMU/UBA (Bundesminiterium für Umwelt und Reaktorsicherheit/ Umweltbundesamt Ed., *Umweltprobenbank des Bundes (Verfahrensrichtlinien)*, Ecomed.-Verlag, 1994 (in press)
[28] G. Scholl, *Landwirtsch. Forsch.*, **26**, 29 (1971)
[29] VDI (Verein Deutscher Ingenieure), *VDI-Handbuch Reinhaltung der Luft*, Düsseldorf (1978)
[30] A. Winter, P. Müller and G. Wagner, *VDI-Berichte*, **901**, 729 (1992)
[31] M. Altmeyer, *Dissertation*, Universität des Saarlandes, Saarbrücken (1993)
[32] J.J. Bromenshenk, in: *The Bioenvironmental Impact of a Coal-fired Power Plant*, E.M. Preston and R.A. Lewis (Eds.), USEPA, 600/3-78-02 (1978)
[33] U. Arndt, W. Nobel and B. Schweizer, *Bioindikatoren, Möglichkeiten, Grenzen und neue Erkenntnisse*, Ulmer, Stuttgart (1987)
[34] G. Wagner, *Ecoinforma*, **4**, 23 (1993)
[35] G. Wagner, *In: Probenehame und Aufschuβ in der Spurenanalyse*, M. Stoeppler Ed., Springer-Verlag, Berlin/Heidelberg/New York, 77-83 (1994)
[36] G. Wagner, *In: Beiträge zur Umweltprobenbank*, M. Stoeppler and H.W. Dürbeck Eds., Jül. Spez., 412, Jülich (1987)
[37] W. Knabe, *In: Monitoring of Air Pollutants by Plants*, L. Steubing and H.-J. Jäger (Eds.), Junk, The Hague, 59-72 (1982)
[38] W.H.O. Ernst, *Swermetallvegetation der Erde*, Fischer Verlag, Stuttgart (1974)
[39] S. Bråkenhjelm, *The National Swedish Environmental Monitoring Programme (PMK)*, Swedish Environmental Protection Agency, Uppsala (1990)

2.7 References

[40] R.A. Lewis, M. Paulus, C. Horras and B. Klein, *MAB-Mitteilungen*, **29**, Bonn (1989)
[41] M. Paulus, B. Horrals, B. Klein and R.A. Lewis, *Final Report UBA 90/108 08001*, (1990)
[42] Northern Council of Ministers, *Guidelines for Integrated Monitoring in the Nordic Countries*, NORD 1988:26 (1988)
[43] R.A. Lewis and B. Klein, *Toxicol. Environ. Chem.*, **27**, 251 (1990)
[44] T.I. Lillestolen, N. Foster and S.A. Wise, *Sci. Total Environ.*, **139/140**, 97 (1993)
[45] V. Krivan and G. Schaldach, *Fresenius' Zeit. Anal. Chem.*, **324**, 158 (1986)
[46] H. Lieth and B. Markert Eds., *Element Concentration Cadasters in Ecosystems. Methods of Assessment and Evaluation*, VCH Weinheim (1990)
[47] A. Wyttenbach, L. Tobler and S. Bajo, *Trees*, 204 (1988)
[48] R.-D. Zimmermann, L. Tobler and S. Bajo, *Trees*, **1988**, 204 (1988)
[49] E. Steinnes, *In: NATO Air pollution Pilot Study, Assessment Methodology and Modelling*, Suppl. to the 3rd follow-up report no. 134, 1-8, (1975-1979)
[50] E. Englund and A. Sparks, *GEO-EAS (Geostatistical Environmental Assessment Software), Programs and User's Guide*, Environmental Monitoring Systems Laboratory, U.S. Environmental Protection Agency, Las Vegas, N.V. (1988)
[51] G. Scherelis and W.D. Blümel, *Karlsruher Manuskripte zur Mathematischen und Theoretischen Wirtschafts- und Sozialgeographie*, Heft 92, Universität Karlsruhe (1988)
[52] G. Wagner, *In: Plants as Biomonitors for Heavy Metal Pollution of Terrestrial Environment*, B. Markert (Ed.), VCH-Publishers, Weinheim/New York, 425 (1993)
[53] G. Wagner, M. Altmeyer, R. Klein, M. Paulus and J. Prengart, *In: Bundesministerium für Umwelt, Naturschutz und Reaktorsicherheit*, Referat N II 2 - 71022/1, Bonn (1993)
[54] W. Knabe, *Forschung und Beratung*, Reihe C37, Landwirtschaftsverlag Münster (1983)
[55] G. Wagner, U. Bagschick, M. Burow, C. Mohl, P. Ostapczuk and M. Stoeppler, *In: Proc. of Int. Conf. Heavy Metals in the Environment*, Athens, 515-518 (1985)
[56] R.-D. Zimmermann, *AFZ 11*, 4-7 (1989)
[57] P. Müller, *Biogeograpie und Raumbewertung*, Wissenschaftliche Buchgesellschaft Darmstadt (1977)
[58] G. Wagner and J. Krüger, *Das Gartenamt*, **31**, 516 (1982)
[59] VDI (Verein Deutscher Ingenieure), *VDI-Handbuch Reinhaltung der Luft*, VDI 3799 Bl.2, Düsseldorf (1991)
[60] W. Erhardt, *VDI-Bericht*, **609**, 701 (1987)
[61] G. Grimmer and A. Hildebrand, *Dt. Lebensmittel-Rundschau*, **8**, 237 (1965)
[62] B. Prinz, G.H.M. Krause and L. Radermacher, *Staub - Reinhalt. Luft*, **50**, 377 (1990)
[63] T.M. Bourgess, R. Webster and A.B. McBratney, *J. Soil Sci.*, **32**, 643 (1981)
[64] G. Miehlich, *Z. Planzenern. Bodenk.*, **139**, 597 (1976)
[65] G. Scherelis, *Stuttgarter Geographische Studien*, Bd. 112, Universität Stuttgart (1988)
[66] A. Paetz and B. Croessmann, in: *Environmental Sampling for Trace Analysis*, B. Markert (Ed.), VCH, Weinheim, New York, Tokyo, 321-334 (1994)
[67] ISO/TC 190, *Soil Quality - Sampling*, ISO/CD 10381 (1993)
[68] E. Heinisch, *Umweltbelastung in Ostdeutschland*, Fallbeispiele: Chlorierte Kohlenwasserwtoffe. Wiss. Buchgesellschaft Darmstadt (1992)
[69] M. Bahadir, *Chemie in unserer Zeit*, **25**(5), 239 (1991)

[70] J. Sprengart and G. Wagner, *In: Bundesministeriums für Umwelt*, Naturschutz und Reaktorensicherheit, Referat N II 2 - 71022/1, Bonn (1993)
[71] J.D. Schladot and F. Backhaus, *In: Progress in Environmental Specimen Banking*, S.A. Wise, R. Zeisler and G.M. Goldstein Eds., NBS Special Publication 740, 184-194 (1988)
[72] P. Müller and G. Wagner, *Report T 86-040*, Bundesministerium für Foschung und Technologie, Bonn (1986)
[73] R. Klein and M. Paulus, *Ecoinforma*, **4**, 49 (1992)
[74] R. Klein, *Sci. Total Environ.*, **139/140**, 203 (1993)
[75] M. Paulus and R. Klein, *In: Bioindikation in Aquatischen Ökosystemen*, G. Gunkel Ed., Fischer-Verlag Jena (in press)
[76] R. Klein and G. Wagner, *MAB-Mitteilungen*, **37**, 117 (1993)
[77] C.A. Edwards and J.R. Lofty, *Biology of Earthworms*, Chapman and Hall, London (1977)
[78] M. Edgren, M. Olsson and L. Reutergardh, *Chemosphere*, **10**(5), 447 (1981)
[79] H. Hecht, *VDI-Berichte*, **901**, 937 (1991)
[80] K.-E. Krüger and R. Kruse, *F+E-Report*, Umweltbundesamt Berlin, Cuxhaven (1984)
[81] F. Tataruch, *VDI-Berichte*, **901**, 925 (1991)
[82] A.J. Niimi, *In: Aquatic Toxicology*, J.O. Nriagu Ed., John Wiley and Sons, New York, Chichester, Brisbane, Toronto, Singapore, 207 (1983)
[83] J. Disser, H. Brunn, A. Nagel and R. Prinzinger, *Ökol. Vögel*, **14**, 173 (1992)
[84] A. Bignert, A. Göthberg, S. Jensen, K. Litsén, T. Odsjö, M. Olsson and L. Reutergardh, *Sci. Total Environ.*, **128**, 121 (1993)
[85] T.B. Bagenal Ed., *Ageing of fish*, Surrey: Unwin Brothers Limited (1974)
[86] R. Klein and K. Netwich, *In: BMU (Bundesministeriums für Umwelt, Naturschutz und Reaktorsicherheit)*, Referat N II 2 - 71022/1, Bonn (1993)
[87] R. Klein and M. Paulus, in: *BMU Referat N II 2 - 71022/1*, Bonn (1993)
[88] D. Busch, *Dissertation*, Universität Bremen (1991)
[89] R. Klein and M. Altmeyer, *In: The Zebra Mussel Dreissena polymorpha*, D. Neumann and H.A. Jenner Eds., Jena: Gustav Fischer Verlag,, Limnologie Aktuell Vol. 4, 255 (1992)
[90] T.F. Nalepa and D.W. Schloesser Eds., *Zebra Mussel: Biology Impacts, and Control*, Lewis Publishers, Boca Raton, Ann. Arbor (1993)
[91] W. De Kock and C.T. Bowmer, *In: Zebra Mussel: Biology, Impacts, and Control*, T.F. Nalepa and D.W. Schloesser Eds., Boca Raton, Ann Arbor, London, Tokyo, Lewis Publishers, 503 (1993)

3.

Quality assurance and quality control of surface water sampling

G.J. Stroomberg[1], I.L. Freriks[1], F. Smedes[2] and W.P. Cofino[1]

[1] Vrije Universiteit, Institute for Environmental Studies, De Boelelaan 1115, 1081 HV, Amsterdam, The Netherlands
[2] Ministry of Transport, Public Works and Water Management, National Institute for Coastal and Marine Management (RIKZ), PO Box 207, 9750 AE Haren, The Netherlands

    River basins and coastal zones generally constitute areas with a high population density. In the course of time, settlements were established among rivers and in coastal zones owing to the favorable living conditions, such as the availability of water for irrigation, industrial and drinking purposes, of fishery products, fertile land, and efficient means of transportation. The increase in population density and economic activities was accompagnied by an increasing emission of a range of substances into aquatic systems, resulting in the contamination thereof. The degree of contamination may impair or, in a number of cases, has impaired functions of river basins and/or coastal areas (e.g. supply of fishery products, drinking water, ecological functioning). The condition of aquatic systems is important environmentally, the economy, and to social well-being. It is the task of present-day water management to arrive at an ecologically rational use of aquatic systems and to conserve water resources in a demanding economic and social context.

    Reliable information on the condition of aquatic systems is a prerequisite for effective and efficient water management. Such information is frequently obtained in surveys or in monitoring programs. The quality of the design, implementation and evaluation of these measurement programs needs to be managed. Much effort has been devoted to quality assurance and quality control of laboratory analysis [e.g. 1-4]. The actual laboratory analysis, however, is preceded by procedures such as sampling, transport, and storage which, as Youden [5] and other authors [6-8] have pointed out, are equally or even more important factors where the reliability of data is concerned.

The impact of sampling on the quality of environmental data has become apparent with the increase in the sensitivity and accuracy of analytical techniques at determining contaminant concentrations in environmental samples. For example, trace metal levels in the oceans have appeared to decrease over the past few decades as a result of improved sampling methods [9]. In Lake Erie and Lake Ontario, surface water samples were taken for trace metal investigations with the contamination control methodology commonly applied in marine research. The concentrations found were much lower than those reported previously, giving rise to the statement that "doubt was cast on most of the published data on trace metal concentrations in the Great Lakes" [10]. This finding demonstrates that in fresh water environments also the level of contamination control should be assessed critically and may be more demanding than often assumed.

Information on aquatic systems is obtained as the result of a chain of activities, the strength of this chain being determined by its weakest link. These activities encompass the assessment of the information needed, the design of the measurement program, planning, sampling and analysis, evaluation of data, and presentation of information. Quality assurance of sampling has to be seen in this context. In the broadest sense, quality assurance and control of sampling should include sampling strategy, planning of the sampling campaign, the collection of the samples, sample handling, preservation, transport, and storage. In this paper, quality assurance and control of the sampling of surface waters is approached from this broad perspective, paying attention to all the items indicated. The scope of this paper is limited to the sampling of surface water since this in itself requires considerable attention to detail. Many of the practices and principles discussed apply, however, to other types of water also, e.g. ground water or precipitation.

3.1 Quality systems for sampling

Quality management in laboratories is accomplished by the introduction of quality systems based on such international standards as EN45001 and/or ISO25. These standards can quite readily be adapted for sampling as indicated briefly in the following. Adapting these standards has the advantage that quality assurance and control of sampling is designed and implemented in a systematic manner.

A quality system for sampling requires that the organization of sampling is well delineated, including clear allocation of responsibilities. A quality manager and technical manager for sampling should be appointed. A procedure for authorization of the release of samples should be established. The quality system should be described in a quality manual, which documents all important sampling functions, procedures, and processes. The quality system should regularly be subjected to quality audits and reviews, i.e. ascertaining whether requirements are being complied with (auditing) and examining the requirements to ensure that the quality system meets the quality objectives (reviewing).

Staff involved in sampling should have specified, appropriate qualifications, training, and skills. Records of training should be kept, persons authorized to take (specific) samples should be identified. Requirements for sampling equipment ought to be defined, the equipment used should be adequate for the tasks as demonstrated by tests on the equipment itself or on the entire sampling procedures. Equipment records should be kept up. Traceability and calibration should be pursued to the greatest extent possible. Methodology for sampling (and in situ field measurements) should be well validated and documented. The facilities used for sampling should enable contamination-controlled sampling and sample work-up and constitute a good, "comfortable" working environment for staff (e.g. provide shelter against wind). An effective documented system for the handling of samples should be present. A systematic record of information of practical relevance to the sampling should be maintained. All information relevant to the validity of the samples should be noted and passed on to the laboratory/client. Procedures should be present for the handling of complaints and anomalies.

EN45001 and ISO25 deal specifically with routine tests and not with research. The design of a sampling program can be regarded as a research activity. To ensure the quality thereof, the approach taken should follow Good Laboratory Practice guidelines, ISO9001, or the Dutch standard NEN3417 which has been established for quality systems in research and development. A research manager supervising the design of the sampling strategy and the implementation thereof needs to be appointed. The qualifications needed by such a manager should be elaborated and documented. A research plan should be prepared and maintained to establish the sampling strategy. The design process should be audited and reviewed. The sampling strategy should be well documented.

In many cases, a laboratory may choose to integrate the quality systems for sampling and analyses.

3.2   Development of a sampling strategy

A sampling strategy is "a predetermined procedure for the selection, withdrawal, preservation, transportation, and preparation of the portions to be removed from a population as samples" [11]. Before designing a sampling strategy, the information needed has to be assessed, as discussed in section 3.2.1. On the basis of a qualitative and quantitative elaboration of the information needed, the sampling strategy can be established as outlined in section 3.2.2. It is apparent that the design of a well founded sampling strategy requires input from different disciplines, e.g. chemistry, hydrology, biology, and statistics. The establishment of a multidisciplinary team is recommended.

3.2.1   Analysis of the information needed

Prior to the development of the sampling strategy it should be clear for which purpose the information is needed and which qualitative and quantitative requirements should be met. Apart from the identity of the substances to be determined, quantitative pronouncements should be made, e.g. the level of confidence required regarding the analyte's identity, the magnitude of type I and type II errors when concentrations in a water body need to be compared to those in a standard, or the minimum temporal trend which should be detected ($x\%$ difference in y years with a probability of $z\%$).

In practice several aims have been identified in the design of sampling programs [12]. The most common are:

Estimating mean concentrations: one of the most basic questions to be asked in environmental field work entails the assessment of the mean concentration of the contaminant in a certain area. The number of samples to be taken in order to describe the mean concentration accurately depends on the variability of the concentrations of the contaminants.

Detecting trends: information on temporal trends and/or spatial distributions may be required for several reasons, e.g. in order to evaluate whether measures taken to reduce contamination have been effective. Superimposed on trends are fluctuations due to natural and/or anthropogenic causes. Examples of natural variations include seasonal cycles of biological activity and, therefore, of nutrient concentrations and seasonal effects in concentrations of contaminants in water related to differences in water discharges. A clear example of anthropogenic causes is provided by concentrations of modern pesticides which rise during the spraying season and decrease thereafter.

Monitoring for compliance: monitoring may be required in order to assess whether environmental quality objectives are met. Examples in Europe include EU-Council Directives 76/160/EEC regarding monitoring the quality of bathing water and 78/659/EEC dealing with fresh water for fish.

Determination of fluxes: rivers play a major role in assimilating or carrying off industrial and municipal waste waters and agricultural run off. The determination of fluxes requires attuned monitoring of chemical and physical parameters.

Early warning monitoring: concentrations of chemicals may be monitored with high frequency in order to safeguard inlets of drinking water production stations from contamination.

Frequently, the assessment of these issues requires discussions between the person responsible for designing the program and the laboratory which performs the work. A layman may be responsible for the design, as discussed by Keith: "It cannot be assumed that the person requesting an analysis will also be able to define the objectives of the analysis properly" [13]. Agreement ought to be reached on what can be expected from the results. The problems intended to be solved should be defined accurately [14]. No uncertainties about the aims of the sampling program should exist once it has been started. Even related questions which are not expected to be answered by the sampling program can be formulated in order to define the boundaries of the usefulness of the data [14].

3.2.2   Design of a sampling strategy

Each type of information need requires a specific sampling strategy [15]. Important aspects of the design of sampling strategy will be discussed consecutively.

3.2.2.1 Representativeness of samples

In the context of sampling, the term "representativeness" is commonly used. Its definition, however, is usually rather vague. Gy provides a theoretical framework for defining representativeness [16]. Gy defines the relative sampling error as the difference between the true value of the sample and that of the batch from which the sample is taken. The sampling error is a random variable defined by the moments: the expected value $m(SE)$, the variance $s^2(SE)$ and the mean square $r^2(SE)$, where $r^2(SE) = m^2(SE) + s^2(SE)$.

In this theory, sampling is accurate when $m(SE)$ is smaller than a certain standard, e.g. the maximum acceptable bias m_0. Sampling is reproducible when $s^2(SE)$ is smaller than a certain standard s_0^2. Sampling is representative when $r^2(SE)$ is both accurate and reproducible, i.e. smaller than a certain standard r_0^2 which may be defined as $r_0^2 = m_0^2 + s_0^2$.

It appears that for sampling of surface waters, SE may be decomposed into three parts:
- spatial error: how well does the sample reflect the spatial characteristics;
- temporal error: how well does the sample reflect the temporal characteristics; and
- sampling procedure: how well does the sample composition reflect the composition of the batch (water mass) it has been taken from at the time and location at hand?

Each part adds to the sampling error in respect of accuracy and variance. The total error is the sum of the sampling error and analytical error. It is clear that inferences from samples to the water body from which the sample has been taken need insight in the magnitude of this total error in order to arrive at meaningful results. In addition, in the design phase the temporal and spatial scales and the limits for the maximal acceptable bias and variance should be established.

3.2.2.2 The sampling site

The first step to undertake is to obtain information on the site(s) to be investigated. This information may include data on the location and magnitude of specific inputs (e.g. rivers, waste waters), specific hydrological features (water discharges, seasonal effects, salinity gradients and turbidity maxima in estuaries,..) and data from previous campaigns. Sometimes such information is not available or is incomplete, in which case it may be useful to conduct a preliminary study in order to establish a sampling program.

A feature to take into account is the (possible) lack of homogeneity of water bodies. When two side-branches of a river meet it can take up to several kilometers before the two joining water masses have completely mixed. A similar phenomenon can occur for the inputs of point sources into rivers. In lakes and seas, the often overlooked effect of patchiness may occur. Depending on the hydrology and the type of water body, a degree of stratification may exist. A relatively highly contaminated surface microlayer may be present at the top of the water column.

The depth of a body of water constitutes another important factor. Resuspension of sediments and thorough mixing of water masses may take place for shallow lakes through the action of strong winds. Fine suspended material may be deposited in the deeper off-shore regions of a lake, whereas coarse sediment may be deposited to a larger degree around the periphery [12]. These effects diminish with increasing depth. When sampling such shallow waters, preceding and ambient weather conditions have to be taken into account and noted.

In estuaries salinity gradients affect the concentrations of contaminants in the water column. The salinity acts as an indicator of the degree of mixing of relatively clean sea water with relatively contaminated fresh water. In addition to dilution of contaminants in the fresh water, chemical, biological, or physical processes may occur which lead to relative enrichment or depletion of contaminants in the water column. Sampling along salinity gradients is a common practice as the profiles of concentrations versus salinity provide information on the processes affecting water-column concentrations, enabling better interpretation of the data [17-19].

3.2.2.3 Determinands

A factor to consider entails the level of detail required. For some objectives, a group parameter (e.g. extractable organic halides) may suffice. This may be the case in early warning monitoring.

The societal use of the analyte is a factor to regard. Pesticides, for instance, are usually applied during the growth of crops, leading to concurrent rises in the concentration levels in surrounding water bodies. This effect will have a seasonal cycle which will call for differing sample densities as the seasons proceed [20].

For some information needs, for instance temporal trend monitoring, sampling of water is not mandatory. In such cases, investigations in environmental compartments such as suspended solids, biota, or sediments may be considered [21]. A range of contaminants is accumulated in these compartments, resulting in concentrations which are higher than in water. In addition, some time-integrating effect takes place reducing short-term temporal variability.

3.2.2.4 Methodology

The information needed and the physico-chemical properties of the analyte must be considered in order to select appropriate techniques for sampling, preservation, and analysis.

Important issues to address include [13]:
- the level of confidence regarding the analyte's identity;
- the analyte levels to be measured, both qualitatively and quantitatively [22];
- the required performance characteristics;
- the degree of confidence needed;
- the degree of method validation necessary; and
- the degree of quality assurance needed.

In the above, the entire set of methodologies for sampling, preservation, storage, and analysis, the materials to be used and the facilities which are required in the field should be taken into account. General guidance is provided by the ISO [23].

A question to be addressed is filtration of the sample. Frequently, measurements are carried out in the context of legislation which requires concentrations of contaminants in total, unfiltered samples or which do not mention at all whether filtration prior to analysis is required [24]. Many contaminants have a high affinity for suspended matter [21,25,26]. The concentrations of such analytes in unfiltered water samples will depend highly on the suspended matter concentration. The short-term between-sample variability of suspended matter concentrations, and consequently of the associated contaminants, can be quite high. Furthermore, the content of suspended matter in the water column may vary substantially with hydrological and weather conditions giving rise to large differences between samples collected in different periods at a location. In unfiltered samples, a redistribution of the contaminants between the water and particulate phases may occur. The presence of suspended particulate matter may also enhance degradation of compounds [27].

The problems indicated above imply that filtration in the field is often preferable. This requires, however, special facilities, additional equipment and chemicals, and highly trained personnel, requirements which are not always easily met. Given the objective of the measurement, a compromise needs to be found between the (possible) negative effects of not filtering samples and the problems associated with filtration.

3.2.2.5 Statistical design

Each objective for measurement requires a sampling strategy with its own statistical considerations. In this paragraph some considerations and subsequent problems that can occur are discussed. It is emphasized that the statistical design should not be approached as a purely mathematical exercise. Statistics should be considered in conjunction with local characteristics and hydrology, environmental processes, and methodological aspects.

Factors which need to be considered when making a statistical design of a sampling program include [28]:
- nonnormality of data;
- nonhomogeneity of variances;
- auto-correlation;
- pronounced seasonal variation;
- numerous outliers or censored data;
- missing data points; and
- strong flow dependence.

In nature, many distributions are log-normal rather than normal [29]. When the acquired data has a nonnormal distribution, biased results will be obtained when a normal distribution is assumed. If a population is known to follow a log-normal distribution, the bias can be reduced by either by transforming the data to a normal distribution or by using appropriate formulae. On the other hand, it has also been reported that when power or exponentially related data are treated as log-normal distributed, bias ranging from 2% up to 57% can occur in extreme circumstances [30]. In any event a population with an unknown distribution should be tested when the application of parametric tests is envisaged. Two possible tests include the W test and the D'Agostino test. The W test is recommended as a powerful test for normality and log-normality for a population with 50 or ferwer
observations. D'Agostino's test can be used for populations with 50 or more observations. Both tests have been described [31] but are not presented here since populations with unknown distributions can be very adequately processed by nonparametric methods, as described further on.

Auto-correlation implies that within certain temporal and/or spatial scales samples are not independent. Taking auto-correlation into account may reduce the number of samples required in order to achieve a certain confidence level or, alternatively, enable a higher degree of confidence with the number of samples selected.

Pronounced seasonal variations may be due to variations in biological activity, e.g. summer-winter cycles of nutrient concentrations, to climatic effects, e.g. seasonal effects on the water discharge of the River Rhine, which in turn has an impact on the concentrations of chemicals, or to temporal patterns in human activities, e.g. the application of pesticides for crop protection in certain periods [20]. The flow-dependence is, in fact, a special form of seasonal variation. An example is found in the River Meuse. During periods of high water levels water discharge will increase. The increased flow is capable of resuspending sedimented matter which, due to adsorption, is enriched with hydrophobic pollutants. It has been reported that during such a sediment wave, 80% of the total annual flux occurs within a month. Consequently, tighter monitoring programs are called for during these sediment waves [32].

Data obtained over an extended period of time often contain anomalies [33]. Data may be missing owing to loss of samples or canceling of sampling cruises. Errors may have occurred so that data need to be regarded as outliers, e.g. contamination of samples, incomplete filling of sample bottles leading to the loss of volatile species, or incorrect preservation procedures. It is useful to anticipate missing data and to plan how to cope with these.

In some cases, data may be encountered which are below the limit of quantification (LOQ). Such data are usually reported as:
- "N.D." (not detected);
- "below LOQ";
- "0";
- " < X" (X being the numerical value of the LOQ);
- some value between 0 and the LOQ (usually half the LOQ); or
- the actual measured value (even if it is below the LOQ, either positive or negative) [31,34].

This last option is based on the assumption that the analytical technique is not biased in the range below the LOQ. In that case any value which is "measured" below the LOQ is in fact being generated in a random fashion, according to the characteristics of the analytical technique. The advantage lies in the fact that linear trends in data around the LOQ are more likely to be detected if the full data set is used, rather than a censored data set [31,34].

The degree to which data below the LOQ can be used is highly dependent on the objective of the measurement. It is recommended to report all data (and therefore not to censor) and to report whether or not a given value is below the LOQ. The confidence level for this range should also be reported. Consequently anyone using the data for further statistical processing will be able to assess whether or not the given result can be used. Options for dealing with "less than" values include [31]:
- Replacing "less than" values with values of, for instance, ½ × LOQ. This approach will render an unbiased mean of the whole population provided that the analytical technique cannot generate negative values. Another condition is that all values generated between 0 and the LOQ follow a normal distribution. If the data set known to be log-normally distributed a transformation to a normal distribution should be done first. It is, however, assumed, in these considerations, that only a small number of actual "less than" results are present.

- Computing the population median instead of the population mean. The population's median is defined as the middle observation of a series of observations ranked from the smallest up to the largest value. The population has to have a symmetric distribution otherwise the median cannot be used as a substitute for the mean. Any observations below the LOQ do not have any value assigned to them. The median can be used even when most of the observations fall below the LOQ. Another advantage is the fact that any extreme values at the "upper end" of the ranked series are also filtered out.
- Calculating a trimmed mean. This trimmed mean is the arithmetic mean of $n \times (1 - 2p)$ observations which remain after discarding the largest $n \times p$ data values and the smallest $n \times p$ data values. When the number of observations below the LOQ is equal to or smaller then $n \times p$, a trimmed mean of the remaining observations can be calculated. Like the median this method is robust against outlier data at the upper and lower ends of the population of observations. It has been suggested that a 15-25% trimmed mean of a population can been considered to be a safe approximation of the true mean.
- "Winsorization" of the data, in which case observations which are below the LOQ are replaced by their next largest value. An equal number of the largest observations is than replaced by their next smallest value. From this new data set a mean and a standard deviation can be calculated. This procedure is known as Winsorization and has the same advantages as using the median or trimming since it is robust against extreme outlying values. But on top of this a standard deviation (s) can be calculated which can be corrected for Winsorization with Eq. (1) and so becomes an approximately unbiased estimate of the true variation.

$$s_w = \frac{s \cdot (n - 1)}{v - 1} \tag{1}$$

In this equation n is the total number of observations and v is the number of observations, the values of which have not been replaced.

A factor to consider when designing or optimizing a sampling program entails segmentation, (e.g. [31]) which in general requires information on the area to be present. Segments are chosen in such a way that each segment represents a more or less homogeneous lot. The segments should still be samples in a statistically valid way but since the variability in a segment is reduced, a smaller number of samples might be sufficient. This method should, however, be checked carefully, as is shown by the following examples.

Green Bay, Lake Michigan, was divided into four segments after an extensive sampling programme. Turbidity was the parameter of interest in this experiment, since it can be measured with relatively high accuracy and precision. It was measured during five sampling cruises at 31 stations. From the acquired data a number of stations was then selected in such a way that 22 stations provided a whole-bay volume weighted mean accurate to within 10% of its value determined, using the full network with 80% confidence. From the obtained data a network of stations was designated which covered the whole lake, according to the variability of the measured turbidity. It was found that using a four segment model or the whole bay model made no real difference. In fact, the four segment model required one extra station in order to provide the same accuracy and confidence level as the whole bay model. A 19 station network with a subjective spacing performed remarkably well compared with the calculated networks. This supported the assumption that sampling networks are less sensitive to configuration than to size. This fact was shown to be theoretically valid provided the spacing stations is not clustered and regions are covered in which levels of the variable sought are least predictable [35].

For Lake Ontario it has been shown that segmentation into an offshore zone, a zone near the shore, and some point sources of contamination would make it possible to reduce the existing monitoring program [36]. It was shown that chlorophyll-*a* levels which were monitored until then by means of 15 annual cruises, could instead be monitored with 4 annual cruises with a reduced sampling program. A condition was, however, that each zone should have one monitoring point for continuous weekly sampling. The 4 sampling cruises could be executed during the middle of each season at extreme high or low chlorophyll-*a* levels. It was concluded that in support of this program every ten years a full sampling cruise program would need to be performed for two consecutive years in order to verify whether the segmentation was still valid [36].

For a meaningful statistical evaluation of data, an appropriate number of samples needs to be taken. Different formulae are available for estimation of this number. However, the calculation of the estimate requires insight into the spatial and temporal variations proper. A preparatory study is recommended if such information is not available. Alternatively, a scheme may be devised in which a first strategy is chosen on the basis of a best estimate, data are gathered and evaluated, whereafter the design of the program is optimized and a new cycle is started.

Several papers provide simple formulae which can be used to estimate the number of samples required [31,37-40]. Some formulae for calculation of the number of measurements for the determination of a mean concentration are [31]:

number of samples n, variance prespecified:

$$n = s^2 * (1 + \frac{2}{n_1})$$

number of samples n, relative error prespecified:

$$n = (Z_{1-\alpha/2} * \frac{\eta}{d_r})^2$$

number of samples n, prespecified margin of error:

$$n = (Z_{1-\alpha/2} * \frac{\sigma}{d})^2$$

In these formulae, n represents the number of measurements needed, n_1 the number of measurements needed to determine σ^2, V the variance specified during the assessment of the information needed, σ^2 the variance of the measurements, d the absolute margin of error, $Z_{1-\alpha/2}$ the standard normal deviation, $\eta = \sigma/\mu$ the coefficient of variation, and d the specified relative error. Reference [31] provides several other formula's for the calculation of n using more sophisticated techniques. It is noted that calculating the minimum number of samples using normal probability statistics gives only a rough indication of the actual number of samples required, and often does not give reasonable assurance the required precision actually will be obtained [41].

An elegant approach to coping with auto-correlation effects in determining the number of samples required for estimation of the time-averaged mean has been described by the group of Kateman [42]. The variance in the difference between the real composition of the lot and the composition of the gross sample, σ_*^2, is calculated using the correlation factor T_x, the number of samples n, the temporal length of the total sampled lot P (e.g. 360 days), the interval between two samples A, $A=P/n$, and p and a, which represent P and A normalised on the correlation factor, i.e. $p=P/T_x$ and $a=A/T_x$. The formula for σ_*^2 is:

$$\sigma_*^2 = \frac{\sigma_x^2}{n}[1 + 2\{\frac{\exp(-a)}{1-\exp(-a)} - \frac{\exp(-a)\cdot(1-\exp(-p))}{n(1-\exp(-a))^2}\}]$$

$$- \frac{2\sigma_x^2}{np}[2n - \frac{1-\exp(-p)}{1-\exp(-a)} + \frac{1-\exp(-p)}{1-\exp(a)}]$$

$$+ \frac{2\sigma_x^2}{p^2}[p - 1 + \exp(-p)] \qquad (2)$$

From σ_*^2 and the known variance of the analytical method used, σ_a^2, the ultimate uncertainty, σ_L^2, can be determined via

$$\sigma_L^2 = \sigma_*^2 + \frac{\sigma_a^2}{n}$$

$$(3)$$

Provided that each sample is analyzed, this value of σ_L^2 can then be used to calculate the error in the determination of the mean value of a component during one year, Δ_*, for any number of samples, n, with

$$\Delta_* = t_n \cdot \sigma_L \tag{4}$$

In this formula, t_n is Student's t for a confidence level of 0.05 and n samples.

Plotting Δ_* against n makes it possible to visualize the effect on the error Δ_* of increasing the number of samples. This has been done in Figure 3-1 where the plot was made for the year average concentration of NH_4^+ in the River Rhine at Bimmen. An assessment of field data rendered an estimate for T_x of 120 days, for σ_a 20.2%, and for σ_x 60.9% of the annual average concentration. As σ_a and σ_x are represented as a percentage of the annual average mean, so is Δ_* also expressed as a percentage of the annual average concentration.

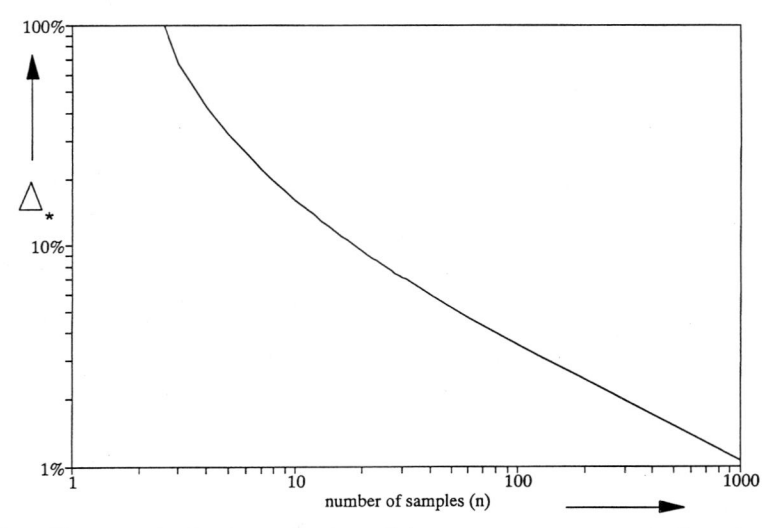

Figure 3-1: The error in the determination of the year-average concentration, Δ_*, (expressed as a percentage of the year-average,) as a function of the number of samples, n, after [42].

From Figure 3-1 it can be concluded that in order to reduce the error from 10% to 5%, the number of samples has to be increased from 18 to 55. Similar calculations have been performed for T_x values of 30 and 10; the results are given in Table 3-1.

Table 3-1: The number of samples required to obtain a 10% error and a 5% error in the determination of the year average concentration for some values of T_x.

	$T_x = 10$	$T_x = 30$	$T_x = 120$
$\Delta_* = 10\%$	33	24	18
$\Delta_* = 5\%$	88	65	55

For the detection of trends in a time series, in general, a method of calculating the number of required samples has been presented [43]. A detailed overview of all options is beyond the scope of this paper. The methods, which are presented below, serve as an illustration of possible approaches. The tests have been assessed and classified according to their usefulness in certain situations [43]. Particular attention has been paid to nonparametric tests, which do not make assumptions about the distributions underlying the data. These tests enable the detection of monotonic trends for nonnormal and truncated data. As a result they are more powerful than parametric tests in situations where there is heterogeneity of variances or nonnormal distribution of errors. For normal distributions, nonparametric tests perform nearly as well as parametric tests. Therefore it was concluded that nonparametric tests are preferable to parametric tests for the analysis of trends in water quality [43].

The choice of the appropriate non-parametric test for detection of monotonous trends can be made according to the scheme depicted in Figure 3-2 [43]. Firstly, the data are freed from any trends, seasonal or otherwise, and correlogram calculations (after Box-Jenkins, 1980) are made (steps 1 to 6). Periodicity can be checked (step 7) by using time-plot series of the residuals (produced at step 4) or the correlograms. When residuals are less dependent after deseasonalization they indicate a periodic process. If periodicity is detected it should be checked for homogeneity of seasonal trends (step 8). Visual examination of the seasonal trend is necessary to guarantee thorough interpretation of the data. When periodicity has been detected, verification of the dependence of the residuals after deseasonalization must be performed (step 9a). When this dependence is proven a test for series with periodicity and persistence must be used (step 12b). If no dependency is detected then a test for cyclic but not persistent series must be applied (step 12a) [44,45].

In the event of a noncyclic series, persistence must be checked with the residuals remaining after detrending (step 9b). When no mutual dependence is established, simple Kendall's or Spearman's tests can be used (step 12d). In the case of mutual dependence and of a Markovian persistence order of 1 (step 10), an adaptation of the Mann-Whitney test by Lettenmaier [46] can be done (step 12c). If the persistence is not of a Markovian order of 1, the data should be submitted to a test adapted for both persistence and periodicity (step 12b). Even when there is no periodicity and the persistence structures are of a larger class, test 12b can still be used [47].

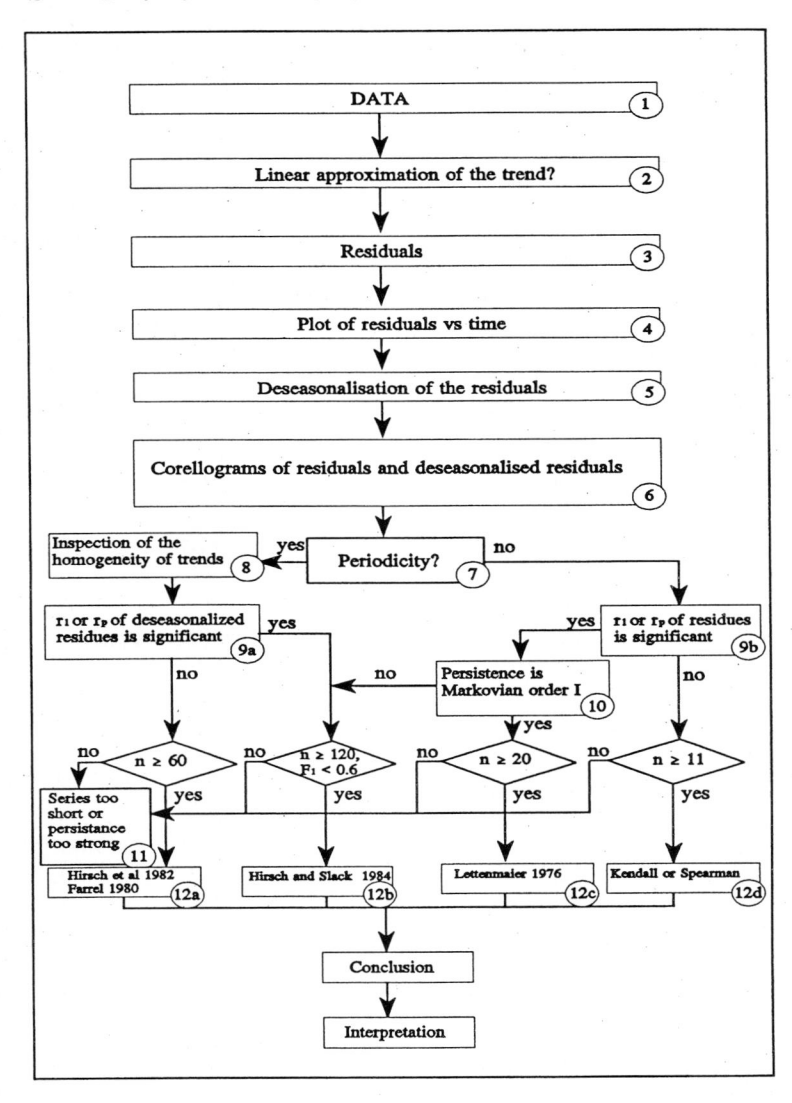

Figure 3-2: A decision tree for determining which non-parametric test is best suited for the detection of a montomous trend, after [44].

The minimum number of observations required in order to perform each of these tests is also displayed in Figure 3-2. It indicates once more the necessity of a properly designed sampling strategy (step 11).

3.3   Documentation and planning

Proper documentation and planning is essential to achieve data with the required level of quality.

In principle, three levels can be discerned for documentation: the design of the program, the information to be provided to staff taking samples, and standard operating procedures.

It is the opinion of the authors that the documentation of the design of the sampling strategy should give, in addition to a description of the program, a clear account of the underlying line of reasoning. The outcome of the assessment of the information need should be given. The specifications which the information should meet ought to be described. Temporal and spatial characteristics and the environmental processes which were taken into account should be stated. The statistical approach needs to be discussed. Particular starting-points and/or assumptions which have been made and any conditions which need to be met for meaningful interpretation of the data (e.g. about distributions) should be recorded. Good documentation in this respect is important as the period between design and evaluation can be quite long. Considerations may be forgotten or even lost as a result of staff turnover.

The sampling strategy needs to be "operationalized" in a sampling plan. This plan should form a good basis for the staff involved in planning, sampling, and analysis to plan and carry out the required activities to the level required. The sampling plan may include:
- the objectives of sampling;
- a concise description of the sampling strategy, providing relevant information for the elaboration and implementation of the actual sampling and analytical programs;
- relevant background information;
- the parameters to be determined;
- a description of the samples to be taken;
- a description of auxiliary information to be collected (e.g. pH, temperature, weather conditions, depth,...);
- type and number of locations to be sampled;
- sampling frequencies;
- the pretreatment of samples in the field (e.g. filtration, preservation);
- methodology; and
- sample coding.

The sampling plan will be elaborated in more detail for the final planning of the sampling cruises (e.g. identification of ships/vans, detailed time schedule, staff responsible for sampling).

The methodology for sampling, sample handling, sample preservation, and analyses may be described in standard operations procedures (SOPs) or protocols for specific procedures (PSPs) [14]. The SOPs refer to standard procedures for operations which are done on a repetitive basis whereas the PSPs are written for special occasions. Obviously a sampling program will consist of both types of procedure, and both should be described in advance.

Prior to its application, the performance of an SOP or a PSP has to be subjected to several tests in order to validate the suitability thereof for the original information needed.

As more people are involved in sample handling and samples are stored over extended periods of time it is obvious that chain of custody procedures are important. Bookkeeping of the history and whereabouts of samples should be kept at a high standard in order to ensure the reliability of the results obtained. From each sample a track record should be kept, consisting not only of information concerning the place and time of sampling but also the sampling and conservation methods used and any anomaly observed at the sampling site or during sampling and storage. The tenability of the sample should be clear, preferably by indicating the day of sampling and the tenability on the sample label. The track record should be available to anyone involved in assessing the quality of the data. The chain of custody procedures are to ensure the identity of a sample from the time of sampling until the final assessment of its data.

Crucial moments in the chain of custody are collection of the sample in its storage bottle (is the bottle correctly coded?) and handling the sample during analytical procedures (do samples get mixed up as they are being transferred into laboratory glassware?). Also handling the data in spreadsheets or other computer programs can sometimes mix up a set of data, especially when a large number of samples is handled. The use of bar coded containers can improve tracking of the samples but here also one should be suspicious of bias. Introducing dummies with known (and possibly varying) concentrations can help to check whether a series of data is still intact or has been shuffled.

An important element in quality control is the design of schemes which provide feedback on the processes concerned. For instance, in laboratories a range of quality control samples is included in the measurement series in order to check the performance. It is difficult to implement such technical feedback mechanisms for sampling procedures as outlined in section 3.4. Therefore, much attention needs to be paid to the "human factor". A comparison may be made with the guidelines for the design of quality control in the laboratory when complex samples or procedures need to be carried out. In this case, a draft version of an ISO/REMCO guide states that the laboratory has to manage the quality of these (research) activities, ensuring that staff with adequate qualifications are involved. The guide emphasizes that skilled, motivated staff should be given sufficient time to carry out the research and recommends that the analysts have a direct but objective interest in the outcome of the analysis and interpretation of the data [48]. These guidelines are also valid for sampling.

3.4 Validation of sampling

It seems that validation of the sampling procedure is often forgotten. In analytical chemistry, validation is defined as determining the suitability of methodology for providing useful analytical data. A method has been validated when the performance characteristics are adequate and when it has been established that the measurement process is in statistical control and produces accurate results [49]. Assessment of accuracy may be accomplished (among other methods by comparison with other (reference) samples or sampling at reference locations (*vide infra*). Assessment of statistical control invokes a time series of data. An important problem for validation of samplers and sampling methods is to distinguish between variation due to temporal or spatial variances within the water body and variation due to operations and conditions to which the sample may be subjected on removal from the water body. For example, if one of two water samplers yields higher analyte concentrations, clear proof must be given whether these higher yields are the result of a temporal variability or are due to analytical anomalies such as contamination.

Sampling performance studies, the equivalent of interlaboratory studies, may be useful in the validation process, but are reported only incidently in the literature. In a large sampling and laboratory intercomparison exercise the analytical variability and the overall variability of eight water quality parameters (total P, total filtered P, soluble reactive P, filtered nitrate + nitrite, ammonia, soluble reactive silica, chloride, and chlorophyll-*a*) has been determined [50]. The analytical variability was determined by using one ship for sampling, and subsequent analysis by three laboratories. The overall variability study was performed by sampling a single water mass simultaneously by three ships anchored 500 m from each other. Each laboratory determined the quality of the water samples taken by their ship in their laboratory according to their own method of analysis. In general, differences in mean levels were due primarily to laboratory procedures and not to different sampling methods (4-L Niskin bottle, 8-L Niskin bottle, and submerged pump), as is illustrated for total phosphorus in Table 3-2.

Table 3-2: Means by laboratory, ship, and depth for total phosphorus in a three ship, three laboratory comparison exercise [50].

Depth	Laboratory	Sampling Method Ship 1: 4-L Niskin	Ship 2: 8-L Niskin	Ship 3: subm. pump	Laboratory Mean
5m	Lab. 1	12.3	12.6	12.5	12.5
	Lab. 2	9.8	9.2	10.0	9.7
	Lab. 3	10.0	10.6	10.4	10.2
	Ship Mean	10.7	10.6	11.0	
24m	Lab. 1	34.4	35.2	34.2	34.6
	Lab. 2	30.4	30.1	31.0	30.5
	Lab. 3	32.2	31.5	32.1	31.9
	Ship Mean	32.3	32.3	32.5	

An extensive statistical analysis of the data was also reported in this study. The results of the statistical analysis and the between-agency differences were compared with estimates of spatial and temporal changes taken from the literature. Laboratory differences for total P were of the magnitude of the decrease of total P estimated to occur over 4.5 to 7 years, depending on the data set used for the estimation of the time trend. The above mentioned impression that laboratory performance has a larger influence on the reported concentration levels than the sampling equipment, has also been given in an early study of trace metals in sea water [51,52]. In this study a comparison of different hydrowires was also included. It was observed that plastic coated stainless steel hydrowires made a slightly, but significantly higher contribution to contamination than polyamide and stainless steel hydrowires [51].

With the improvement in laboratory performance the impact of sampling on data quality becomes apparent. This is illustrated by a recent intercomparison of samplers with different capacities (Ruttner (2 L), van Dorn (5 L), Niskin (2, 8 and 30 L) and Go-Flo (20 L)) carried out in the Mediterranean around the Italian peninsula. The trace metals lead, cadmium, and copper were assessed in this investigation [53]. The samples were pressure filtered on board. Measurements were carried out on board 24 hours after acidification at pH 2, and in the laboratory after frozen storage. It was observed that the volume of the sampling devices had a large effect on sample contamination, the effects becoming lower at larger volumes. It was concluded that when samplers with a volume larger than 8 L are used the problem of sample contamination could be considered under control. It was also observed that contamination levels decreased during the sampling trip due to improved cleaning of the sampling devices. These improvements were probably among others brought about by the rapid feedback: the participants received the analytical results prior to the next sampling occasion, enabling them to adjust methodologies.

3.5 Execution of the sampling plan

In this section important aspects of the execution of the sampling plan will be discussed. Processes which influence the quality of a sample are treated in section 3.5.1. Equipment, chemicals, and cleaning procedures are addressed in section 3.5.2. Section 3.5.3. deals with filtration. Finally, transportation and storage are discussed in section 3.5.4. As all factors that are known to effect the quality of the sample have to be considered, the supervision of sampling operations by an experienced chemist has been regarded as extremely important [54].

3.5.1 Processes affecting sample quality

Sample quality may be affected by physical, chemical, and/or biological processes.

Physical processes which change the properties of surface water during the sampling procedure include:
- malfunctioning of the equipment;
- degassing;
- photophysical degradation;
- precipitation;
- damaging of suspended matter; and
- whirling of sediment and suspended matter.

It might be argued whether sorption processes (contamination, adsorption and absorption) are physical or chemical processes. In this chapter the arbitrary choice has been made to consider these phenomena as chemical processes.

Malfunctioning of the sampling equipment seems an obvious item, but unexpected events (such as failure of locking mechanisms used for sampling at greater depth [55]) may easily remain unnoticed. In an intercomparison exercise of different sampling devices it was observed that the overall salinity average of grab type samplers was slightly higher than for peristaltic pump samplers. One of the possible explanations for this observation was that grab samplers might have been contaminated by adjacent waters a few meters above the desired sampling depth might have occured [56].

Losses of dissolved gases may take place in specific circumstances such as cooling of the water sample below 5° C or depressurization of samples taken at great depths.

Sampling protocols often prescribe minimal exposure of the sample to light. This is a proper practice, although photophysical degradation during the actual sampling will generally be negligible. Even the photolysis half lives of compounds that are sensitive towards photodegradation (for example the dithiocarbamate metam-Na, the phenolic herbicide dinoseb, and the carbamates aldicarb and propoxur) are given in the order of days [57-59].

Submersing the sampling equipment in shallow water may cause whirling of sedimented material. This will yield unrepresentative samples containing too high and irreproducible amounts of suspended matter. This problem can be easily circumvented by carefully examining the water depth prior to sampling, and careful introduction followed by acclimatization of the sampling equipment. If a pump is used for water sampling, the flow rate should be low enough to rule out the possibility of whirling of the sediment. This can be achieved by sampling isokinetically, e.g. sampling without changing the water velocity as it enters the sampler intake [60]. In the case of automated samplers and/or samplers using a pump, precipitation of suspended matter in dead volumes of the sampling equipment (pump, tubing) may occur during sampling. There is little literature available on this topic. When using pumps for surface water sampling, the flow rate of the sampling apparatus must be sufficiently high to ensure total recovery of suspended matter. A flow rate of 1 m^3 and/or a linear velocity of 1 m/s seems appropriate for sampling both suspended matter and dissolved contaminants [24]. Unpublished research by the Dutch Ministry of Transport, Public Works and Water Management has shown that the use of a pump for surface water sampling only slightly damages the alkyl chains of phyto- and zooplankton [61]. Particle size distributions, as determined by a Coulter counter, change slightly due to the use of a pump for sampling [62,63]. This effect can be minimized by using low sampling flow rates [62,63].

Properties of surface water change during the sampling procedure due to *chemical processes* such as:
- chemical degradation;
- redistribution of analytes;
- adsorption and absorption; and
- contamination (e.g. desorption).

Little is known on the chemical degradation of analytes during sampling. Sorption phenomena may be reduced by a suitable choice of materials, as discussed later.

The importance of an adequate degree of contamination control cannot be overemphasized [6,64]. The level of contamination introduced by the sampling vessel has led to the statement that many of the analytical values to be found in the older literature merely immortalize the extent of shipboard contamination' [65].

For sampling of trace metals in water using a boat it has been advised not to use the sampler at depths of less than 10 m [10] because of the contamination plume around the boat. Sampling of water samples for organotin analyses has been performed from a rubber boat, in order to avoid contamination due to anti-fouling paints from the ship [66]. Sampling by air [37,67] has been carried out occasionally. Lead contamination arising from the use of leaded fuel has been reported [67].

An initially low concentration of a specific analyte in a sample may be increased as a result of previous sampling actions. This specific form of contamination is called "memory-effect", "carry-over", or "cross-contamination". The risk of both contamination and cross-contamination of surface water samples increases with increasing surface area of the sampling equipment in direct contact with the sampled water. This has been illustrated by a comparative study of different types of sampler; a vacuumpump with a metering chamber (and thus a relatively large inner surface) resulted in more cross-contamination than a peristaltic pump [68].

Many aspects of forementioned risks of contamination during sampling also apply to risks for storage, as discussed later.

The *biological processes* that change the properties of surface water during the sampling procedure generally concern biological degradation. Fungi, for example, are able to degrade PAH and chlorinated aromatic compounds aerobically [69]. The amount of time an analyte remains in the sampling equipment is generally very short. It normally will take at least several days before noticeable biological degradation can be observed (e.g. [69,70]). In most cases, biological degradation during the actual sampling will be negligible in comparison with biological degradation during transport and storage.

3.5.2 Equipment, chemicals and cleaning procedures

Contamination of samples can have many sources. It is obvious that these sources should be removed or have their influence minimized. Some of these sources are related to the sampling site itself, others depend on the materials, chemicals, and procedures used.

3.5.2.1 Materials

Commonly used materials are (borosilicate) glass, (high/low density linear) polyethylene, polypropylene, and PTFE (the so-called Teflon). Each type of material has its drawbacks and advantages. Glass, as an inorganic material, is well suited for use with organic contaminants. Leaching and absorption of those compounds is limited. This is contrasted by its behavior in contact with inorganics. As glass is made of silicates, residual active sites capable of binding ions are always present. Leaching of trace metals is also known; this makes glass inappropriate for handling trace metal samples in general. Mercury constitutes an exception: for this metal glass is required owing to the volatility of mercury species.

Polyethylene is a relatively inexpensive material. It is frequently used for trace metal investigations. It is less suitable for organic compounds, owing to sorption and leaching problems, and for mercury, owing to its porosity. Polypropylene is an alternative to polyethylene. It is reported to absorb more contaminants than polyethylene and should be used with caution. Teflon is known for its low level of contamination and favorable sorption characteristics for both metals and organic compounds, although adsorption of neutral organic compounds is known. Teflon is more expensive than other materials and susceptible to diffusion of gases; this lessens its usefulness as tubing material [71].

For some of the most frequently used materials the results of leaching experiments on metal impurities are given in Table 3-3, after [72]. An extensive list of materials to be used for sample bottles and chemicals for preservation is given by the ISO [73].

Metals are sometimes incorporated into the design of sample collectors. Commonly used metals are aluminum and stainless steel, since these metals are less prone to corrosion. Coating of the metal surfaces with inert organic layers is an option to consider.

Table 3-3: Trace elements in laboratory ware materials, based on [72]. Concentration range: mg/g.

Material	100 - 10	10 - 0.1	0.1 - 0.01	0.01 - 0.001
Polyethylene and polypropylene	Na, Zn, Ca, (Al, Ti)*	K, Br, Fe, Pb, Cl, Si, Sr	Mn, Al, Sn, Se, I	Cu, Sb, Co, Hg
Poly(vinyl chloride)	Na, Ca, Al, (Sn)**	Br, Pb, Sn, Cd,	As, Sb Zn, Mg	-
Teflon	K, Na	Cl, Na, Al, W,	Fe, Cu, Mn, Cr, Ni	Cs, Co
Polycarbonate	Cl, Br	Al, Fe	Co, Cr, Cu, Mn, Ni, Pb	-
Glass	Al, K, Mg, Mn, Sr	Fe, Pb, B, Zn Cu, Rb, Ti, Ga, Cr, Zn	Sb, Rb, La, Au, As, Co	Sc, Tl, U, Y, In
Silica	-	Cl, Fe, K	Br, Ni, Cu, Sb, Cr	Sb, Se, Th, Mo, Cd, Mn, Co, As, Cs, Ag

* Sometimes used as catalysts.   ** Sometimes used as stabilizer.

Especially as sampling devices become more and more sophisticated, and more different types of material are in contact with the sample, proper measures should be taken to ensure that the device, including the containers, does not alter sample integrity. Leaching and sorption tests can be performed before using any material in order the assess its usefulness.

3.5.2.2 Chemicals

Chemicals used for cleaning and preservation should be of high grade. Acids used for acidifying samples, such as nitric and hydrochloric acids, can be distilled just as organic solvents. A quartz still is necessary to reduce contamination from the labware. Distillation might be more cost-effective then buying high grade reagents. If proper quality checks are in place it should be possible to reduce reagent costs because they can be "upgraded" in one's own laboratory.

The same considerations apply to any water used. The metal content can be minimized by following the three-stage procedure described below:
- "distilling" by reverse osmosis (RO);
- redistillation in a quartz still; and
- treatment in a Milli-Q system.

Repeated recirculation through the Milli-Q system will result in even higher purity. These steps are effective for metal contamination but might be less effective for some organic components. The Milli-Q system is usually equipped with a active carbon filter as a final stage; this can introduce hydrocarbons. Distillation in a glass or quartz still, discarding of the first and last few milliliters seems a more appropriate approach.

Using materials and chemicals selected/cleaned and/or upgraded in this way makes it possible to diminish and control the level of bias. If, however, some bias cannot be excluded it should at least be estimated. The estimates of the level of contamination should reported with the analytical data.

3.5.2.3　Sampling devices

About a hundred models of portable (semi)automatic sampling devices are reported to exist [9]. Each device has been developed (or modified) for different types of contaminant, sampling location, and sampling method. A sampling device can be anything from a plain glass bottle to a remotely controlled submersible sampling vehicle [75]. The choice of sampling device is, of course, predetermined by factors already mentioned. A few of the most common ones will be dealt with later by describing their advantages and their drawbacks.

Criteria for the selection or design of the sampling device for trace metals in sea water have been described in literature [76,77]. With a few adaptations these criteria can easily be extended to fresh water samples and organic contaminants:

- A sampler with optimum properties should be built from a plastic material which has a low trace metal/organic contaminant content, resists vigorous attacks by chemicals during the cleaning operation, and shows no significant memory effect due to adsorption/ desorption processes. The surface of the material should be neither porous nor rough.
- The sampler size should enable handling by one person. The retrievable volume should be large enough to obtain suitable numbers and volumes of subsamples for different parameters and enable the flushing of storage bottles before filling.
- The samples should preferably come into contact with a small selection of (suitable) materials only. The number of metal parts of the construction should be kept to a minimum and be protected from direct contact with the water.
- The design should allow for in situ preservation of the sample (i.e. acidification for metals).
- The sampler should be strongly built and of an uncomplicated design, all parts should be exchangeable.
- To avoid the inside being coated with the contaminated surface micro layer, the sampler should pass the water surface in a closed state.
- For simultaneous sampling at several depths and for the reduction of the necessary station time, the sampler should be suitable for serial operation.
- The sampler should be operated using coated hydrowire (e.g. polyethylene; an alternative is the use of Kevlar cables) and Teflon coated messengers for releasing or closing operations.
- To avoid losses of fast sinking particulate matter, the dead volume below the drain cock should be kept to a minimum.

A quite different approach entailed the deployment of scuba divers for sampling of marine waters [78]. The divers were able to sample depth intervals as small as 2 - 3 cm. The intermediate mixing zone (which exists between the surface layer which is less saline and the undiluted seawater) was their target. Because of the difference in salinity between those two bodies of water a visible precipitation layer of organic material is present. The divers were able to selectively sample this layer directly making sure not to contaminate it by facing the direction of the current.

Comparison with pretreated Teflon samplers showed that samples obtained by the divers produced the same low levels of contamination. The limitations of scuba divers are, however, obvious.

A few commonly used sample devices are discussed below; the descriptions given provide an outline of their working mechanisms. A more extensive description of a large number of sampling devices is given in several articles and manuals [9,37,40,55]. Contradictions exist between authors concerning the exact names of some of the samplers. For clearness' sake, and the fact that it has found the most support by other authors, a decision was made about nomenclature [40]. Some of the devices are schematically depicted in Figure 3-3.

Sampling bottle: the simplest sampling "device" is the plain sampling bottle available in many shapes and sizes and made of a variety of materials. Its size is usually dependent on analytical procedures. When sampling is carried out properly, little or no contamination is to be expected during the sampling operation. By putting the bottle up side down in the water prior to filling and inverting it to fill it, contamination from the surface microlayer can be avoided.

Sampling bucket: another favored sampling "device" is the plastic or stainless steel bucket. A sample of up to several liters may be taken at once, homogenized and split into subsamples. Care is needed to avoid contamination, e.g. by air or by mixing of the surface microlayer with the final sample.

Ruttner bottle: the Ruttner bottle is an early sampling device made up of a cylinder with plate valves on each end. The valves are connected via a metal rod through the cylinder. Releasing is achieved via a messenger which makes the cylinder drop a few centimeters on the lower valve and the upper valve drop on to the top end of the cylinder. It can only be operated in an upright position. The sampler is open when it passes the surface microlayer.

Van Dorn bottle: another simple sampling device is the Van Dorn bottle, which consists of a cylinder stoppered at both ends by rubber hemispheres or balls. The hemispheres are connected to each other via a rubber string passing through the sample chamber. It is lowered down with both ends held open because the stoppers are attached to a "lock". When the required depth has been reached a messenger is sent down to release the lock and stoppers. The rubber string pulls both stoppers into position, thus closing the sampler. It enables sampling at predetermined depths and can be used in both horizontal and vertical orientations, the horizontal position being useful when sampling close to the sea or river bed [55]. This sampler is also open when it passes the surface.

Go-Flo bottle: the Go-Flo bottle is also a cylinder-type sampling device. Here the valves are balls which have been bored through. The balls are positioned at the ends of the cylinder in such a way that they can be rotated. The cylinder can be lowered in the closed state and at the required depth the balls are rotated 90° to enable flushing and filling of the device. Thereafter the balls are again rotated 90° to the closed position and the device can be pulled up.

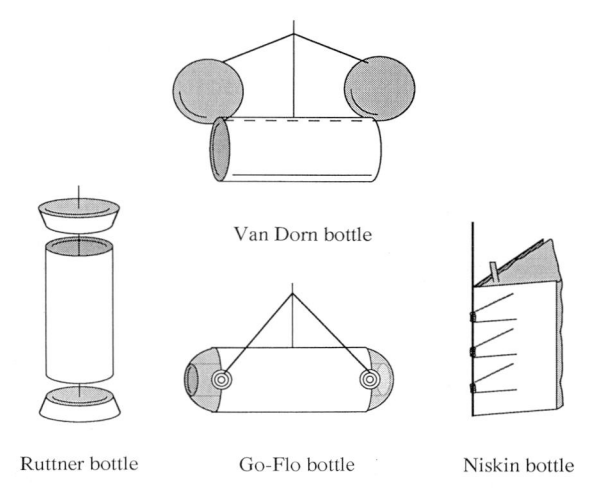

Figure 3-3: Schematic diagrams of a Ruttner bottle, a Van Dorn bottle, a Go-Flo bottle and, a Niskin bottle (after [37,40]).

Niskin bottle: the Niskin bottle, or bag sampler, consists of two hinged aluminum plates with between them a polyethylene bag connected to both. After being lowered to an appropriate depth a messenger unlocks the plates which are pulled from each other by a spring which opens the two hinged plates like a book. The bag fills itself via an inlet tube which is initially closed by a protecting plastic casing. The activating messenger releases a device which cuts the cover of the inlet tube. When filling is complete a clamp closes the inlet tube and the sampler can be pulled up.

The operation of the Niskin and Go-Flo bottles follow the closed-open-closed principle, i.e. they pass through the water surface in a closed state. This in contrast with the Ruttner and Van Dorn bottles which become "coated" with the contaminated suspended matter of the surface microlayer as they are lowered. The successful operation of these samplers is related to the reliability of the mechanism used to release the stoppers or valves [55]. The operation of messengers should somehow be controlled to prevent sampling at the wrong depths.

The samplers mentioned up to now are all grab samplers, which collect a predetermined sample volume at once. The following samplers collect a variable sample volume which makes them suitable for a different mode of operation.

Sample pumps: pumping systems are gaining importance. Peristaltic pumps are favored since they do not expose the sample to any mechanical parts of the pump. Any tubing used should, of course, consist of inert material. Pumping speed has been shown to be one of the critical parameters. On the one hand sedimentation in the tubing should be avoided and the residence time should be kept to a minimum. Also, the formation of a biofilm can be minimized by maintaining a high flow velocity. On the other hand a high flow velocity might influence sample composition. For instance particles of suspended matter can break up into smaller fragments which have a larger specific surface area. This could change partitioning conditions for those organic compounds which are readily adsorbed by suspended matter. Volatile organic compounds may also be lost from the liquid phase because of the low pressure generated in the system.

At high flow velocities differential sampling of suspended matter may occur. When the flow velocity at the inlet differs too much from the velocity of the sampled water, suspended matter particles will be collected in a biased way. Because of their inertia they are capable of missing the inlet as their own speed differs from the flow velocity at that point. This can be avoided by pumping "iso-kinetically", which means that the flow velocity at the inlet point is always maintained at the same velocity as that of the sampled water. This however requires a more sophisticated sampler design.

Sampling fishes/sampling torpedoes: particularly when sampling is done from rather large vessels the use of a so called sampling fish or sampling torpedo can be very helpful. Contamination can be avoided by using a crane to keep the sampling device away from the vessel and by ensuring that the samples are taken upstream of the ship and/or upwind. The torpedo can be fitted with several measuring devices for on-line measurements such as pH, temperature and conductivity. Ion-selective electrodes can also be employed. These variables can be used as guidance parameters in order to collect information on stratification and/or to find point sources of contamination. Samples can be collected either by filling of containers mounted on the device or by pumping the sample on board via tubing.

Solid phase extraction systems: solid phase extraction (or SPE) has, due to its capacity to reduce sample handling prior to analysis, proven to be a sampling technique that is gaining in importance in the field of environmental chemistry. The relative ease with which samples are taken, stored, transported, and subsequently worked up make SPE a valuable tool in surface water sampling. SPE is based on the principles of liquid chromatography in which compounds are retained by interactions with the stationary (solid) phase. The choice of stationary phases for SPE is, therefore, almost as wide as for liquid chromatography, and includes sorbents based on reversed phase, normal phase, adsorption, ion-exchange, or size exclusion techniques [79]. From this series the solid phase with the most universal application is octadecyl- or C_{18}-modified silica. Its hydrophobic interaction with a wide variety of (even slightly) hydrophobic compounds makes it suitable for a large number of aquatic contaminants. Besides the traditional column configuration, C_{18}-modified silica has also been incorporated in filter disks which have even better extraction characteristics. The silica particles are "trapped" in a network of PTFE fibrils which make up the (Empore) filter disk. Washing the filter with a small volume of a suitable solvent (such as methanol) suffices for elution of the contaminants [80]. SPE systems can be used both in "off-line" and "on-line" extraction systems. Off-line application (in the field) of SPE has the advantage of reducing the sample volume from 1-2 litres down to 5 mL of organic solvent. SPE solid phases such as XAD-2 or C_{18}-modified silica are known to preserve trapped hydrocarbons over a period of up to 100 days in the presence of oleophilic bacteria [81]. "On-line" systems, however, make it possible to reduce the handling of samples to a minimum. A (river) water sample of 30 mL can be preconcentrated, cleaned-up and analyzed within a single system. The system, using an SPE column packed with polymer PLRP-S (styrene-divinylbenzene), is capable of detecting the large majority of 54 test compounds in Rhine water sample, at the 1-5 mg/L level [82-84].

Time-integrated samplers: samplers which sample a body of water over an extended period of time are called time integrated samplers. The content of the sampler will represent a time-averaged mean of the composition of the bulk under investigation over a specified period of time, thus making it possible to assess the quality of the bulk over a specified time period. Temporarily extreme concentrations will pass by almost unnoticed, since the measurement will only render an account of the average conditions during the sampling period. This technique may be useful for determining fluxes of contaminants over a fixed period of time with a minimum of number of required analyses. A few examples of time integrated samplers are treated below.

An easy to use and simple to manufacture passive sampler is Pisces [85]. It consists of two pieces of pipe soldered together in a T-shape. The "foot" of the T is sealed with a Teflon disk whereas the "arms" are sealed with polyethylene membranes 1 mm thick. After filling the sampler with hexane (200 mL) it was suspended in the water column from a float. In time PCBs became concentrated in the hexane. Calibration of the PCB uptake was possible so that time-integrated concentration levels could be calculated. Although the sampler design still needed some improvement it was possible to use it to identify PCB sources along the Black River (NY). Because the PCBs were collected in hexane, sample storage, transportation, and laboratory procedures were simplified and, therefore, costs were reduced. It should be noted however that this method is only suited for collecting dissolved PCBs. PCBs associated with particles or dissolved organic matter could not be extracted [85].

Several publications highlight the possibilities of dialysis as a tool for investigating the presence of bioavailable contaminants next to dissolved contaminants. As a cut-off pore size 1000 D is recommended, since it seems that passive uptake of hydrophobic organic compounds by organisms occurs at particle sizes less than this. By using dialysis bags filled with hexane it becomes possible to sample a site over prolonged periods of time before the sampler becomes saturated. Because of the concentrating capacity of the system it is possible to sample sites with contaminant concentration far below the detection limits of conventional methods [86,87]. An interesting application of dialysis which, at the same time, mimics biotic conditions, is the use of model lipids as the receiving solvent. Purified lipid extracts from aquatic organisms as well as a single lipid, triolein, were employed. About 1 g of recieving lipid was placed in about 80 cm of dialysis tubing (polyethylene tubing: 2.6 cm wide, 76.5 μm film thickness, exposure area ~ 360 cm^2). The lipid was squeezed down the length of the tubing after which the tube was sealed and placed in an exposure tank or an outdoor pond enclosure. It was shown for two model tetrachlorobiphenyls that their lipid:water partitioning coefficient corresponded closely with their K_{ow}s (octanol:water) [88].

Uncoated capillary columns have been employed to concentrate organic compounds from solution. Simply passing the aqueous samples through the colums enabled efficient trapping on the capillary walls. Column dimensions varied from 15-30 m in length and 0.31-1.6 mm internal diameter. Plastic columns, in particular, were reported to remove the organics ((chlorinated) mono aromatic compounds and some aliphatic compounds in water) from the solution extremely efficiently. Subsequent flushing of the columns with a suitable organic solvent enabled the collection and concentration of the organic compounds. Further research on field samples was still needed [89,90].

Another approach to time-integrated sampling is the use of a sparger [91]. By running a stream of air through a column of water volatile hydrophobic chemicals can be extracted in situ and collected on appropriate column material. Only truly dissolved analytes available for extraction by the air stream will be concentrated. The tested analytes were mercury, some chlorobenzenes, chlorostyrene, and six PCB congeners. This method has, so far, only been shown to be useful for semiquantitative work as a survey for the presence of contaminants. Quantitative analysis has been considered feasible but remains to be examined.

A new development is the application of a tubular silicone membrane through which extractant (such as hexane) is pumped. The sample is led through a second tube surrounding the membrane, enabling exchange of analytes between the sample and the extrant. The extrant is concentrated and transferred to a gas chromatograph for analysis. This technique still requires further development but could in future enable on-line monitoring of waste streams [92]. A fully developed and commercially available system for in situ time-integrated sampling is the Infiltrex sampler (previously known as the Seastar sampler). It is a self-contained instrument consisting of a small pump, a filter mount and a solid phase column. A microprocessor controls the pumping regime and the whole system is powered either by batteries or an external supply. Several materials are available for the column packing, including 8-hydroxyquinoline attached on a XE-305 resin for the concentration of transition metal ions and an Amberlite XAD-2 resin is suitable for hydrophobic organic contaminants. The system is usually used over periods of three to four days. Results have been reported for DDT metabolites, cyclodiene pesticides and PCBs [93], and three (large) PAHs [94]. The reported concentrations are relatively low because the system is only capable of collecting the dissolved fraction of the contaminants.

3.5.2.4 Cleaning procedures

Depending on the material used and the type of contaminant under investigation different cleaning procedures must be followed. When several samples are taken on a field trip with the same sampling device, precautionary action should be undertaken in under to prevent cross contamination. Especially when contaminant levels differ significantly, the level of cross-contamination should not be underestimated. In between sample operations rigorous cleaning of the sampling device should be considered.

After cleaning, the sample apparatus (and the container) should be packaged in order to prevent contamination during transportation to the sampling site. Usually wrapping up in polyethylene bags will suit this purpose well. At the sampling site both the sampling device and the container should be rinsed with water of the sampled area before use. The materials used for cleaning depend both on the materials from which the sampling device is constructed and the contaminant under investigation.

For trace metals several articles have been published in which cleaning procedures for sampling devices and containers have been reviewed [10,56,77]. A detailed description of an eight step cleaning procedure in order to minimize trace metal contamination is given below. In view of the low levels of trace metal concentrations subsequently measured it seems appropriate to consider this procedure as state of the art.

- Degrease the bottle in a soap bath for 24 h, rinse well with reverse osmosis (RO) "distilled" water, and shake off excess water.
- Fill with or soak in reagent grade acetone for 1 h, drain off acetone and allow to air dry, rinse with RO-water, and shake off excess water.

- Fill with or soak in reagent grade concentrated hydrochloric acid for 1 h, rinse with RO-water, and shake off excess water.
- Fill with or soak in reagent grade concentrated nitric acid for 1 h, rinse with RO-water, and shake off excess water.
- Fill with or soak in reagent grade 6 M nitric acid for 72 h, rinse with RO-water, and shake off excess water.
- Soak bottle in warm (40-50°C) 2 M reagent grade nitric acid for 72 h, rinse with RO-water, and shake off excess water.
- Rinse inside with 0.5% reagent grade nitric acid then rinse with quartz-distilled water.
- Place piece(s) in a heavy polyethylene bag rinsed outside and inside with quartz-distilled water.

Once new bottles had been put through this procedure only the last three steps were repeated. When bottles were exposed to high metal concentrations the last five steps were executed. All labware involved was treated in a similar manner. Since this method was developed for very low trace metal levels, adaptations are possible when higher concentrations are encountered.

Standard procedures include cleaning with water and detergent, rinsing with distilled or deionized water, oven drying at 105^0 C for 2 h, rinsing with the extraction solvent to be used after cooling of the bottle and drying with purified air or nitrogen. For bottles which have already been used it is recommended in addition to extract with acetone, rinse with hexane and dry as described [48].

3.5.2.5 Considerations in the field

The information need may be satisfied by different approaches. The basic issue is that the considered or selected approach is thoroughly evaluated and validated.

Taking samples on the waterside or from permanent structures such as bridges enables precise positioning of the samples so that on return exactly the same spot can be sampled again. Bridges make it possible to collect samples from the whole cross section of the water. Sample sites selected this way are often rather chosen for reasons of convenience and cost, however, instead of on considerations of representability. It should be assessed whether the sampling scheme which results from logistical considerations enables the required inferences to be made.

Surveys of rivers may be carried out with relatively simple sampling techniques such as grab sampling. Grab samples taken at the surface of a body of water may be convenient when speed is wanted but may be less appropriate when a stratified site is sampled. In such cases sampling of individual strata might lead to better information.

As an example, a comparison was made between grab sampling and integrated (stream) depth sampling [95]. Four stream sites, representing a range of hydrologic characteristics, were selected within the Kentucky River basin. A range of flow conditions was sampled over a period of approximately two years. The integrated depth samples were composed of subsamples taken at evenly spaced depths. The amount of water in the subsamples was proportional to the stream velocity. The grab samples were taken in mid stream during the collection of the integrated samples. It was shown that for dissolved constituents there were no differences between the two methods, but that grab sampling rendered lower concentrations for constituents associated with suspended solids [95].

In some cases final decisions about the locations and/or depth where the samples are to be taken can be based on the results of rapid on-line screening techniques. For example, a nephelometer or CTD can be used to assess vertical gradients. The information on the gradient can be employed to establish the depths at which samples are taken.

As discussed several times, control of contamination during sampling is important. When using a ship, caution should be taken not to cross sections of the area to be sampled. On a flowing river this will be less of a problem since water upstream of the vessel is free of any contamination caused by the vessel. On stagnant waters the sampling route should be planned in such a way that no sampling sites are crossed by the sampling vessel before sampling. The wind direction should also be taken into account since the area downwind of the vessel could become contaminated by exhaust fumes [64]. Therefore, sampling from a vessel should be done in such a way that the vessel is downstream and downwind of the sampling site. When small boats, powered by outboard motors, are used it can be worthwhile to shut the engine off several meters before reaching a site and paddle the last few meters in order not to disturb the site. On large vessels the use of telescopic devices or cranes is advised in order to take samples up to several meters away from the vessel in order to reduce risks of contamination. Because of the pollution plume of the sampling vessel it is recommended that samples should be taken at depths of at least ten meters [10,96]. Obviously this is somewhat impracticable for shallow waters. In those cases samples could still be taken at a minimum ten meters from the vessel, preferably upwind.

It is recommended to wear polyethylene gloves in order to reduce the risk of contamination of the sample when handling sampling equipment and samples. This also provides protection for the personnel involved in taking the samples since pollution levels are not always known. Smoking at the sampling site should be prohibited in order to prevent contamination by organics from smoke. This also applies to any engine activity near the sampling site since exhaust fumes provide a ready source of hydrocarbons and even lead levels are known to rise due to the use of leaded fuel [56]. An example to this effect was shown in California were investigators used a helicopter to sample some 170 lakes in the High Sierra. when the sites were sampled again using careful procedures to prevent contamination it was shown that lead levels were reduced by a factor of about 300 whereas the levels of other investigated metals, K, Ca, and Sr remained the same [67].

Facilities with clean rooms or laminar flow benches are essential when very low levels of pollutants are investigated. It was shown for Lake Erie and Lake Ontario that metal concentrations were lower than reported in previous studies due to the considerable precautions that were taken, including high quality chemicals and materials and the use of a portable clean room.

3.5.3 Filtration

Suspended matter (SPM) can accumulate hydrophobic organic contaminants and trace metals, sometimes concentrating them by a factor of a thousand or more. The concentrations in an unfiltered sample are in such cases determined heavily by the SPM content. The between-sample variability of the latter may be high, resulting in a similarly high variability for the contaminants of interest. Therefore, separate analyses of SPM and filtrate or centrifugate may be worthwhile.

The use of filters can introduce contamination, or reduce concentration levels by adsorption. Proper testing of the filters, and the filtering device, is called for. Filtration is prone to contamination so that special facilities need to be present when filtration is carried out in the field.

Filtration can be done either under gravity or by applying vacuum or gas pressure. Using gravity will take far too much time when, e.g., a 0.45 μm filter is used. Vacuum filtration or filtration by gas pressure is, therefore, desired. Both these methods can be used in on-line or off-line modes [52].

Also for filters and filter devices appropriate cleaning procedures should be observed. Procedures for filtering samples with very low concentration levels of metals have been described. They consist of:
- using filter membranes, made of polycarbonate, with 4 μm pore size;
- leaching them for one week in 20% ultra pure nitric acid prior to the sampling trip;
- keeping them soaked in Milli-Q water until used in the field; and
- handling them using Teflon-coated tweezers only.

Since chlorinated biphenyls (CBs) tend to be associated with suspended matter (SPM), removal of SPM is desired when levels of dissolved CBs are to be determined. The level of sorption increases with the degree of CB chlorination. It was observed that upon filtration all dissolved CBs adsorbed by both glass fiber and membrane filters [97]. The level of adsorption ranged from 50 - 100%. It was concluded that only CBs associated with dissolved organic matter (DOM such as humic acids, proteins, and fulvic acids) were able to pass the filters. It was not determined whether the DOM were adsorbed by the filters, as was shown in other studies.

Glass fiber filters (Whatman type GF/C, 0.45 μm mesh size) adsorb dissolved organic carbon (DOC). The adsorption amounted to 10-20% of the total DOC content [98]. Before filtering, the filters had been precombusted at 450°C for 4 h to reduce any sample contamination. From this result it was concluded that samples for DOC analyses should be taken at a late stage of the filtering process. In this way absorption sites on the filter become saturated and losses can be prevented.

Sorption of PAH by glass fiber prefilters (0.45 μm) has been reported [99]. Four filters (two types from two manufacturers) were tested, and resulted in absorption percentages ranging from 0.0-25.3% for 13 PAHs. Besides binding to the filters, PAHs such as benzo(a)pyrene and pyrene, have been shown to bind readily to DOM [100]. It was concluded that differences in affinities of benzo(a)pyrene, pyrene and PCBs could be explained by different proportions of hydrophobic acids and the degree of aromaticity of the DOM.

An alternative to filtration is centrifugation. Comparison of filtration and centrifugation has been carried out for organic contaminants [97]. CB concentration patterns in centrifugates resembled the patterns found in SPM and differed from those found in filtrates. This observation was explained by the presence in the sample of particles with a density equal to or lower than water, consequently raising concentration levels of bound CBs. Centrifugation has the advantage that it enables the collection of large amounts of suspended matter for analysis.

Clogging of filters alters the effective filter pore size in an unpredictable and unmeasurable manner [99]. An alternative approach is the use of a combined filter arrangement in which several filters of ever decreasing pore size are stacked on each other. A combination of filters has been described for the filtration of soil leachate, which is rich in particulate matter. A 10

μm nylon filter, a 0.75-1.5 μm cellulose membrane filter and a 0.45 μm cellulose membrane filter were stacked and in this way prevented immediate clogging. A commercial tangential flow system is available. This system has been used to collect large amounts of estuarine and marine waters for the preparation of reference materials [101].

3.5.4 Transportation and storage

Transport of samples to the laboratory should be carried out as soon as possible. Adequate conditions and/or preservation methods should be employed.

Analyte concentrations can change during storage because of:
- adsorption of the analyte to storage vessel or suspended material;
- biodegradation;
- photophysical or other reactions;
- thermal lability;
- (photo-)chemical reactivity;
- evaporation;
- degassing;
- precipitation; and
- (cross-)contamination.

The factors involved in sorption losses during storage may be classified into four categories [102]:
- the analyte, especially its chemical form and concentration;
- the solution, the presence of acids (pH), dissolved material (e.g., salinity, hardness), complexing agents, dissolved gases (especially oxygen), suspended matter (competitor in the sorption process), and microorganisms;
- the container, its chemical composition, surface roughness, surface cleanliness and the specific surface. The history of containers (e.g., age, method of cleaning, previous samples, exposure to heat) is of importance because it may be of direct influence on the type and number of active sites for sorption; and
- external factors such as temperature, contact time, light and occurrence of agitation

Under normal surface water conditions (6 < pH < 8.5) hydrolyzable metals tend to be adsorbed onto surfaces [103]. The analytes can be adsorbed on (or absorbed by) dissolved organic material [100], storage vessels [102,104,105], particulate matter [104], and biological material [25,99,106]. Losses of trace metals from distilled water and artificial sea water by sorption on various container surfaces have been observed. The significant loss of 30% of Pb (within 24 h) during storage of unacidified, pre-filtered, low-salinity samples was explained by adsorption on the polyethylene container wall [105]. A special form of adsorption is ion-exchange [40], which is observed using glass containers [107]. It has been stated that for trace metals preservation of water samples in the laboratory should be preferred to preservation on location, unless the samples cannot be transported to the laboratory within a few hours [108]. This observation is probably due to sample contamination in the field, modern facilities enable contamination-controlled sampling. The continuous decrease of trace metal concentrations in sea water [55,109], achieved by increasing the volume:inner-surface ratio and careful choosing the container materials [55], has been one of the most striking examples of successful minimization of contamination. But even using 50 L precleaned polyethlene bottles it remains difficult to avoid trace metal contamination of sea water samples, especially by Zn and Pb [52,103].

Comparison of storage of water samples in polyethylene, borosilicate glass, and PTFE containers showed PAH losses well over 60% after 30 min due to adsorption on the surface of the polyethylene container [104]. Interactions of organic contaminants and trace metals with particulate matter in unfiltered water during storage can also be a problem for surface water analysis. Reliable storage of filtered sea water samples for butyltin analysis can be achieved by acidifying at pH=2 and storage in the dark in Pyrex bottles, whereas water samples with a high suspended matter content stored in the same way show stability problems [110]. The Cu concentration of filtered unacidified surface water samples remained constant, whereas unfiltered unacidified samples showed significant redistribution of Cu between particulate and dissolved phase [105]. PAHs are very sensitive to losses during storage because of adsorption by most surfaces, including suspended particulate matter [104]. Modern pesticides are sensitive to degradation and reversible biosorption during storage. Degradation even occurs via nonmetabolic processes. It has been shown using ^{14}C labeling, equilibrium, and kinetic studies that malathion is degraded to water-soluble products in the presence of dead cells and isolated microbial cell walls [106]. A study on the influence of storage and shipping conditions (time, temperature) on TOC levels in water samples [111], showed that storage time and temperature are not critical for river water with high TOC (220 μg/L). For water containing low level TOC (169 μg/L) an increase of approximately 30% (46 μg/L) due to contamination was observed when the water was bottled in amber glass bottles [111].

Cooling of the samples to 4° C [40,112] or freezing to -20° C [113,114] is a generally accepted preservation technique. For thermal labile compounds in particular immediate cooling is important. Freezing samples slows down chemical, biological, and physical processes (e.g. migration of analytes through the sample) and is one of the most universal preservation methods. Slowing down migration by freezing also diminishes the risk of adsorption of the analytes by the storage vessels. Specific care should be taken towards volatile components, however, since cooling of a sample also causes degassing (especially below 5°C) [40].

Freezing sea water samples in high-density polyethylene containers enabled the conservation of nitrate and phosphate levels for up to six months [113]. Before use containers were soaked in 10% HCl solution for 48 h and rinsed three times with Milli-RO/Milli-Q water. Samples were frozen at -40°C in a blast freezer within three hours of sampling and stored at -20°C. Concentrations of phosphate in filtered sea water stored at -20 °C steadily decreased in samples stored longer than four months. Although it was shown that the observed loss was not due to the effects of surface adsorption, no other explanation could be given [113].

Shock freezing has been recommended for storage of heavy metal samples [115]. The procedure consists in collecting a (sub)sample in a 20 mL polypropylene syringe, freezing it in liquid nitrogen, and storing it at -10°C. No further preservation measures were considered necessary. This does, however, require container materials which can manage these conditions. Upon thawing of the samples stratification will occur and proper mixing of the samples will be necessary. Freezing can change sample composition especially when suspended matter has not been removed. Cells of, for instance, algae or bacteria will undergo lysis because of the formation of ice crystals inside them, resulting in dispersal of the contents of the cells in the sample. Therefore it is recommended to filter samples before they are frozen.

Another preservation strategy entails the addition of preservation agents. It is common practice to preserve samples for trace metal determination by acidification to pH<2 (ISO,1994). A separate procedure is necessary for mercury. In view of the volatility of the metallic form, mercury has to be stored in its ionic form. This can be accomplished by addition of a sufficient amount of $K_2Cr_2O_7$ (0.05% (w/v) of the total sample) in order to oxidize the mercury present. A yellow color indicates that enough $K_2Cr_2O_7$ is present to keep the mercury in its ionic state, otherwise the colour will be the green/blue of chromate. The dichromate should be added to the bottles prior to the sampling trip. When chromium is also under investigation extra care should be taken to prevent cross-contamination.

When considering persistent organic contaminants one should keep in mind that persistent does not mean indestructible. For example, PAHs are reported to be sensitive to biodegradation especially when stored over extended periods of time [104,116]. Consequently for PAH samples a number of preservation measures has been suggested. Addition of organic solvents such as 40% (v/v) acetonitrile was proposed in order to reduce adsorption by container surfaces. Since organic solvents are usually volatile, inflammable, or even toxic a new approach was devised [104]. It involves the addition of surfactants at concentrations which enable the formation of micelles. Tests were done with anionic (cetylpyridium chloride), nonionic (Brij-35), and cationic (sodium lauryl sulfate) surfactants. The stabilizing effect was the same as was shown by 40% acetonitrile. The stability of aqueous indeno[1,2,3-*cd*]pyrene (one of the less soluble PAHs) was shown to be independent of the surfactant used.

The addition of 1% formaldehyde (v/v) has been applied in order to inhibit biological activity [116]. Samples of raw sewage remained stable over a period of 56 days stored at 15°C. Especially the low molecular weight PAHs, acenaphthylene, acenaphthene, fluorene, phenanthrene, and anthracene and two of the high molecular weight PAHs, benzo[*k*]fluoranthene and indeno[1,2,3-*cd*]pyrene, shown to be susceptible to biodegradation, had their stability improved.

The stability of volatile organic compounds can be enhanced by the addition of sodium bisulfate ($NaHSO_4$) until pH = 2.0 and storing samples at 4°C [117]. The low pH was shown to prevent dehydrohalogenation of chlorinated organic compounds and also reduce the degradation of aromatic compounds. The most dramatic examples were two of the least stable aromatics, ethylbenzene and styrene, which had their stabilities increased from around 14 days without acidification up to 122 days with the addition of sodium bisulfate. The stability of five (EPA target) ketones was also improved.

Cross-contamination has been shown to occur with samples containing high concentrations of volatile organic compounds [118]. Sample containers of borosilicate glass with Teflon-lined silicone disks in screw caps were checked by filling them with neat dichloromethane, *trans*-1,2-dichloroethylene, chloroform, and 1,2-dichloroethane. Storage of these containers proximal to containers filled with water resulted in low levels of cross-contamination. Storing saturated aqueous solutions of these compounds next to water did not result in any significant levels of contamination. Nevertheless it is a good practice to keep low and high level contaminant samples separated in order to reduce the risks over prolonged storage periods.

3.6 Quality control practices

In the previous paragraphs, elements of a quality system for sampling have been described. Quality systems also include feedback mechanisms for two reasons [119].
1. Feedback is used to maintain a specified level of performance. When deviating data are observed, the laboratory seeks the source of deviations and takes corrective actions.

2. Feedback provides information on the sampling process, e.g. on long-term variability. This information is important for assessing the overall performance, to define or evaluate criteria for proper functioning of sampling, and to identify weak points which need or could be improved.

Feedback is obtained through a program of measurements. For analytical laboratories, the concept of quality control of analytical data has been introduced, which is described as the set of procedures undertaken by the laboratory for continuous monitoring of operations and results in order to decide whether the results are reliable enough to be released; QCAD primarily monitors the batchwise accuracy of results on quality control materials, and precision on independent replicate analysis of test materials. Feedback is also achieved by registering observed quality problems and complaints. Such a register enables those responsible for sampling to identify weaknesses which need improvement. In this paragraph, four areas pertinent to quality control of sampling will be discussed: quality control samples, reference locations, special sampling trips and plausibility checks on the data.

3.6.1 Quality control samples

Contamination control is an important issue for sampling. Not surprisingly, much attention has been paid to the verification of contamination control. So-called sampler blanks, filter blanks, field or procedural blanks, trip or bottle blanks, and field-spiked blanks have been described in the literature [10,37]. If water samples are filtered on location, it is advisable to prepare a daily filter blank by passing a sample of ultrapure distilled water through one of the prewashed filters in the filtration apparatus, and then preserving, storing and analyzing the water in the same way as the actual water samples [37]. Trip or bottle blanks are defined as "samples of analyte-free media taken from the laboratory to the sampling site and returned to the laboratory unopened". Trip blanks consisting of pure water and preservation agents provide information on the level of contamination control arising from the sampling bottle and the chemicals used for preservation. Sampler blanks are prepared by "ultrapure distilled water poured into or permitted to pass through the sampler" [37] and can be used to detect possible sample contamination due to the sampling equipment. Field blanks are "samples of analyte-free media similar to the sample matrix that are transferred from one vessel to another or exposed to the sampling environment at the sampling site". Field (or procedural) blanks serve to determine the contamination level due to the sampling bottle, filtration equipment, additives and environment all together. If field blanks are not available, the use of a simulated or synthetic field blank is advised [13].

When analytes are very sensitive towards such processes as degradation, volatilization or adsorption, spiking of the sample is an often invoked quality assurance measure in the laboratory. Spiked samples prepared in the laboratory do not give information on analyte losses during transport and storage. Therefore [10,37], field-spiked samples are prepared at the sampling site by adding a fixed and precisely known amount of the analyte to a portion of a single water sample. The spikes enable the determination of the recovery of analytes after transport, storage, and analysis. It has been advised to obtain field-spiked samples at least once at each sampling point and at three different levels of the parameter of interest, within the concentration range capability of the analytical methodology employed [37]. It has been recommended that the level of the spike should be near the level of the investigated contaminant in order to prevent bias [13]. If the concentration level of a spike is too high the recovery of the spike might not represent the native contaminant since adsorption sites in the matrix could become saturated and the reported recoveries would be too high.

Spiked field-blanks have been recommended for correction for eventual losses of the analyte during the sampling procedure. It is noted, however, that recoveries found in spiked- field blanks may be quite different from those obtained for spiked real water samples because of the absence of sample matrix.

Duplicate samples (splits) are obtained in the field by splitting one sample into two or more identical subsamples [37]. This type of control sample can be used "to obtain the magnitude of errors owing to contamination, random and systematic errors, and any other variabilities which are introduced from the time of sampling until the time the samples arrive at the laboratory" [37].

As far as both the number of blanks and field spikes is concerned, 5 - 10% of the total number of samples is considered appropriate, provided that the method used is quite stable [14].

3.6.2 Reference locations

The use of reference locations (or control sites) has been recommended in order to furnish information on background levels [9]. As an area is investigated and data becomes available a comparison must made with similar areas so that it becomes clear whether or not the determined contaminant levels are elevated.

It is also possible to choose control sites in very undisturbed areas which are relatively uncontaminated. These areas can also serve as a control on the sampling procedure and to identify possible sources of contamination. These areas must, however, have stable contaminant levels. This approach was taken during a baseline study of contaminants in sea water organized by ICES [120]. Several stations were identified, one or more of which were to be sampled by the participants in the study. The water masses at the locations and depths concerned were known to be very stable and homogeneous, implying that the different data should be comparable despite the different times at which the samples were obtained.

In fresh water and coastal zone environments, it is difficult to find watermasses with comparable characteristics. Sampling strategies in which stations are sampled by different organizations in a certain period may, however, be useful.

3.6.3 Special sampling trips

A laboratory can organize a sampling trip conducted by a technician, after which it is repeated by another staff member completely independently. Alternatively, sampling may be carried out by two staff members, one of whom is merely making observations [121].

3.6.4 Plausibility checks

In addition to data on quality control samples, i.e. checks built into the system, plausibility checks can be carried out as tool for verification. Plausibility checks invoke expert judgement with respect to the hydrological, chemical, and biological features of the water system and of the characteristics of the measurement techniques (i.e. sampling and analysis). Over time, more insight is obtained in both, implying that the checks become better founded over time. Options include:

3.6.4.1 Inspection of data by graphical means

Graphical plots are very illustrative, e.g. plots in time, in space or frequency plots. Problems can be manifested as unexpected events, patterns, etc. Examples include:
- an observation which deviates strongly from measurements conducted before and after it ("outlier");
- a constant signal, with or without very little noise;
- a strongly fluctuating series of data; and
- deviations from expectations.

An example of the last is provided by metal-salinity relationships observed in estuarine and coastal areas [19]. When data deviate from the expected relationship, an investigation into the causes should be carried out.

3.6.4.2 Comparison with reference data

The reference data may entail data from the same area obtained in the same season in different projects, data on other, comparable water systems, and/or "historical" data, i.e. data on the water system collected in previous years. When the data do not comply with expectations, a more detailed investigation needs to take place [122].

It is emphasized, however, that noncompliance with expectations does not necessarily imply that a problem has occurred. In many cases, no obvious reason for a deviating result is found. In such cases, the data should be accepted, although they can be flagged as being suspect.

3.7 References

[1] CEN/CENELEC, *General criteria for the operation of testing laboratories*, European Standard EN 45001, Brussels, Belgium (1989)

[2] International Organization for Standardization, *General requirements for the technical competence of testing laboratories*, ISO/IEC Guide 25, Geneva, Switzerland (1982)

[3] F.A.J.M. Vijverberg and W.P. Cofino, *Control procedures: Good laboratory practice and quality assurance*, Techniques in marine environment science, International Council for the Exploration of the Sea, Copenhagen, Denmark (1987)

[4] W.P. Cofino, in: *Environmental analysis - Techniques, applications and quality assurance*, D. Barceló (Ed.), Elsevier Science Publishers, Amsterdam, The Netherlands, 359 (1993)

[5] B. Kratochvil, D. Wallace and J.K. Taylor, *Anal. Chem.*, **56**, 116R (1984)

[6] G.E. Batley and D. Gardner, *Water Res.*, **11**, 745 (1977)

[7] D. MacDougall, *Anal. Chem.*, **52**, 2242 (1980)

[8] M. Valcárcel and A. Rios, *Anal. Chem.*, **65**, 781A (1993)

[9] L.H. Keith (Ed.), *Principles of Environment Sampling*, ACS Professional Reference Book, American Chemical Society (1988)

[10] J.O. Niagru, G. Lawson, H.K.T. Wong and J.M. Azcue, *J. Great Lake Res.*, **19**, 175 (1993)

[11] Anon, *Nomenclature for Sampling in Analytical Chemistry*, Commission on Analytical Nomenclature, 1194-1208 (1990)

[12] J.C. Ellis and R.F. Lacey, in: *River Pollution Control*, M.J. Stiff (Ed.), Ellis Horwood Ltd., Chichester, United Kingdom (1980)
[13] L.H. Keith, R.A. Libby, W. Crummer, J.K. Taylor, J. Deegan Jr. and G. Wentler, *Anal. Chem.*, **55**, 2210 (1983)
[14] J.K. Taylor, *Principles of Quality Assurance of Chemical Measurements*, US National Bureau of Standards, Report NBSIR 85.3105 (1985)
[15] ISO, International Standard 5667/1, Geneva, Switzerland
[16] P. Gy, *Heterogeneity, Sampling - Homogenization*, Elsevier Science Publishers, Amsterdam, The Netherlands (1991)
[17] GESAMP-IMO/FAO/UNESCO/WMO/WHO/IAEA/UN/UNEP Joint Group of Experts on the Scientific Aspects of Marine Pollution, *GESAMP Reports and Studies*, **32**, UNESCO, Paris pp.172 (1987)
[18] P. Balls, W. Cofino, D. Schmidt, G. Topping, S. Wilson and P. Yeats, *ICES J. Mar. Sci.*, **50**, 435 (1993)
[19] W.P. Cofino, *Helgoländer Meeresunters*, **43**, 295 (1993)
[20] R.P. Richards and D.B. Baker, *Environ. Toxicol. Chem.*, **12**, 16 (1993)
[21] P.M. Chapman, G.P. Romberg and G.A. Vigers, *J. Water Pollut. Control Feder.*, **54**, 2 (1982)
[22] W.G. de Ruig and H. van der Hoet, *108th Annual Assoc. Off. Anal. Chem. Int. Meeting*, Portland, Oregon, USA (1994)
[23] ISO, International Standard 5667 series (1994)
[24] F. Smedes, *Int. J. Environ. Anal. Chem.*, **54**, 1 (1994)
[25] B.O. Jacobsen, N. Nyholm, B.M. Pedersen, O. Poulsen and P. Østfeld, *Water Res.*, **10**, 1505 (1993)
[26] X. Wang and C.P.L. Grady Jr., *Water Res.*, **28**, 1247 (1994)
[27] P. Switzer, *Water Resources Res.*, **15**, 1712 (1979)
[28] R.E. Kwiatkowski, *Environ. Monitor. Assessment*, **17**, 253 (1991)
[29] A.L. Koch, *J. Theor. Biol.*, **12**, 276 (1966)
[30] M.C. Newman, *Environ. Toxicol. Chem.*, **12**, 1129 (1993)
[31] R.O. Gilbert, *Statistical Methods for Environmental Pollution Monitoring*, Van Nostrand Reinhold, New York (1987)
[32] M.H. van der Weijden, *Waterkwaliteitsmetingen in de Maas bij Eijsden tijdens een periode van hoog water*, RIZA, Werkdocument 92.042X (1992)
[33] B.A. Bodo, *Environ. Toxicol. Chem.*, **10**, 1105 (1991)
[34] P.S. Porter, *Statistical Analyses of Water Quality Data Affected by Limits of Detection*, Thesis, Colorado State University, Colorado, USA (1986)
[35] B.M. Lesht, *J. Great Lake Res.*, **14**, 325 (1988)
[36] R.E. Kwiatkowski, *J. Great Lake Res.*, **4**, 19 (1978)
[37] World Meteorological Organization, *Manual on water-quality monitoring*, Operational Hydrology report No.27, WMO, Geneva, Switzerland (1988)
[38] R.C. Ward, J.C. Loftis, K.S. Nielsen and R.D. Anderson, *J. Water Pollut. Control Fed.*, **51**, 2292 (1979)
[39] D.P. Lettenmaier, *Water Resources Res.*, **14**, 884 (1978)
[40] J.M. Krajca (Ed.), *Water Sampling*, Ellis Horwood Ltd., Chichester, United Kingdom (1989)

[41]   L.G. Blackwood, *Environ. Sci. Technol.*, **25**, 1666 (1991)
[42]   G. Kateman and L. Buydens, *Quality Control in Chemical Analysis*, 2nd Edition, John Wiley & Sons, Inc. (1993)
[43]   D. Berryman, B. Bobee, D. Cluis and J. Hammerli, *Water Resources Bull.*, **24**, 545 (1988)
[44]   R. Farrel, *Methods for classifying changes in environmental conditions*, Tech. Report, VRF-EPA7.4-FR80-1, Ann Arbor, Michigan, USA (1980)
[45]   R.M. Hirsch, J.R. Slack and R.A. Smith, *Water Resources Res.*, **18**, 107 (1982)
[46]   D.P. Lettenmayer, *Water Resources Res.*, **12**, 1037 (1976)
[47]   R.M. Hirsch and J.R. Slack, *Water Resources Res.*, **20**, 727 (1984)
[48]   ISO, International Standard/REMCO N271, Draft, Geneva, Switzerland (1993)
[49]   J.K. Taylor, *Anal. Chem.*, **600A** (1983)
[50]   S.R. Esterby and P.E. Bertram, *J. Great Lake Res.*, **19**, 400 (1993)
[51]   J.M. Bewers and H.L. Windom, *Mar. Chem.*, **11**, 71 (1982)
[52]   J.M. Bewers, P.A. Yeats, S. Westerlund, B. Magnusson, D. Schmidt, H. Zehle, S.S. Berman, A. Mykytiuk, J.C. Duinker, R.F. Nolting, R.G. Smith and H.L. Windom, *Mar. Pollut. Bull.*, **16**, 277 (1985)
[53]   C.G. Capodaglio, P. Toscano, G. Cescon, G. Scarponi and H. Muntau, in: *Analytical Quality Control and Assessment Studies in the Mediterranean Basin*, AQUACON MedBas (1993)
[54]   L.J. Ottendorfer, *Proc. Anal. Div. Chem. Soc.*, **15**, 53 (1978)
[55]   IUPAC, *Sampling water - a comparison of water samplers*, Draft document, Working Group on Analytical Quality Assurance, IUPAC Commission V2 (1992)
[56]   C.S. Wong, K. Kremling, J.P. Riley, W.K. Johnson, V. Stukas, P.G. Berrang, P. Erickson, D. Thomas, H. Petersen and B. Imber, in: *Trace Metals in Seawater*, C.S. Wong, E. Boyle, K.W. Bruland, J.D. Burton and E.D. Goldberg (Eds.), Plenum Press, New York, USA (1983)
[57]   H.G.K. Ordelman, P.C.M. van Noort, J.M. van Steenwijk, T.E.M. ten Hulscher, M.A. Beek, J. Botterweg, R. Faasen, P.C.M. Frintop and H.G. Evers, *Watersystem reconnoitring for dithiocarbamates*, RIZA Report 93.025, Haren, The Netherlands (1993)
[58]   H.G.K. Ordelman, P.C.M. van Noort, J.M. van Steenwijk, J.E.M. Beurskens, R. Faasen, M.A. Beek and H.G. Evers, *Watersystem reconnoitring for phenolherbicides*, RIZA Report 94.004, Haren, The Netherlands (1994)
[59]   H.G.K. Ordelman, P.B.M. Stortelder, T.E.M. ten Hulscher, F.H. Wagemaker, J.M. van Steenwijk, J. Botterweg, P.C.M. Frintrop and H.G. Evers, *Watersystem reconnoitring for carbametes*, RIZA Report 93.010, Haren, The Netherlands (1993)
[60]   D.T.E. Hunt and A.L. Wilson, *The Chemical Analysis of Water*, 2nd Edition, The Royal Society of Chemistry, Cambridge, United Kingdom (1986)
[61]   L.P.M.J. Wetsteyn, *Phyto- and zooplankton before and after passage through the Good Measurement Pump*, Internal Report GWIO-90.16044, Rijkswaterstaat, The Hague, The Netherlands (1990)
[62]   G. den Hartog and G. Spronk, *Pump for the sampling of surface water (in Dutch)*, Internal Report GWIO-91.16011, Rijkswaterstaat, The Hague, The Netherlands (1991)

[63] G. den Hartog and G. Spronk, *Testing for a Good Measurement Pump (in Dutch)*, Internal Report GWIO-90.16044, Rijkswaterstaat, The Hague, The Netherlands (1990)
[64] L. Mart, *Fresenius Z. Anal. Chem.*, **296**, 350 (1979)
[65] M.B. Yunker, F.A. McLaughlin, R.W. Macdonald, W.J. Cretney, B.R. Fowler and T.A. Smyth, *Anal. Chem.*, **61**, 1333 (1989)
[66] Ph. Quevauviller and O.F.X. Donard, *Appl. Organometal. Chem.*, **4**, 353 (1990)
[67] C.C. Paterson and D.M. Settle, in: *Accuracy in Trace Analysis: Sampling, Sample Handling, Analysis*, P.D. LaFleur (Ed.), NBS Special Publication 422, Washington, USA (1976)
[68] J.W. Owens, E.S. Gladney and W.D. Purtymun, *Anal. Lett.*, **16**, 253 (1980)
[69] B.E. Logan, B.C. Alleman, G.L. Amy and R.L. Gilbertson, *Water Res.*, **28**, 1533 (1994)
[70] H.G.K. Ordelman, P.C.M. van Noort, J.M. van Steenwijk, T.E.M. ten Hulscher, M.A. Beek, J. Botterweg, R. Faasen, P.C.M. Frintop and H.G. Evers, *Watersystem reconnoitring for dithiocarbamates*, RIZA Report 93.025, Haren, The Netherlands (1993)
[71] N.T. Loux, A.W. Garrison and C.R. Chafin, *Intern. J. Environ. Anal. Chem.*, **38**, 231 (1990)
[72] L. Kosta, *Talanta*, **29**, 985 (1982)
[73] ISO, International Standard 5667-3, Geneva, Switzerland (1994)
[74] L. Kosta, *Talanta*, **29**, 985 (1982)
[75] J.V. Klump, R. Paddock and D. Lovaldo, *J. Great Lake Res.*, **18**, 309 (1992)
[76] L. Brugmann, E. Geyer and R. Kay, *Mar. Chem.*, **21**, 91 (1987)
[77] P. Freimann, D. Schmidt and K. Schomaker, *Mar. Chem.*, **14**, 43 (1983)
[78] G. Kniewald, M. Kwokal and M. Branica, *Mar. Chem.*, **22**, 343 (1987)
[79] M. Zief and K. Ruthann, *Solid Phase Extraction for Sample Preparation*, J.T. Baker Inc. (1988)
[80] E.R. Brouwer, H. Lingeman, U.A.Th. Brinkman, *Chromatographia*, **29**, 415 (1990)
[81] D.R. Green and D. Le Pape, *Anal. Chem.*, **59**, 699 (1987)
[82] E.R. Brouwer, I. Liska, R.B. Geerdink, P.C.M. Fintrop, W.H. Mulder, H. Lingeman and U.A.Th. Brinkman, *Chromatographia*, **32**, 445 (1991)
[83] E.R. Brouwer, D.J. Van Iperen, I. Liska, H. Lingeman and U.A.Th. Brinkman, *Intern. J. Environ. Anal. Chem.*, **47**, 257 (1992)
[84] I. Liska, E.R. Brouwer, A.G.L. Ostheimer, H. Lingeman and U.A.Th. Brinkman, *Intern. J. Environ. Anal. Chem.*, **47**, 267 (1992)
[85] S. Litten, B. Mead and J. Hassett, *J. Environ. Toxicol. Chem.*, **12**, 639 (1993)
[86] G.D. Johnson, *Environ. Sci. Technol.*, **25** (1991)
[87] A. Södergen, *Environ. Sci. Technol.*, **21**, 855 (1987)
[88] J.N. Huckins, M.W. Tubergen and G.K. Manuweera, *Chemosphere*, **20**, 533 (1990)
[89] A. Zlatkis, S. Weisner and L. Ghaoui, *Chromatographia*, **21**, 19 (1980)
[90] A. Zlatkis and R.P.J. Ranatunga, *Anal. Chem.*, **62**, 2471 (1990)
[91] J.W. Sproule, W.Y. Shiu, D. Mackay, W.H. Schroeder, R.W. Russel and F.A.P.C. Gobas, *Environ. Toxicol. Chem.*, **10**, 9 (1991)
[92] R.G. Melcher and P.L. Morabito, *Anal. Chem.*, **62**, 2183 (1990)
[93] D.R. Green, J.K. Stull and T.C. Heesen, *Mar. Pollut. Bull.*, **17**, 324 (1986)

[94]   A.P. Murray, B.J. Richardson and C.F. Gibbs, *Mar. Pollut. Bull.*, **22**, 595 (1986)
[95]   G.R. Martin, J.L. Smoot and K.D. White, *Water Environ. Res.*, **64**, 866 (1992)
[96]   S.S. Berman and P.A. Yeats, *CRC Crit. Rev. Anal. Chem.*, **16**, 1 (1985)
[97]   J.H. Hermans, F. Smedes, J.W. Hofstraat and W.P. Cofino, *Environ. Sci. Technol.*, **26**, 2028 (1992)
[98]   A.R. Abdel-Moati, *Water Res.*, **24**, 763 (1990)
[99]   D.H. Bauw, P.G.M. de Wilde, G.A. Rood and Th.G. Aalbers, *Chemosphere*, **22**, 716 (1991)
[100] J. Kukkonen and A. Oikari, *Water Res.*, **25**, 455 (1991)
[101] S.S. Berman, R.E. Sturgeon, J.A.H. Desaulniers and A.P. Mykytink, *Mar. Pollut. Bull.*, **14**, 69 (1983)
[102] R. Massee, F.J.M.J. Maessen and J.J.M. De Goeij, *Anal. Chim. Acta*, **127**, 181 (1981)
[103] J.G. Fabris, K.A. Smith, J.E. Atack and A.L. Kilpatrick, *Water Res.*, **20**, 1692 (1986)
[104] A. López García, E. Blanco González, J.I. García Alonso and A. Sanz Medel, *Anal. Chim. Acta*, **264**, 241 (1992)
[105] J.A. Campbell, M.J. Gardner and A.M. Gunn, *Anal. Chim. Acta*, **176**, 193 (1985)
[106] M. Tsezos and J.P. Bell, *Water Res.*, **25**, 1039 (1991)
[107] D.E. Robertson, *Anal. Chim. Acta*, **42**, 533 (1968)
[108] J.W. Owens, E.S. Gladney and W.D. Purtymun, *Anal. Lett.*, **16**, 253 (1980)
[109] A.G. Kelly, I. Cruz and D.E. Wells, *Anal. Chim. Acta*, **276**, 2 (1993)
[110] Ph. Quevauviller and O.F.X. Donard, *Fresenius J. Anal. Chem.*, **339**, 6 (1991)
[111] R. Otson, D.T. Williams, P.D. Bothwell, R.S. McCullough and R.A. Tate, *Bull. Environ. Contam. Toxicol.*, **23**, 311 (1979)
[112] L.H. Keith (Ed.), *Compilation of EPA's Sampling and Analysis Methods*, Lewis Publishers, Chelsea, USA (1992)
[113] L.A. Clementson and S.E. Wayte, *Water Res.*, **26**, 1171 (1992)
[114] S.H. Harrison, P.D. LaFLeur and W. Zoller (Eds.), in: *Accuracy in Trace Analysis: Sampling, Sample Handling, Analysis*, NBS Special Publication 422, Washington, USA (1976)
[115] H. Scheuermann and H. Hartkamp, *Fresenius J. Anal. Chem.*, **315**, 430 (1983)
[116] N.D. Bedding, A.E. McIntyre, J.N. Lester and R. Perry, *J. Chromatrogr. Sci.*, **26**, 606 (1988)
[117] M.P. Maskarinec, L.H. Johnson, S.K. Holladay, R.L. Moody, C.K. Bayne and R.A. Jenkins, *Environ. Sci. Technol.*, **24**, 1665 (1990)
[118] S.P. Levine, M.A. Puskar, P.P. Dymerski, B.J. Warner and C.S. Friedman, *Environ. Sci. Technol.*, **17**, 125 (1983)
[119] W.P. Cofino, *Encyclopedia for Anal. Sci.*, in press
[120] ICES, *A review of measurements of trace metals in coastal and shelf sea water samples collected by ICES and JMP laboratories during 1985-1987*, Report No.178, Int. Council Expl. Sea, Copenhagen, Denmark (1988)
[121] NKO/STERIN/STERLAB, Dutch Accreditation Body, *Supplementary criteria for sampling*, draft document (1995)
[122] NKO/STERIN/STERLAB, Dutch Accreditation Body, *Guidance for interpretation of accreditation criteria for environmental laboratories* (1989)

4.

Quality assurance of pre-determination steps for dissolved nutrients in marine samples

A. Aminot
IFREMER, B.P. 70, F-29280 Plouzane, France

The term nutrients denotes the compounds which contain the indispensable nutritive elements for phytoplankton growth, i.e., nitrogen, phosphorus and silicon. These elements may be present in the marine environment in various chemical species, including both inorganic (mainly nitrate, nitrite, ammonia, orthophosphate, and silicate) and organic. Some of these compounds may be found in both dissolved and particulate phases. This chapter focuses on dissolved nutrients, with particular attention being devoted to those most often determined, the inorganic forms.

Nutrients are important determinands in both oceanographic research and environmental monitoring. Their concentration and behavior have been studied in a wide variety of programs describing element cycling in connection with biological activity (from local to world scale) eutrophication studies and modeling assessment of continental input effects or simply as markers of water bodies and mixing. Because of the wide interest of nutrients, there is concern about the quality assurance of their determination throughout the entire marine domain, from the open ocean to the smallest estuary. Consequently, much work has been done on the determination of inorganic nutrients in sea water samples over the past five decades and procedures have now reached a satisfactory stable level of performance.

The basic protocols, used by a large majority of analysts, can be found in numerous manuals on sea water analysis [1-6]. But there is no point in measuring a determinand concentration in a sample if the integrity of the sample has not been carefully preserved, from the water body to the laboratory benchtop. The determination itself is simply one step in the whole procedure including operations such as sampling, subsampling, filtration and storage. However, these pre-determination steps have received less attention, and only scarce information is available in most manuals, except in two dealing specifically with these topics [4,5].

Whereas determination problems can be assessed through intercomparison exercises (see for instance [7]), it is particularly difficult to check the reliability of the preceding steps. It is worth remembering that the entire protocol, from sampling to the final result of a determination, should be considered as belonging to the analytical chain and consequently under the analyst's responsibility. None of the steps should be thought of minor importance and ignored: the chain is as weak as its weakest link.

The aim of this chapter is firstly to focus on the basic principle which should be borne in mind to ensure quality at each of the pre-determination steps; secondly to present the state of the art of practices in use for preservation of sample integrity; and thirdly to guide analysts confronted with specific problems.

4.1 Outline of the context

4.1.1 Introductory remarks

Before dealing with the successive steps through which a water aliquot is taken from its original medium to the nutrient determination procedure, we should summarize the working context which plays an important role in fixing the suitable pre-determination protocol. Nautical means, water characteristics, and processes involving nutrients are main criteria in this context. Given the complex relationship between these criteria, thorough knowledge of potential sources of errors at sea is a key factor for adaptation of the procedure with respect to quality assurance principles. Presence of skilled analysts on board for sampling and related operations is, therefore, a pre-condition for the production of reliable data. It is essential that the analysts be aware of the specificity of nutrients. Indeed, since these compounds are intimately associated with all life processes they are (or should be assumed to be) present everywhere, at all times in the close vicinity of the samples, with the consequent risk of contamination. Additionally, any compound containing a nutritive element can be converted into another through normal bacterial activity in natural waters. The risks of contamination and modification of concentrations are constant and require analysts' unwavering vigilance for potential sources of sample alteration.

4.1.2 Nautical means

Several types of ship are necessary to study the various marine environments. Open sea studies require large oceanographic vessels, with working equipment very similar to those of onshore laboratories. On board analysts can find optimum conditions for quality assurance. Satisfactory conditions are also generally encountered on board the smaller ships belonging to marine institutes, used for coastal and large estuary studies. However, in these nearshore areas, various other types of small ships, such as fishing boats or boats lent by official bodies involved in monitoring (e.g. the navy or harbour authorities) are occasionally or regularly used. Most of these units are not adequately equipped for water sampling and handling, which makes it difficult to apply quality assurance rules. Finally, the least satisfactory working conditions are encountered when sampling has to be done using outboard motor boats, typically rubber rafts, in shallow areas close to the shore, at low tide or in small estuaries. In poor working conditions, the analyst will often have to accept the alternative of *minimizing the risks of low quality data* rather than trying to attain the highest possible level of quality. However, it should never be forgotten that, whatever the quality of equipment on the ship, working at sea presents greater difficulties than working in onshore laboratory conditions. Human capabilities are more or less reduced by the ship's instability and this becomes, in many instances, the actual limiting factor for quality assurance.

4.1.3　　Water characteristics

From the open sea to upstream sections of estuaries, the marine analyst will encounter very different types of water, whose treatment will vary accordingly. Oceanic surface water is characterized by the lowest nutrient concentration levels in a high salinity medium (34-38 PSS) very poor in suspended matter (generally less than 0.5 mg per litre). At the other extreme, in estuarine waters high concentrations for nutrients are found in a turbid medium where salinities range from almost zero to about 30 PSS. Between estuarine and oceanic waters lies the domain of coastal waters with relatively high salinity, a more or less detectable continental impact, and variable turbidity influenced by river plume extension, tide and wave induced resuspension, and phytoplankton blooms. Additionally, nutrient concentrations may vary from undetectable levels in summer to high ones in winter. Depending on the various characteristics of the water to be analyzed for nutrients, it appears clear that the treatment of the samples will involve specific care or operations. In other words, the relative importance of precautions and procedures applied to samples may differ from one type of water to another.

4.1.4　　Processes altering nutrient concentrations

4.1.4.1　　General comments

Some knowledge of the natural processes which can alter nutrient concentrations in a sample may be useful to guide rational application of pre-analytical treatments. It must be clear that as long as no treatment has been applied to a sample, this sample potentially exhibits the same biological activity as the water body it comes from. However, differences in the relative magnitude of the processes are introduced by modification of the water's physical environment (turbulence, light, temperature, bottle walls...). Dissolved nutrients can be removed from or released into the water sample through the various processes described hereafter and summarized in Table 4-1. Growing microorganisms (phytoplankton, bacteria) incorporate dissolved nutrients into their biomass; this conversion from non-living to living (and from dissolved to particulate) material is called assimilation. The energy needed for this process comes either from light (photosynthesis) or from chemical reactions. In the latter case, some inorganic or organic forms of nutrients are transformed into other compounds which are released into the water. It should be noted that several of the processes involving nutrients can take place simultaneously and that, because of their different rates, *variation of nutrients in an unpreserved sample is rarely predictable*. Additional information about the nitrogen cycle in the sea is available in the literature (for instance [8,9]).

4.1.4.2 Processes

Photosynthesis is the main mechanism for primary production in sea water. It enables microalgae (phytoplankton) growth from dissolved inorganic carbon and nutrients according to the following equation [10]:

$$106CO_2 + 16NO_3^- + HPO_4^{2-} + 122H_2O + 18H^+ => C_{106}H_{263}O_{110}N_{16}P + 138O_2$$

Light is necessary as the energy source. Nitrate, nitrite, or ammonia can be used as nitrogen donors, with a decrease in energy requirement from oxidized to reduced forms. Additionally, silicate is simultaneously incorporated by diatoms into their skeleton.

Mineralization (or regeneration) of detrital and dissolved organic matter by heterotrophic bacteria is, in most instances, the main internal source of inorganic nutrients. Mineralization of living-derived matter is the reverse reaction of that representing photosynthesis (see above) [10]. It supplies energy and essential elements to the cells for biomass production while excess nutrients, notably ammonia and phosphate, are released into the water. In specific cases, when the organic substrate contains no N or P, like hydrocarbons or carbohydrates, bacteria take up dissolved inorganic nutrients from the surrounding water for their biomass synthesis. Plastic bottles may constitute a typical nutrient-free source of organic matter. After cell lysis, silicate is also recycled by dissolution.

Table 4-1: Summary of biological processes altering nutrient concentration.
Abbreviations: DIN = Dissolved Inorganic Nitrogen (nitrate, nitrite and ammonia); DON = Dissolved Organic Nitrogen; DOP = Dissolved Organic Phosphorus.

Process	Loss	Gain	Remark
Photosynthesis	DIN, PO$_4$, Si		Light necessary
Mineralization (1)	DON, DOP	NH$_4$, PO$_4$	
Mineralization (2)	DIN, PO$_4$		If excess or organic C
Excretion		DON, NH$_4$, PO$_4$	
Cell lysis		DON, DOP, (PO$_4$)	
Nitrification	NH$_4$	NPO$_2$	Nitritation
Nitrification	NO$_2$	NO$_3$	Nitratation
Denitrification	NO$_3$	(NH$_4$, NO$_2$)	In anoxic conditions

Cell lysis, which occurs after cell death, releases mainly organic nutrients. However, if more inorganic nutrients than needed have been accumulated ("luxury consumption"), they will also be released (particularly phosphate).

Excretion of microorganisms can be of several types. Living phytoplankton cells are known to release soluble organic nutrients (amino acids, ...) by exudation. Other microplankton cells, such as heterotrophic microflagellates or ciliates, excrete ammonia and phosphate. Larger zooplankton organisms, such as copepods, abundant after microalgal blooms, also excrete particulate nutrients (faecal pellets) which can be decomposed by bacteria and redissolved.

Nitrification converts ammonia into nitrite (nitritation), then nitrite into nitrate (nitratation). Two specific bacterial species are involved in this process which takes place only in oxygenated waters. Because of their low yield, these processes produce a small amount of biomass [11].

Denitrification is a general term for the reduction of nitrate which takes place in anoxic waters and can be undertaken by different types of heterotrophic bacteria. Nitrate is essentially used as an oxygen donor and is converted into gases, e.g., N_2 and N_2O. Side reactions can produce ammonia from N-rich organic substrate while some fermentative species can produce nitrite [11,12].

4.2   Sampling

4.2.1   Definition

Sampling can be defined as the operation which consists in taking an aliquot of the water body under study for subsequent analysis. In most instances, this water aliquot will again be divided into subaliquots deposited in specific vials for specific determinations (see Section 4.3). Sampling involves two major devices, the ship and the water sampler.

4.2.2   The ship

The ship is a source of contamination of the surrounding surface water. While this problem has been clearly identified for organic compounds and trace metals, it is often underestimated or ignored for nutrients. Several kinds of effluent are discharged in surface or sub-surface sea water: cooling water, domestic-type waste (toilet and kitchen waters) and bilge water. Domestic wastes unquestionably represent the major source of contamination by compounds containing nitrogen and phosphorus (ammonia, phosphate, organic nitrogen, especially urea, and organic phosphorus). Cooling water represents a different type of risk. Since it is pumped from a certain level below the surface and discharged at the sea surface, perturbation of water body concentrations is to be feared, depending on the cooling water flow and on the local vertical nutrient gradient (generally related to that of salinity) in the upper water layer. Bilge water may be considered as generating similar modifications. Although large modern vessels collect certain effluents in thanks, assurance should be given that no uncontrolled emptying could occur at crucial moments during the cruise, especially in station. However, since most ships currently in use have no special equipment, great care must be taken to ensure that sampling is not performed on the same side as effluents are discharged. If outlets are present on both sides of the ship, the analyst should obtain information about the nature of the waters discharged on the sampling side and whether they can be interrupted in station.

Another source of perturbation of surface water is the turbulent mixing generated by the propellers to position or to keep the ship in station. It is important to wait until stabilization is restored, even at the price of slight drift. The above sources of nonrepresentativness of a sample are inherent to normal ship operation and are therefore permanent. Additional sources can be identified as a result of occasional use of various maintenance products. Some antirust products comprise almost pure phosphoric acid, 1 kg of which corresponds to a mean contamination of 0.05 μmol/l of phosphate in 2.10^8 m^3 (in other words a 20 cm layer over an area of 1 km^2). Many floor cleaning products are rich in ammonia. Attention must be paid to possible use of such products by the crew during the cruise. These types of insidious random contamination can pass unnoticed since they are not discharged through the outlets. Moreover, the products, which are rinsed off the deck with a water gush, trickle down the hull to the sea surface and represent a potential source of contamination for a long time after their use.

Finally the hydrowire, the device needed for deep immersion of water samplers, may be considered free from interference in nutrient sampling.

The general conclusion from the examination of the adverse effect of the ship on nutrient concentration, is that the larger the ship, the greater the risk of perturbation; pneumatic boats are the "cleanest" units. Consequently, from small estuaries towards open sea, because of the decrease in nutrient concentrations, coupled with the increase in ship size, relative perturbation increases rapidly and the importance of the impact of the ship has to be taken into account accordingly.

4.2.3 The water sampler

For nutrients, it is extremely unusual to collect marine water samples directly in the vial where they will be stored until analysis. Intermediate water sampler (also called sampling bottles) are necessary equipment. Several types of water samplers are currently in use in the marine domain (for a more extensive description, see [5]). Their purpose is to isolate from the water body under study the aliquot of water which will be brought up on board as a representative part of that water body for subsequent determination of its characteristics. Depending on the sampling depth, the water can remain in the sampler for up to several hours. No modification of the concentration of the determinands of interest should therefore be generated by the sampler itself during this period of time, either as a result of exchanges with the sampler wall or any chemical process converting one nutritive form of an element into another (see section 4.1.4).

Modern water samplers are generally made of plastic and most have been used at sea by numerous workers for a long time. From this wide experience, it appears that no special problem has been encountered with conventional water samplers as a result of their lack of chemical inertness towards nutrients [13]. Although not in common use, glass samplers used in the past (and possibly still in use) can leach silicate into the sample. Since this process depends on several factors (salinity, pH, temperature), users should check their sampler if silicate is to be determined.

Despite rare interactions with nutrients, due attention must be paid to water samplers regarding their handling and maintenance. New water samplers should preferably be cleaned in the same way as subsample bottles: Stefansson and Olafsson recommend rinsing or soaking in diluted (about 1 mol/l) hydrochloric acid, followed by thorough rinsing with deionized water, especially when phosphate and silicate are concerned [14]; Koroleff recommends brushing with a nutrient-free detergent, followed by thorough rinsing [15]. This treatment has to be repeated from time to time, more frequently when biologically rich or turbid waters are sampled.

When not in use, the water samplers should be carefully drained and stored closed, in a clean place. On board, it is recommended they be maintained in an upright position in adequate holders, which avoids contact with contaminated materials and facilitates subsampling. The samplers should be opened immediately prior to immersion for sampling and kept closed the rest of the time. When handling open samplers to attach them to the hydrowire, great care should be taken not to grasp them by the aperture or to inadvertently touch the internal side of the sampler or the lids. It has in fact been demonstrated that fingerprints are a severe source of nutrient contamination [16].

During the descent to their sampling depth, water samplers, other than those used on the surface, are extensively rinsed with sea water. In order to rinse the surface sampler, it is good practice to immerse the samplers a few meters deeper than necessary and then to return it to the sampling depth.

On board small rubber boats, sampling is often somewhat different from on large vessels. Because of the limited space, water sampler holders can not usually be used. Users should still verify that the samplers are placed in satisfying conditions. Immersion of the sampler is often done by hand for surface water collection and using a thin rope for sub-surface water, but the normal precautions pertain to those types of sampling. More questionable is the use of a bucket, sometimes mentioned as an alternative for collection of surface samples [13]. Indeed, the bucket should not be recommended for several reasons: i) the immersion depth is not fixed and strong surface gradients may affect the representativeness of the sample; ii) the surface film is collected, including all floating particles; iii) when pulling up the bucket on board, water running down along the rope falls down into the sample and contaminates it; and iv) in the absence of a bottom stopcock, correct subsampling for nutrients cannot be achieved.

4.3   Subsampling

4.3.1   Definition

Once the water sampler is brought on board, its content is divided into several separate aliquots collected in the appropriate bottles for specific determinations; this step constitutes the subsampling operation. The smaller aliquots now obtained are in fact actually called "samples" (not subsamples) according to the usual analyst terminology.

4.3.2   The sample bottles

The bottles in which the subsamples are collected, actually called "sample bottles", have to meet certain requirements in respect of the following general rules:
- the material must be chemically inert towards nutrients for the amount of time the sample remains in the bottle;
- the bottle must not be altered by storage techniques (e.g., acid or deepfreezing);
- the closure must be leakproof, and of a type which guarantees the absence of contamination risk.

The question whether it is preferable to collect a unique sample for all nutrients or separate samples for each nutrient is frequently asked. Several considerations guide the choice: the nutrients to be determined, the sample bottle material, the storage procedure and time, the order and grouping of analyses, etc. A unique sample is satisfactory if i) the container wall does not interact with the nutrients of interest; ii) the sample pre-treatment and the storage method are adequate for all these nutrients; and iii) all determinations occur within a short period of time (unless it is certain that withdrawing successive aliquots cannot alter concentrations). If specific constraints do not enable quality assurance rules to be applied, several nutrient samples must be drawn.

4.3.2.1   The bottle material

Plastic bottles are suitable for all nutrients if the samples are stored for a short time, i.e., a few hours, or are rapidly frozen. Polyethylene and polypropylene seem satisfactory, but the latter suffers badly on multiple deep-freezing. It is known that plastics present a surface suitable for bacterial growth, which is why they are sometimes not recommended for nutrient sample bottles. Grasshoff described an unexplained decrease in the amount of nitrate in sea water stored several days at room temperature in polyethylene bottles [17]. Similar experiments performed in our laboratory with aged, filtered (1.2 μm) sea water confirmed the loss of nitrate and phosphate in polyethylene bottles while almost no loss was observed in glass bottles. From these results there is evidence that biodegradation of organic carbon leached from the plastic walls took place in the sample, nitrate and phosphate being used as the only sources of nutritive elements. The hypothesis of strong adsorption of phosphate ions on plastic bottle walls, especially polyethylene, must be rejected considering the results of some authors [18] and our own experiments (unpublished): each time the biological activity is inhibited, by use of either mercuric chloride or heat treatment, the phosphate concentration remains stable in plastic bottles.

Glass is often recommended, except for silicate determination since this is gradually soluble in sea water and therefore contaminates the sample. Recent work has shown that phosphate is also leached from glass during long storage [19]. Thus, glass is completely suitable for nitrogenous compounds only, except when borosilicate glass is used [3, 19]. Plain glass may be used for phosphate over very short storage times or if the samples are frozen, since freezing prevents glass dissolution. However, when frozen, glass bottles can break if the sample salinity is low and/or if insufficient room is left for ice expansion. Glass is also the recommended material for storage of samples for the determination of organic nutrients, since it can be adequately cleaned.

4.3.2.2   The bottle closure

The closure is an important, often neglected, component of the sample bottle. Its role is firstly, to isolate the bottle contents from the surrounding atmosphere and, secondly, to prevent loss of water; it must therefore be absolutely leakproof under normal tightening. Screw caps are the most appropriate systems for either plastic or glass bottles. Ground glass stoppers must be strictly avoided. Leakproof closures proposed by manufacturers are either two-piece or one-piece type (Figure 4-1). With two-piece closures, the seal may be obtained by two means: either a flat, soft liner is placed in the cap, or a plastic (polyethylene) insert is introduced into the bottle mouth before the cap is screwed on. One-piece closures are lineless screwcaps; a seal ring, molded inside the closure, fits tightly against the beveled inner edge of the bottle neck. *Only good quality one-piece closures should be used.* Two-piece models exhibit numerous disadvantages which increase contamination risk. Internal components (liner or insert) are difficult to handle and to clean. The liner should be removed from the cap for satisfactory cleaning or rinsing, otherwise water can be trapped between cap and liner then expelled when the cap is screwed on to the bottle neck. In addition, the liner tends to escape from the cap and fall on to the ship deck at the crucial moment of subsample collection. Closures with an insert are particularly unsuitable since the insert can neither be removed nor placed into the bottle neck without contact with the fingers. When removed for filling the bottle, the insert must be held in the hand or placed somewhere and is therefore liable to severe contamination.

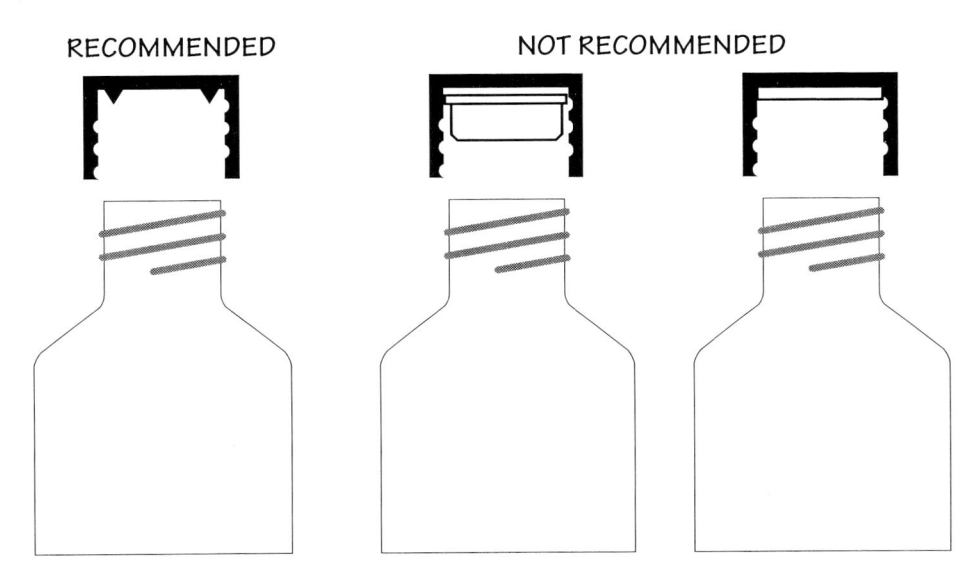

Figure 4-1:   Bottle closure types: one-piece, linerless, caps (left) are recommended.

4.3.2.3   Bottle cleaning

Before the first use of new bottles for the determination of inorganic nutrients, it is recommended that they be cleaned with a nutrient-free detergent or with hydrochloric acid (1 mol/l) and rinsed thoroughly with deionized water. The cleaning procedure is again regularly applied after several uses, more frequently when unfiltered or rich waters are sampled regularly, and especially when the samples are not frozen. This will remove bacterial films growing on the bottle wall, therefore preventing nutrient consumption or release. Brushing is recommended for more effective cleaning [15]. Once the nutrients have been determined in the sample, the bottle must be emptied, bottle and cap thoroughly rinsed with deionized water, drained by shaking (preferably dried in a clean oven) and stored capped. *Never store sample bottles containing sea water* after all determinations have been performed. For organic nutrients, the glass bottles must be cleaned in such a way that residual organic compounds are decomposed by the process. Pyrolysis at 450-500 °C or treatment with sulfochromic acid are recommended. Caps should be thoroughly rinsed with deionized water and dried, and bottles stored capped.

4.3.3 The subsampling procedure

Although simple, subsampling has to be done very carefully since the (sub)sample has to be representative of the entire contents of the water sampled, i.e., in fact, of the water body. At this stage, problems may appear in areas of strong surface stratification, such as in estuaries, since the content of the sampler can exhibit a vertical gradient of concentration. When such a situation is suspected, the only solution is to homogenize the content of the sampler by inverting it several times (after a little water has been removed to enable efficient mixing). If that operation is omitted, successive subsamples drawn off from the sampler will originate from different water layers and the concentration of the determinands will no longer be related to each other. Note that mixing will alter the concentration of dissolved gases; consequently a specially adapted procedure is necessary if oxygen is to be determined.

Subsampling is then undertaken without delay. The samples are drawn in the right order with respect to the risk of change in concentrations (gases, pH, nutrients..., for details, see references [1-6]). For consistency, *samples for salinity and nutrients must be successively drawn off* because of the strong relationship between these determinands in coastal and estuarine waters. Before a sample bottle is filled, it must be rinsed at least twice, each time with about one tenth of the bottle capacity. The cap is screwed on, the bottle shaken, and the water poured into the cap for complete rinsing of the thread. Then the sample is collected (no more than 3/4 of the bottle capacity), and the cap firmly tightened.

Regarding contamination, experience has shown that it is not infrequent, especially under rough sea conditions, that the operator touches the stopcock orifice without being aware of the fact. In order to minimize the risk of such contamination, a simple solution consists in fitting the stopcock with a piece (10-15 cm) of soft plastic tube. Wearing gloves may be an alternative, but this is frequently not applicable in practice. It must be stressed that atmospheric contamination, especially by ammonia, remains possible on the deck from the ship's smoke or from a smoker. Contamination during subsampling is typically of a random nature and should particularly be kept in mind in areas where low nutrient concentrations are encountered and in summer, when nutrients have been consumed by phytoplankton.

4.4 Filtration

4.4.1 Definition

Filtration consists in screening the water sample to separate dissolved substances from particles. But, because of the continuous nature of the size spectrum of particles in an aquatic medium, the application of that definition requires that an operational limit be fixed, somewhat arbitrarily, between the two types of material. A limit value of around 0.5 μm is commonly accepted in oceanography [3]. The wide use of commercial filtration membranes with a nominal pore size of 0.45 μm has almost imposed that figure as a norm. It is interesting to note that almost all living bacteria are retained by such membranes and, consequently, that bacteria are implicitly considered as the smallest particles. All material passing though the filter is therefore assumed to be dissolved.

4.4.2 Is filtration necessary?

Bearing the above definition in mind, when the substances under consideration are dissolved, the water should be filtered through 0.45 µm filter in order to remove all particulate matter before the determination of the compounds. However, beyond the definition itself, the main question is: does suspended matter interfere, in the post-sampling protocol, in such a way that erroneous concentrations will result?

The necessity for filtration can finally be summarized as follows: filtration is necessary when the magnitude of interference from particles in the determination of dissolved compounds exceeds what is considered acceptable for the purpose. Following that approach, filtration is carried out in order *to remove potentially interfering substances* but not for separating substances into different size classes. Methods for particle removal may therefore vary as a function of the type of interference expected, using different pore size filtration or replacing filtration by centrifugation whenever applicable. Removal of most microalgae, for instance, will undoubtedly improve nutrient stability. Potential interference from particles in nutrient determination is summarized in Table 4-2. Additionally, it is worth remembering that each handling of a sample exposes it to contamination risks and should be done with this consideration in mind: filtration should therefore be carried out only if it is unavoidable. Open sea waters, for instance, are preferably not filtered since filtration can introduce greater errors than those it aims to correct.

Table 4-2: Potential interferences from particles in nutrients determination.

Step	Process	Nutrients mainly involved
Storage	Exchange with non-living particles	Phosphate
	Removal/release by living organisms	All
	Release from rupture of dead cells	Organic forms, phosphate
Analysis	Reaction under analytical conditions	Phosphate
	Adsorption of the reaction product	All
	Turbidity blank	All

4.4.3 Vacuum or pressure filtration?

Filtration can be performed under vacuum or under pressure. The vacuum technique is more widely used although pressure filtration has been shown to be more convenient for highly productive waters (high density of microalgae) since it produces less cell rupture [20]. Organic nutrients and phosphate are mainly affected by this kind of contamination. Vacuum filtration can however be used in most cases, provided that the pressure difference accross the membrane does not exceed about 0.3 bar [3]. Plankton-rich water must be treated with the greatest care regarding this point.

4.4.4 Filtration unit

Since the filtered water is the filtration by-product of interest in the present context, the complete unit, i.e., filter-holder and collecting vial, should be considered as a whole in terms of a source of contamination of the samples. Indeed, the filtered water is rarely directly collected in the sample bottle itself, but in a collecting vial fitted to the filter-holder, from where it is poured into the sample bottle (Figure 4-2). Consequently, systems in which the holder is fitted to the vial using a cork plugged into the vial neck are unsuitable: firstly, fingerprint contamination of the vial neck (either direct contact or from the cork) are transferred to the sample when it is poured into the sample bottle; secondly, residual water (from the previous sample or from any mishandling) trapped in the space between the cork and the upper edge of the vial will undoubtedly contaminate the sample when the vial is disconnected.

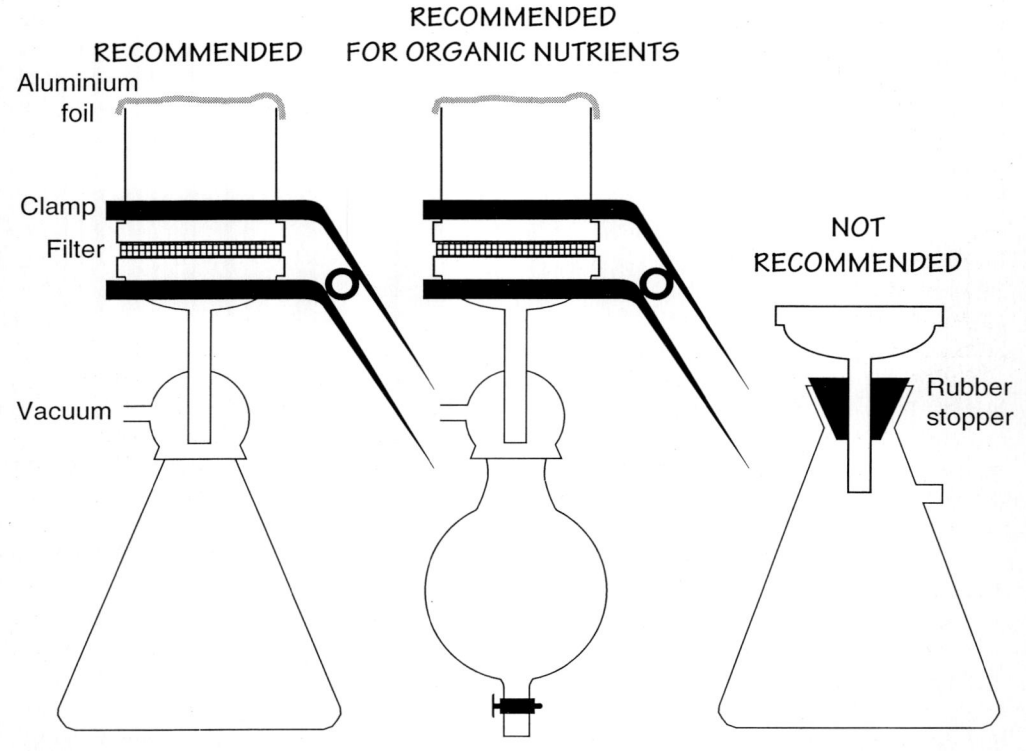

Figure 4-2: Filtration unit types: units fitted to the collecting vial using a cork plugged into the vial neck are unsuitable (right).

Conventional sintered glass, or Buchner funnels, and even some two-piece modern filter-holders, which exhibit the above problem, are therefore not suitable for nutrients. Additionally, sintered glass can clog when used for natural waters and during subsequent rinsing, and cleaning difficulties will result; conventional funnels use inappropriate folded paper filters.

Among glass equipment only those with an *inverted ground joint fitting* (male ground joint on the collecting vial) are adequate. Modern plastic units with the filter-holder screwed on top of the collecting vial are also quite suitable. In these units, the preferable filter-holder system is a two-piece model in which the membrane disk filter is pinched between the base and the funnel. Glass holders with a removable stainless steel grid instead of a sintered glass base may be preferred for cleaning reasons.

4.4.5   Filters

According to the above considerations, it appears that membrane filters should be preferred to any other type (e.g., folded paper filters) since they exhibit the smallest contact area with the sample and are easily handled using flat-end tweezers. The diameter of the membrane will be chosen according to the loading capacity of the filter and the expected suspended matter content of the waters: 47 and 25 mm are the current, convenient sizes commercially available. Manufacturers supply a large choice of filters differing in structure, material and pore size.

The structure type (agglomerated fibres, tortuous, or regular pores) is less important although it determines the filtration rate and maximum loading before clogging.

The pore size is the second criterion determining filtration rate and loading capacity. It is preferable to use a filter with the largest pore size, as long as particles passing through the filter have no adverse effect on nutrient storage or determination; pore size up to several micrometers may be quite convenient.

The type of material used is the main criterion since the filter must neither adsorb dissolved nutrients nor contaminate the sample. Classical membranes are made of either glass fiber or various kinds of polymer, mainly cellulose esters and polycarbonate. Contamination can have two origins: one is the presence of the nutrient as an impurity in or on the material, the other is the filter matrix itself (intrinsic contamination). Data available about potential alteration of nutrient concentrations from filters is scarce: the currently cited study by Marvin and co-workers [21] compared paper filters, glass wool, sintered glass, and two types of membrane for gain or loss of nitrate, nitrite, ammonia, and phosphate, in fresh and sea water. Concerning membranes, they found that Millipore AA and Gelman GA4 (cellulose esters) are satisfactory for nitrite and phosphate in fresh water and for ammonia in sea water; Gelman GA4 is also appropriate for nitrite and phosphate in sea water. But, as reported by Grasshoff [17], Koroleff has found that *interference is not always reproducible*: analysts must therefore remain vigilant for potential alteration of the samples from filters, check their own materials and apply sufficient rinsing when appropriate. Only systematic errors generated by intrinsic contamination are known with certainty; obviously, glass fiber membranes and membranes containing cellulose nitrate should not be used for subsequent determination of silicate and nitrate, respectively. Indeed, no rinsing can effectively remove such interference. Estimates made in our laboratory (unpublished data) indicated contamination of 0.1 - 0.3 μmol/l silicate from Whatman GF/C and GF/F and nitrate from Millipore HA and GS. Beside contamination, loss of nutrients by adsorption must also be suspected; phosphate and ammonia, in particular, are known to exhibit such behavior in fresh water. Glass fiber filters do adsorb a small quantity of ammonia from demineralized water (unpublished laboratory data), which could be a problem with low salinity estuarine water. Glass fiber filters have also been shown to induce the release of dissolved free amino acids from organisms [22]; this should be kept in mind during organic nutrient studies. Finally, experience shows that glass fiber filters (despite minor disadvantages, and with a high filtration rate and negligible contamination levels after short rinsing) and polycarbonate filters may be recommended, provided the above limitations are taken into account. Because of these various constraints, filtration may lead to an increase of the number of subsamples if specific membranes must be used for subsequent determination of certain nutrients.

4.4.6 Alternative to filtration

Given the problems generated by filtration, especially the risk of concentration alteration, it may be convenient to replace it by alternative methods, provided they are as simple and as reliable. Two methods are suggested: centrifugation and pre-filtration.

4.4.6.1 Centrifugation

Centrifugation is an elegant way of removing particles since the only additional vessel in contact with the sample is the (easy to clean) centrifuge tube. Precautions have to be taken when the sample is transferred from sample bottle to tube, and *vice-versa*. The sample bottle containing the turbid water can be re-used to collect the centrifuged sample, provided it is correctly rinsed with the centrifuged water to remove remaining particles. Although centrifugation is limited to samples of a few dozen millilitres, this volume is generally sufficient for nutrient determination. Comparison of filtration and centrifugation (2500xg, 5 min) for phosphate determination in turbid samples from a river plume has shown the reliability of the latter technique [16]. It should be noted that centrifugation can be applied either to freshly sampled water or (if suspended matter does not alter the concentrations during storage) to stored samples just before analysis. In the latter case, centrifugation is considered to be preferable since the stored volume is generally small and more exposed to alteration by handling. Direct *centrifugation of the samples in their own bottle* even constitutes an advantageous improvement since it avoids transfer, hence subsequent potential contamination [16].

4.4.6.2 Pre-filtration

Pre-filtration is a filtration technique using a larger-pored filter than conventional filtration; this enables a high flow rate through the membrane. One can thus use an *on-line filter-holder* which is fitted to the water sampler stopcock, using a soft plastic tube, for direct collection of the sample into the sample bottle and immediate analysis or storage. Pre-filtration is meant to remove large living organisms which are likely to modify nutrient concentrations at any later step in the protocol. Nylon plankton nets, available in a large range of pore sizes from a few micrometers, are convenient. Pre-filtration at around 50 μm removes zooplankton, algal fragments and part of the microalgal population. Pre-filtration through smaller pores can remove a substantial part of the microalgal population. This will improve concentration reliability in both unpreserved and preserved samples by reducing metabolic activity or release from cell lysis [23]. In our laboratory, 10 μm polypropylene membranes, from Gelman, have been employed for many years and appear quite satisfactory. Mounted in a 47 mm on-line Millipore holder (low dead volume), they are rinsed by passage of a few dozen millilitres of the water before collection of the subsample in a bottle. If only a small volume of sample (50-100 ml) is necessary, many samples can be filtered through the same membrane before it has to be changed. With turbid or plankton-rich waters, it may be necessary to change the membrane more frequently, and preferably to have a couple of ready-to-use filter-holders prepared. Good practice consists in having several holders with various pore size filters and using the smallest convenient size, depending on the expected particulate matter content of the water, this factor determining the clogging rate.

It should, however, be remembered that pre-filtration does not replace filtration since, even at 10 μm, it does not eliminate nanoplankton (2-10 μm) and picoplankton (<2 μm), which represent a significant part of the biological activity of some waters.

4.5 Storage and preservation

4.5.1 Introductory remarks and definition

Because of their biological activity, the concentrations of nutrients in the sample bottle can change after collection (see section 4.1.4). The greater the biological activity and productivity of the water, the greater the expected change will be; spring and summer eutrophic coastal waters are, therefore, the most exposed to such change. Exchange with inert particles can also alter concentrations (especially of phosphate). However, there is no way of foreseeing which will be the dominant processes and their magnitude. Consequently, it is widely admitted that nutrient determination should be carried out without delay, i.e., within a few hours of collection. Unfortunately, application of this good practice is limited to large vessels with well equipped laboratories and scientific staff devoted to nutrient determination using autoanalysers. In most other cases (small ships, coastal and estuarine monitoring or multi-parameter cruises with a high station concentration and small teams) the nutrient samples must be stored for various periods of time (hence preserved from concentration changes) until analysis at the shore laboratory. Even when analyses are performed on-board, a preservation method must be foreseen in case of breakdown of the analytical equipment.

According to the variety of water characteristics and nutrients under consideration, it is stated that no general procedure can be recommended for sample preservation during storage. This overall negative assertion should not discourage analysts from developing a rational approach to their storage problems. But before dealing with procedures, it is useful to give a definition of storage which occurs when a sample, one isolated from its water body, is not "immediately" analyzed. In this context, "immediately" is understood as being the short period of time (say, a few minutes) required to bring the sample to the laboratory and start the analysis. According to that general definition, water sampled in the deep ocean, for example, is considered as being stored in the water-sampler up to one hour or more until it is brought up on deck and subsampled. This type of specific storage problem related to the primary sample is out of the scope of this review, which will be limited to storage and preservation of subsamples.

Because changes of nutrient concentration are mainly of biological origin, every procedure that tends *to reduce the micro-organism activity* of the sample will improve the concentration's stability. Two categories of treatment can be applied, either alone or in combination: i) reducing the biological activity by physical treatment (cooling, filtration, darkness); and ii) inhibiting or killing the organisms by use of chemicals.

Depending on the storage time, the water characteristics, the ship facilities and other eventual, local constraints, the analyst may choose the most adequate treatment, according to the following guidelines, separated into short-term and long-term procedures. For a more extensive review of the literature dealing with storage of nutrient samples, readers can refer to [5].

4.5.2 Short-term preservation

The first operation which can be done without delay is to remove part of the suspended matter by on-line *pre-filtering* of the sample (see section 4.4.6.2.): this is easy and not time-consuming, and will slow both physical and biological exchange processes. The second operation is immediately to place the sample *away from light*; this will prevent intense phytoplanktonic assimilation (but affects neither respiration nor bacterial activity). The third important point is *to cool the sample* down to 0-5 °C in order to slow metabolic processes.

Application of these three treatments will enable the storage time to be increased from a few hours to one or two days in most instances. This procedure is easily applicable, even on board rubber boats, provided an ice-box, with enough cooling capacity, is available.

If samples are filtered through a membrane with small pores ($\sim$ 1μm), almost all small planktonic and bacterial cells will be removed, hence storage time can be significantly increased.

4.5.3 Long-term preservation

4.5.3.1 General comments

Logistical reasons generally lead laboratories to group their analyses in series, requiring long-term storage of samples. Because of the instability of nutrients, "long-term" should be understood as longer than one day and up to several months.

Numerous authors have worked on the stabilization of nutrient concentrations by the addition of organic or inorganic poisons to the samples, or by refrigeration, or by both treatments together. Some work has produced inconclusive or contradictory results. From the review of Kirkwood [24], it may be concluded that a large share of the apparent confusion can be attributed to two main factors: firstly, calibration (long-term reliability) problems may have biased the results; and secondly, the various researchers seldom compare "like with like". Poisoning with acids, chloroform, and mercuric chloride (the most popular poison in current use [24]), and freezing will be discussed.

4.5.3.2 Poisoning

4.5.3.2.1 Acidification

Acidification should be used with great wariness since it exhibits the following disadvantages: i) the severe matrix modification requires that the determination protocol be reconsidered, in particular the reagent combination, to re-optimize the reaction conditions; ii) nitrite can be lost as dinitrogen as a result of its reaction with ammonia or primary amines (Van Slyke reaction); and iii) hydrolysis of condensed or organic forms of phosphorus can occur [25]. The use of acidification for preservation of nitrate is neither well documented nor expressly mentioned as a possible alternative in manuals, and consequently should be avoided. Acidification should therefore be limited to filtered samples specifically designed for subsequent silicate determination, at a concentration of 9 mmol/l of sulfuric acid [26].

4.5.3.2.2 Chloroform

Chloroform has been investigated by many researchers, and from their data it may be concluded that it is unsatisfactory. It has been shown to cause a release of phosphorus from microalgae and therefore must not be used for unfiltered samples [23,27]. While apparently encouraging results were obtained with filtered waters, several more recent works reported poor preservation [24]. A strong argument against chloroform has been documented by researchers who found, in sea water samples, several strains of bacteria able to use chloroform as a carbon source along with simultaneous nitrate consumption [28]. *Chloroform must therefore be avoided* as a preservative for nutrients.

4.5.3.2.3 Mercuric chloride

Mercuric chloride for sample preservation was mentioned in the literature as early as 1920. A very large range of concentrations added to the samples has been reported: 1 to 500 mg/l [24]. Since the killing effect of mercuric ions is attributed to bonding with enzyme sulfydryl groups, efficiency of low concentrations can be drastically reduced in highly productive waters. Although there is no widespread agreement on the amount that should be added to the samples, Kirkwood suggests that a concentration of 20 mg/l or more seems to be able to guarantee comfortable excess in most instances [24]. But, as for every preservation method, removal of suspended material before poisoning, by filtration or pre-filtration, will invariably improve the reliability of mercury preservation. This precaution is indispensable for plankton-rich waters, not only to ensure mercury efficiency but also to prevent contamination from nutrient release after organism death and subsequent cell rupture. Within the above limitations, mercuric chloride appears to be an efficient poison, but it exhibits the following disadvantages: i) it strongly interferes in the indophenol blue method for ammonia determination [29], sometimes preventing measurement at the levels present in natural water; ii) it can interfere in the cadmium column reduction method for nitrate determination [17,24]; and iii) it exhibits strong human toxicity. It should be noted that accurate correction for interferences may be difficult because, as previously stated, the concentration of free mercury(II) in solution is unknown.

4.5.3.2.4 Summary of poisoning

In summary, poisons can be used as preservatives for nutrient samples as follows:
- acidification (pH~2.5) is suitable for silicate in filtered samples,
- mercuric chloride (20-40 mg/l) is satisfactory for all nutrients provided water turbidity is low and interference in determination methods (especially ammonia and nitrate are concerned) is under control; and
- chloroform is unreliable and must be avoided.

Apart from the restrictive uses of poisons specifically due to nutrient problems, analysts must be aware of potential contamination risks towards other determinands. Volatile organic poisons can invalidate data from measurement of dissolved organic carbon or of trace organic compounds in samples treated simultaneously. Working with saturated mercuric chloride solutions in the vicinity of colleagues hunting picomoles of mercury is perceived as a serious risk of contamination.

4.5.3.3 Freezing

4.5.3.3.1 Usefulness

Freezing as a preservative technique for nutrient samples appeared in the 1950s when freezers progressively became common laboratory equipment. One of the main advantages of freezing over poisoning is that it reduces contamination risks, because handling is minimized and no extraneous product is added into the sample. As the matrix is unchanged, reaction conditions for the determination are identical to those applied to fresh samples. The basic principles for application of freezing are firstly that the rate of chemical reactions (hence organism metabolism) is slowed by the temperature, and, secondly, that reduction of molecular collisions due to solidification will virtually halt most reactions. An argument against freezing was raised by Wangersky and Zika [30] on the basis of work showing that some reactions of biochemical systems are enhanced in the frozen state (-18 °C) when compared with the liquid state at 0 °C. However, in these studies the reactions were studied in aqueous solutions exclusively composed of the few reactive compounds at concentrations 10^3 to 10^5 times greater than those of nutrients in marine waters. It is thus extremely improbable that biochemical-type reactions involving nutrients can occur in frozen marine waters because of the very low concentration levels of the potentially reacting substances.

4.5.3.3.2 Mechanism

To apply a strict freezing procedure some knowledge of the behavior of sea water during cooling is useful. The formation of sea ice in cold seas, already well studied [31,32], can serve as a basis for describing sample freezing. Firstly, salts in solution lower the freezing point as follows:

Salinity PSS	0	10	20	30	35
Freezing point °C	0	-0.5	-1.1	-1.6	-1.9

Secondly, as freezing takes place, the formation of ice crystals, consisting of pure ice without any salt, enriches the remaining water in salts, hence lowering its freezing point. Both the amount of ice and the brine concentration increase simultaneously, and progressively, given that there is a single temperature at which the ice and salt concentration are in equilibrium. When saturation points are reached, salts start precipitating at an almost constant temperature: sodium sulphate crystallizes at -8°C and sodium chloride at about -22°C. At the latter temperature the whole mass becomes frozen solid. In a sample bottle placed in a freezer, the water is cooled from the wall on which ice therefore starts crystallizing, rejecting brine toward the inner and upper parts of the bottle. The content of the bottle therefore becomes very heterogeneous. While solidification is progressive in high salinity waters, it can occur very quickly in fresh and low salinity waters, because of supersaturation and the absence of salt regulation. The third point concerning water freezing is that the density of ice is about 10 % lower than that of water, which leads to a corresponding increase in the volume of the frozen sample.

Consequences of the mechanism of sample freezing can be seen in several respects. An insufficiently low storage temperature enables nonsolidified, highly concentrated brines to persist; in these certain chemical reactions can slowly occur. Calcium phosphates, including apatites which are in a metastable state in sea water [10], are likely to be formed and contribute to phosphate disappearance [33], since they will probably not redissolve after sample thawing. Silicate polymerization/depolymerization may also be affected (see section 4.5.3.3.4). If bottles are too full, the loss of a little brine, pushed by ice pressure through a leak in the cap, will lead to a significant decrease in nutrient concentrations. Because of phase separation and subsequent heterogeneity of the frozen sample, the sample must be entirely thawed and mixed before aliquots are taken for analysis.

Many workers have investigated the stabilization of nutrient concentrations by freezing (see the cited manuals [4,5]). A review of that literature shows that working conditions differ largely in terms of nutrients tested, water origins, and characteristics (oceanic, coastal or estuarine, surface or deep water, salinity and nutrient concentrations), water treatment (fine, coarse or no filtration), sample bottle types (tubes, bottles or bags, in glass or various plastics, from 15 ml to 1 l), storage time (days to years), thawing method (warm or hot water, ambient or hot air). This wide combination of conditions, although somewhat puzzling for rigorous comparison of the data, leads to the conclusion that freezing is a reliable preservation method, provided some application rules and limitations are respected; these are described below.

4.5.3.3.3   Operational

The following conditions should enable reliable preservation by freezing, bearing in mind that thawing requires as much care as freezing itself.
- Sample bottles must resist pressure due to ice expansion: below a salinity of 5 PSS, glass bottles can break, depending on their shape and water volume (reduce filling percentage as salinity decreases).
- Because of ice expansion, bottles must not be filled too full: 3/4 is a maximum; this will leave enough room for ice and prevent breakage.
- Sample bottles must be frozen in an upright position; this will avoid loss of sample under freezing, hence significant negative errors.
- Samples should preferably be "clarified" by pre-filtration, filtration, or centrifugation.
- Samples must be cooled as soon as possible after collection, by placing them in a normal deep freezer or by quick freezing in a dry ice-alcohol bath if available.
- For entire solidification during storage, deep freezing is required, i.e. the temperature should not be higher than -22 °C to -23 °C.
- Storage should preferably not exceed a few months.
- Samples must be thawed in an upright position; if a water bath is used, care must be taken that no external water comes into the neck thread (strong risk of contamination when cap is unscrewed.
- Because of the heterogeneity which results from freezing, samples must be entirely thawed and thoroughly mixed before an aliquot is taken for analysis.
- Thawing must be done without excessive heating and the samples must be regularly shaken in order that they remain cool after ice has completely melted.
- The thawing step must not exceed a few hours and samples should be analyzed very shortly after thawing.

4.5.3.3.4 Specific problems

Two nutrients, phosphate and silicate, can present specific problems. Phosphate can decrease after several months storage [33], a phenomenon which is attributed to precipitation, as suggested above. Silicate presents a particular problem, since it polymerizes on freezing, this effect being more pronounced as salinity decreases [34-37]. In so far as is possible, it is therefore preferable *to avoid freezing silicate samples*. Nevertheless, silicate can be determined in frozen samples, provided total depolymerization takes place before determination. Four factors can affect measured silicate: salinity, silicate concentration, storage time, and time thawed before determination. For a silicate/salinity ratio <2 μmol/l per PSS, a storage time <2 months, and <1 h thaw time, few problems will be encountered. For longer storage time and higher Si/S ratio, thaw time must be increased accordingly; a 24 h thaw time is required if Si/S$=3$ and storage time is 6 months or Si/S$=10$ and storage time is 1 month (for more extensive information, refer to [37]).

4.5.4 Should each analyst develop storage studies?

It is often recommended that analysts should develop storage studies, appropriate to their water types. This suggestion appears rather exaggerated since, as there is no definition of a water type, it could lead to excessive work with poor output. Indeed, because of the multi-factorial origin of the variations of nutrient concentrations in a sample, the reliability of such studies is intimately dependent on the nutrient laboratory background, which largely exceeds pure analytical competence. Laboratories should rigorously apply a conventional procedure well within its reliable application range. Nevertheless, it is good quality assurance practice to compare, from time to time, concentrations measured in fresh and stored duplicate samples of waters of various characteristics. This can additionally help laboratories checking their entire analytical protocol. Extensive studies must only be developed if unconventional storage methods are used.

4.6 From sampling to analysis: typical schemes

Depending on working conditions and context, various on board schemes can be proposed for optimizing handling and preservation of nutrient samples, from the water body to analysis.

- On board ships with analytical equipment
  i) sampling - subsampling (with pre-filtration) - (short-term storage) - analysis
  ii) sampling - subsampling - (short-term storage) - filtration or centrifugation - (short-term storage-) - analysis

- On board ships without analytical equipment
  i) sampling - subsampling (with pre-filtration) - long-term storage (poisoning or freezing)
  ii) sampling - subsampling - (short-term storage) - filtration or centrifugation - long-term storage (poisoning or freezing)

- On board small boats (typically rubber rafts)
    i) sampling - subsampling with pre-filtration - short-term storage (darkness and cooling)
    ii) sampling - subsampling with pre-filtration - long-term storage by poisoning.

Acknowledgement

I am grateful to Roger Kérouel for his helpful comments.

4.7 References

[1]   S.R. Carlberg, *ICES Coop. Res. Rep. Ser. A*, **29**, 1 (1972)
[2]   E.G. Goldberg, *A Guide to Marine Pollution*, New York: Gordon and Breach Sci. Publ. (1972)
[3]   J.D.H. Strickland and T.R. Parsons, *Bull. Fish. Res. Bd. Can.*, **167**, 1 (1972)
[4]   A. Aminot and M. Chaussepied, *Manuel des Analyses Chimiques en Milieu Marin.* Brest: CNEXO/Documentation (1983)
[5]   K. Grasshoff, M. Ehrhardt and K. Kremling, *Methods of Seawater Analysis, 2nd ed.*, Weinheim; Verlag Chemie (1983)
[6]   T.R. Parsons, Y. Maita and C.M. Lalli, *A Manual of Chemical and Biological Methods for Seawater Analysis,* Oxford: Pergamon Press (1984)
[7]   D. Kirkwood, A. Aminot and M. Pertilä, *ICES Coop. Res. Rep.,* **174**, 1 (1991)
[8]   A. Hattori, *J. Oceanogr. Soc. Japan*, **38**, 245 (1982)
[9]   E.J. Carpenter and D.G. Capone, *Nitrogen in the Marine Environment,* New York Academic Press (1983)
[10] W. Stumm and J.J. Morgan, *Aquatic Chemistry*, New York: John Wiley & Sons (1981)
[11] G. Martin, *Le problème de l'Azote dans les Eaux*, Paris: Technique et Documentation
[12] A. Hattori, in: *Nitrogen in the Marine Environment:* E.J. Carpenter and D.G. Capone Eds., New York: Academic Press, 191 (1983)
[13] J.P. Riley, K. Grasshoff and A. Voipio, in: *Guide to Marine Pollution:* E.G. Goldberg Ed., New York: Gordon and Breach Sci. Publ., 81 (1972)
[14] U. Stefansson and J. Olafsson, *ICES Inf. Tech. Meth. Sea Wat. Anal.,* **3** (1970)
[15] F. Koroleff, *Methods of Sawater Analysis, 2nd ed.:* K. Grasshoff, M. Ehrhardt and K. Kremling Eds., Weinheim: Verlag Chemie, 125 (1983)
[16] R. Kérouel and A. Aminot, *Mar. Environ. Res.* **22**, 19 (1987)
[17] K. Grasshoff, in: *Methods of Seawater Analysis, 2nd ed.:* K. Grasshoff, M. Ehrhardt and K. Kremling Eds., Weinheim: Verlag Chemie, 21 (1983)
[18] D.F. Krawczyk and M.W. Allen, *EPA Environ. Monitor. Ser.* 600, 4-74, 004, 180-193 (1974)
[19] A. Aminot, R. Kérouel, D. Kirkwood, J. Etoubleau and P. Cambon, *Intern. J. Environ. Anal. Chem.* **49**, 125 (1992)
[20] G. Liebezeit, *Dissertation*, Univ. Kiel (1980)
[21] K.T. Marvin, R.R. Proctor Jr. and R.A. Neal, *Limnol. Oceanogr.*, **17**, 777 (1972)
[22] J.A. Fuhrman and T.M. Bell, *Mar. Ecol. Prog. Ser.*, **25**, 13 (1985)
[23] G.P. Fitzgerald and S.L. Faust, *Limnol. Oceanogr.*, **12**, 332 (1967)

[24] D.S. Kirkwood, *Mar. Chem,* **38**, 151 (1992)
[25] D. Jenkins, in: *Trace Inorganics in Water,* R.G. Gould Ed., Am. Chem. Soc. Publ. **16**, 265 (1968)
[26] F. Koroleff, in: *Methods of Seawater Analysis, 2nd ed.:* K. Grasshoff, M. Ehrhardt and K. Kremling Eds., Weinheim: Verlag Chemie, 174 (1983)
[27] P.G.W. Jones, *J. Cons. Int. Explor. Mer,* **28**, 3 (1963)
[28] J. Castellvi and M. Cano, *Rapp. Comm. Int. Mer Médit.,* **28**(8), 59 (1983)
[29] F. Koroleff, in: *Methods of Seawater Analysis, 2nd ed.:* K. Grasshoff, M. Ehrhardt and K. Kremling Eds., Weinheim: Verlag Chemie, 150 (1983)
[30] P.J. Wangersky and R.G. Zika, *The Analysis of Organic Compounds in Sea Water*, Report NRCC-16566, Halifax, NS, Canada: MACSP, Atlantic Regional Laboratory, National Research Council of Canada (1977)
[31] C.A.M. King, *Introduction to Physical and Biological Oceanography,* London: Edward Arnold (1975)
[32] P. Groen, *The Waters of the Sea,* London: D. van Nostrand Company Ldt. (1967)
[33] L.A. Clementson and S.E. Wayte, *Wat. Res.* **26**(9), 1171 (1992)
[34] J. Kobayashi, in: *Chemical Environment in the Aquatic Habitat:* H.L. Golterman and R.S. Clymo Eds., Amsterdam N.V. Noord-Hollandsche Utgerus Maatschappij, 41 (1968)
[35] J.D. Burton, T.M. Leatherhead and P.S. Liss, *Limnol. Oceanogr.,* **15**(3), 473 (1970)
[36] R.W. Macdonald and F.A. McLaughlin, *Wat. Res.,* **16**, 95 (1982)
[37] R.W. Macdonald, F.A. McLaughlin and C.S. Wong, *Limnol. Oceanogr.* **31**(5), 1139 (1986)

5.

Quality assurance of sediment sampling

M. Perttilä[1] and B. Pedersen[2]
[1] Finnish Institute of Marine Research, PO Box 33, FIN-00931 Helsinki, Finland
[2] National Environmental Research Institute, PO Box 358, DK-4000 Roskilde, Denmark

Quality assurance (QA) is the total system of activities required to guarantee the appropriate quality of the product, and is, therefore, an integrated part of all sampling programs. The sample to be analyzed has to be taken adequately, otherwise all efforts in the laboratory for good quality control are futile, since the obtained analytical result, however accurate, can not be used for the purpose intended.

Below are listed the most important items which should be included and documented in a comprehensive quality assurance plan for sediment sampling to ensure that the data generated has the appropriate quality (modified after [1]):

1) description of the project including the objectives and the quality of the data needed (precision, accuracy, completness, representativeness and sensitivity);
2) project organization and responsibility including the designation of a Quality Assurance (QA) coordinator;
3) site selection;
4) sampling period frequency and number of replicates;
5) sampling procedures for each major measurement, including subsampling;
6) storage and sample handling procedures including type of container, cleaning, procedures, labeling;
7) internal quality control programme (replicates, split samples, blanks, etc.);
8) education and learning programme and malfunction procedures; and
9) performance and system audits and a report to management.

Many good and comprehensive books, guidelines, and articles have been published covering the more general aspects and planning of environmental sampling and quality assurance (items 1,2,8 and 9 above) [2-5]. Therefore, mainly the more technical aspects of the implementation of the QA system (items 3-7 above) in relation to sediment sampling will be dealt with here.

The more general sections not discussed here are by no means less important. Education, project organization, and the audit system, to verify that the QA/QC plan has been carried out according to the protocols, are very essential parts of a sampling QA/QC system to ensure that the data has the appropriate quality. An effective audit system combined with detailed and specific written protocols for the different procedures are in fact the key elements in an effective QA/QC sediment sampling system.

Quality control of sediment sampling is in some respects easier than that of sea water sampling, e.g. fewer contamination problems owing to the rather high natural concentrations in sediments. However, sediment sampling is usually more difficult and also more time-consuming than water sampling. This stresses the importance of adequate training of the persons responsible for the sampling as an important part of the quality assurance procedure.

5.1 Sampling objectives

The quality assurance procedure is highly dependent on how the data will be used. In consequence, the objectives of the project should be well defined and clearly stated, as these will determine the sampling period frequency, the number of replicates to be taken, sampling sites, and subsampling procedures. The user of the data should, in cooperation with the sampling personnel and the analysts, also define the quality of data necessary, e.g. precision and accuracy.

There can be several reasons for undertaking sediment sampling; these include pollution monitoring, net sedimentation and mass balance calculations, sediment transportation investigations, other sediment process studies, and health and environmental risk assessments in polluted areas. Each of these objectives may pose different requirements on the quality of the data.

5.2 Sampling site selection

5.2.1 General considerations

As stated earlier, the aim of sediment sampling will have a large influence on sampling site selection.

Sediments can provide an excellent archive enabling study of the past and present environmental pollution. Environmental monitoring of the geographical distribution and time trends of the concentration of a trace contaminant in the sediments is, therefore, often the aim where sediment sampling is a part of the program. The choice of sampling sites is, however, of crucial importance, especially in time-trend monitoring programs, where the sampling should be repeated at pre-defined intervals at exactly the same positions (sites).

The site should be representative of the area and there should be steady sedimentation conditions. Because of varying hydrographic conditions such as internal currents, the net sedimentation rate may vary significantly from site to site. The following aspects should be considered when selecting sampling sites:

- hydrography (currents, vertical structure of water mass);
- structure of the sedimentation basin;
- benthic communities (bioturbation);
- rate of sedimentation;
- sediment structure (overall topography and the fine structure of the sediment surface; and
- available historical data and other information.

Some of these aspects are further discussed below.

In addition, in some highly polluted areas (e.g. in regions where dumping of military wastes is known to have occurred) special precautions must be taken for clean-up procedures in cases of accidental lifting of hazardous materials on deck.

5.2.2 Transport processes and hydrography

Compounds are transported to the sediments from the water column by means of scavenging adsorption on to or incorporation into precipitating particles. In the absence of strong currents, the most polluted sediment areas are, therefore, generally found in the vicinity of a pollution source, and the concentrations will decrease with increasing distance from the source.

The first sedimentation area is usually not the final sink for the sediments. The sediment is resuspended through the activity of winds and waves, and through macroscale currents transported gradually to the final accumulation areas.

The general hydrography in the area is also important. The stratification of the water column may prevent the vertical cycling of the water mass and the penetration of atmospheric oxygen to the bottom. The redox conditions at the bottom may, therefore, vary with time. Continuous monitoring in such areas will usually not be meaningful because:

- precipitation/dissolution properties vary with the redox potential for certain compounds (e.g. iron and phosphate);
- the benthic communities may change as a consequence of oxygen availability, which will affect the bioturbation of the sediment surface and hence the concentrations found in a sample; and
- the diagenetic processes within the sediment vary with the redox potential, e.g. through varying solubility.

5.2.3 Sea bed structure (topography)

The sea bed topography could be very irregular both in coastal and estuarine areas as well as in sea areas like the Baltic Sea. One sedimentation basin is often divided into smaller sedimentation "holes", separated by sharp edges. The whole sea bed area should therefore be mapped, e.g through echo-sounding, before the sampling positions are selected. The navigation systems of many ships usually allow coupling of navigation data with echogram data, so that the sea bed topography can be described visually in a three-dimensional picture (see Fig. 5-1). The most promising areas can be selected from such echogram data. However, to select the most suitable sampling position, data about the depth of the soft sediment is also necessary. Scanning of the area using a low-frequency (<10kHz) sounding can give this information. This technique not only gives a signal from the sediment surface but, because it penetrates the surface, also gives a signal from other, deeper-lying reflector surfaces.

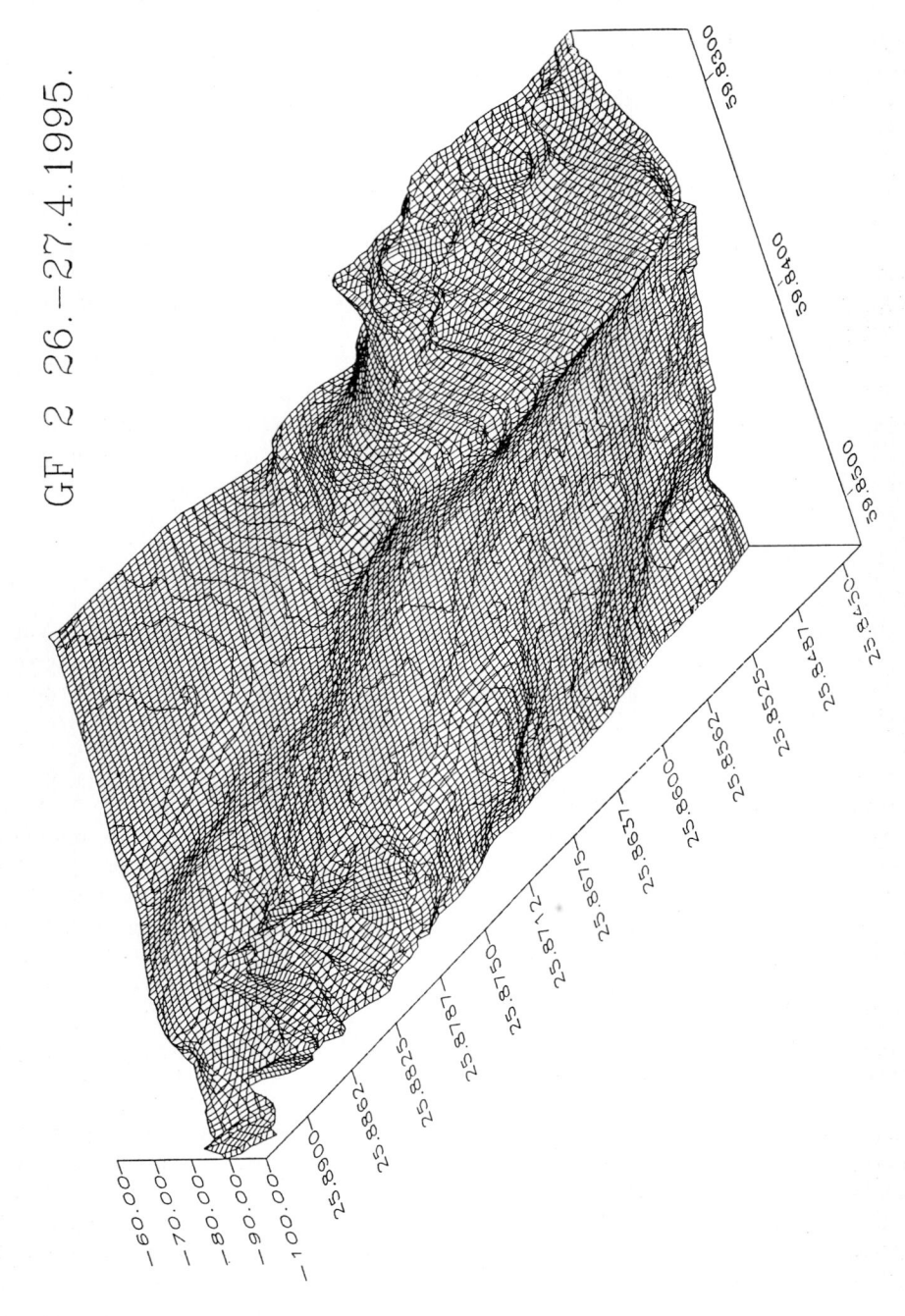

Figure 5-1: Sea bed topography

To a certain extent, other qualities of the sediment bed can also be deduced from a low-frequency echogram, e.g. the occurrence of gas bubbles, which may mix the sediment layers. Also the sample itself may be mixed by the expanding bubbles when it is lifted towards the surface.

The possibility of mechanical disturbance of the sea bed in shallow areas (< 200 m) should also be controlled before any sampling. This disturbance could arise from fishery bottom trawling. The most powerful method for detection of the fine-structure of the sediment surface is the side-scan sonar. An experienced side-scan sonar user can easily detect, e.g., the signs of bottom trawling in the echograms. Again, it is essential to be able to combine the positions of the side-scan sonar pictures with the navigation data of the ship to ensure proper site selection.

Preferably, all three scanning methods described above should be applied to ensure appropriate site selection and thereby minimize the costs which result from improper sampling.

Despite all mapping techniques, the final test for selection of a site is provided by the visual inspection of a test core. The sediment core description (texture, estimates of grain size, color, odor, etc.) should be documented for future comparison. Changes in sediment structure and sedimentation conditions do occur under exceptional conditions, and it is worth while to make sure visually that no changes have taken place.

5.2.4 Ship positioning

One of the prerequisites for quality is the documented repeatability of the results. This also includes sampling position. Accurate positioning of sediment sampling sites is important in any program, particularly when the sampling is to be subsequently repeated, as in time-trend monitoring programs [6] In the earlier days of marine research it was not possible to be certain of having sampled at exactly the same position as earlier, e.g. when using the decca navigation network and especially when taking the difficulties of keeping the ship in position in rough weather into account.

The global positioning system (GPS) using satellite signals enables positioning to better than ± 20m. This precision may not always be necessary, e.g. in open-sea water sampling. However, it is indeed mandatory in time trend sediment monitoring in areas with a complicated sea bed topography, as, e.g., in the Baltic Sea, where apparently similar sedimentation basins may be separated by sharp underwater mountains. The separation can effect the concentrations in the subareas as the near-sea bed current system, and hence the rates of sedimentation/ resuspension processes may vary in the different areas. Benthic communities are usually very stable and may therefore also differ when going from one basin to another. This will also have an influence on the concentration levels in the different subareas as the benthic community will effect the sediment concentrations through their bioturbation and irrigation activities [7].

Several parallel samples must often be taken during a stay at a station. This sets strict requirements on the ability of the ship to hold the selected position exactly for a prolonged period of time. Any drifting should be closely followed. This can be done through modern computer-aided navigation, and even improved by use of the taut-wire technique in which a weight is lowered to the bottom and the wire is kept tight by means of an automatic tension-compensating unit. Small deviations in the angle of the wire with respect to the vertical line are sensed by a circular sensor. The sensor signals are passed to the navigation computer which then gives orders to the propulsion mechanism for corrective measures. In depths less than 400 m this technique enables the ship to maintain position within a radius of a few meters.

5.2.5 Sampling period frequency

The objectives of sediment sampling will greatly influence sampling period frequency and the following elements should at least be taken into account in the planning phase:

- net mass accumulation rate and linear sedimentation rate;
- subsampling thickness to get the necessary amount of material for the analyses;
- sediment surface mixing depth (bioturbation);
- concentrations of the pollutants;
- expected change in contaminant concentration from year to year;
- analytical reproducibility.

The sensitivity of the sediments in respect of the detection of changes in the environment with time (e.g. concentrations of pollutants) depends on the rate of sedimentation, the mixing depth of the sediment, primarily as a result of bioturbation, and analytical reproducibility [8].

Generally, the rate of sedimentation varies from the very low oceanic values (order of magnitude 0.1 mm/y) to high values (several mm/y) in eutrophic lakes and semi-enclosed seas such as the Baltic Sea [9].

The rate of sedimentation and changes in sedimentation conditions can be determined and studied by means of the vertical distribution of the ^{210}Pb isotope (see, e.g. [8]). The observed ^{210}Pb profile can also be used for evaluation of the suitability of the site for monitoring purposes [8].

The surface mixing depth could be several centimeters corresponding to several decades, depending on the sedimentation rate. Consequently, the most suitable areas for sediment monitoring are those with reasonably rapid and stable sedimentation and very little (or no) bioturbation or resuspension. A full core should, therefore, always be investigated at the beginning of a monitoring survey even though only sediment from the surface is generally used in environmental monitoring. This can also give very valuable information when assessing the data, as the data from the bottom of the core, if from pre-industrial time, can be used as a site-specific background value.

5.2.6 Replicate samples

Demonstration of the precision of the sampling, the variability, and that the samples are representative of conditions in the area under investigation should be a mandatory part of a QA/QC sampling plan [1]. There could be a large spatial accumulation and hence concentration variability, even within one single net sedimentation basin, as a consequence of currents and internal waves, bottom structure and topography.

The variability of the concentration in a restricted sampling area is usually randomly distributed. The number of replicates necessary for a decision or a statement to be made depends, therefore, on the confidence level required. This increases when the number of samples is increased. But the number of replicate samples cannot be increased too much, as this will make the program too expensive. The optimum number of samples can, however, be calculated if the natural variation and the analytical costs are known.

If the natural variation is not known, a preliminary or exploratory sampling programme can provide useful information for the design of the final sampling program. There exist many good textbooks about the statistical design of a sampling programme (see, e.g. [5]). One of these (and preferably a statistician) should always be consulted in the planning phase of any sampling program to ensure that the data will have the required level of confidence. Field replicate samples should not be mixed with laboratory replicate samples which give an estimate of the analytical variability only. This is further discussed under section "Split samples".

5.3 Sampling

5.3.1 Sampling devices

"There are many factors which need to be considered in the selection of a suitable sampling device. These factors include the sampling plan, the type of sampling platform available (vessel, small boat, etc.), location and access of the sampling site, physical character of the sediments, the number of sites to be sampled, weather, the staff, and the cost" [6].

Only two of the most common sediment samplers will be shortly described here. These are the box corer and the gravity corer. The Kajak corer [10] has sediment penetrating properties similar to those of the gravity corer. Different types of grab sampler device are also often used. For further information about sampling devices see, e.g., Mudroch and MacKnight [6].

The box corer has the advantage that, because of its large surface area, it is usually only necessary to lower the sampler once; further subsampling can be performed on deck. The best results are obtained on very soft sea beds, and especially when only the sediment surface, i.e. the most recently deposited layers, needs to be studied. On the other hand, this type of sampler easily disturbs the sediment surface if lowered too fast. The gravity corer penetrates sufficiently deep, even if the sea bed consistency is somewhat harder, and collects a column of the subsurface sediment. Blomqvist [11] has reported an *in situ* study on the reliability of core sampling, showing the effect of the sampling tube diameter on the compression of the sample core.

This demonstrates that the choice of sampling device can affect the results, and hence the sampling accuracy, in a systematic way. A comparison of different gravity core samplers has also confirmed that the compression of a core sample is dependent on the tube inner diameter [12].

The ordinary corer has the disadvantage that because of its inherent asymmetry it often hits the sediment surface at an angle. To overcome this drawback, a double corer has been developed. Neither box corer nor gravity corer gives satisfactory results with very sandy sea beds. In such circumstances, a drilling corer, in which the corer tube is then pressed slowly into the sediment by means of hydraulic pressure, might prove useful. Mechanical back and forth turning of the corer tube can also be applied to facilitate the penetration. However, it should be noted that hard sea beds are usually erosion sea beds and consequently usually not suitable for sediment sampling, e.g. for monitoring purposes, although sampling in these areas could be of interest for other purposes.

5.3.2 Sampling procedure

The actual sampling procedure is considerably more demanding for sediment work than for water chemistry studies. It needs a lot of experience to "feel" the correct rate of lowering of a sampler, and to raise the sampler from the sediment without changing the integrity of the sample, e.g. losing the sediment and/or disturbing the surface layer. Therefore, only a designated and well-trained person should be allowed to handle the winch; this underlines the importance of training as part of an effective QA programme.

The sampler should always be kept in an upright position to avoid mixing of the surface layer. On board, irrespective of the method of sampling, the first step should always be visual inspection of the sediment. A core should be rejected if there are any irregularities in the exterior of the sample. The core surface, especially, can be disturbed. This occurs if the sampler hits the sediment surface at an angle, or if there have been problems during the raising of the sampler. Severe bubble formation - due to the pressure difference or occurrence of gas -may also necessitate rejection of the sample and/or the sampling site.

5.3.3 Subsampling

A sediment core as a whole is seldom of interest for any sediment study. In order to ensure the suitability of the core, i.e. its structure, texture, and intactness, it must be subsampled. The core is cut down to suitably thin slices (1 mm to 10 mm, depending on purpose, sedimentation rate, etc.). Each slice is stored in a separate container. In many routine monitoring programs, only the sample from the sediment surface is of interest, but a subsample from the deeper sediment core should also be analyzed to provide an actual "background" concentration from the site.

Subsampling is the procedure during which the sample is exposed to the highest risk of contamination. However, the risk is not so great as when sea water samples are handled, because of the usually much higher concentrations of trace elements in the sediments. The equipment must, of course, be cleaned thoroughly. The sea water from the area can usually be used. The cutting and other instruments should also be of a material which do not compromise the sample. For many purposes, polyethylene and polystyrene plastics provide the most practical materials, as these are easy to handle and keep clean. Stainless steel or glass equipment can be used when subsampling for trace organic analysis. The cutting of a sediment core usually starts from the top where the highest concentrations are generally found and proceeds down the core; there is, therefore, a risk of contamination from the previous subsamples of a core. The contamination can partly be avoided by rigorous and regular washing of equipment whith sea water between each cuts. The partial mixing of subsamples cannot always be avoided as the usually very soft sediment surface layer, containing over 90% water, can easily slide down the inside of the core tube. If mixing has occurred, the only solution is to discard the sample.

The cutting instrument should be suitably calibrated, in order to ensure accurate and regular subsample thickness.

5.3.4 Sample treatment

Proper sample handling between sampling and analysis is of crucial importance to the quality of the data. The samples can become contaminated, mixed up, or in other ways compromised if not handled correctly. It is, therefore, essential that the procedures covering these aspects are adequate and properly documented and followed as a part of the overall QA/QC sampling system. Below the most important procedures are discussed further.

5.3.5 Drying, grinding and sieving

Depending on the further use of the sediment sample it might be necessary to dry, sieve, or grind the sediment. Three different methods are commonly used to dry a wet sediment: air drying, oven drying and freeze drying. All three methods can alter the structure and composition of the sediment in different aspects such as loss of volatile compounds, and oxidation and aggregation processes. Freeze drying is, however, the most useful drying method, because of the lower risk of losing the more volatile elements and providing a powdery material instead of hard clay aggregates [13]. Freeze-dried samples can, therefore, generally be used for the analysis of many of the less volatile organic compounds and of inorganic compounds/ elements, although some authors have reported loss of mercury following freeze drying [14].

There are different ways of separating sediments into various particle size fractions: wet sieving of wet sediments, and dry or wet sieving of dry sediments. The choice of the method selected depends on the purpose of the fractionation and the further use of the results. Mudroch and MacKnight [6] and Loring and Rantala [13] described the different techniques in more detail, and Barbanti et al. [15] have recently investigated the effects of the different procedures involved in grain size fractionation, e.g. freeze drying, sonication, and desorption, to obtain a better understanding of the magnitude of the changes caused by different analysis protocols. Even small changes in a sieving procedure, e.g. maintaining the net of sieves perfectly horizontal, might have an effect on the outcome of the sieving [16].

Some analyses, e.g. X-ray-fluorescence and PIXE (proton induced X-ray emission), require very uniform and small grain sizeand it will be necessary to grind a sediment sample to the desired particle size. Different equipment is available [6].

There is always a possibility of contamination or loss of samples during the drying, sieving, and grinding processes; to ensure the quality of the data this should be tested by processing different test samples, e.g. blank, reference, and replicate samples, before a new procedure is introduced.

5.3.6 Handling atmosphere/room facilities

Air contamination by dust and oil should be avoided, otherwise no special atmosphere is usually required for the subsampling of the sediments, e.g. for environmental monitoring purposes, to continue with our hypothetical example. However, redox measurements, an important background parameter, require special handling to maintain oxygen-free conditions in order to avoid the effects of atmospheric oxygen. The sample slice has to be stored in an airtight and oxygen-free container. An easy way to do this is by using a cutting instrument where the container is included. The redox measurement should take place as soon as possible and always within less than 10 minutes.

Also for special purposes, e.g. in speciation studies of the chemical forms of trace elements in pore water, the redox conditions must be maintained. This requires a controlled atmosphere, e.g. the use of atmosphere (glove) boxes, where the slicing and other manipulations can take place.

5.4 Sample storage

Tightly sealed sample containers must always be used to avoid any further contamination or evaporation.

For samples meant for nutrient and trace element analysis, polyethylene or polystyrene containers are the most suitable for storage. Disposable plastic containers can often be used as these do not usually contain detectable amounts of nutrients and trace metals. This should, however, always be tested, e.g. by the use of a reference material or a blank sample.

Petri dishes are frequently used for storage, especially for trace metal and nutrient samples, but these containers have the disadvantage of not closing tightly and have to be taped tightly if used.

For samples meant for trace organic analysis, glass or PTFE containers should be used and the bottles or jars should be closed tightly with a piece of aluminum foil under the cap to protect the sample from any plastic or rubber material in the cap. The shiny side should not be in contact with the sample because it is coated with a lubricant [3].

Sample containers for trace organic analysis and preferably also any aluminum foil to be used, should, depending on the compounds of interest, be washed with a suitable organic solvent mixture, e.g. 50:50 hexane:acetone, and dried carefully to avoid contamination.

Each sample container must be carefully labeled to avoid any confusion. There should be clear instructions about the labeling and it should be tested under field conditions to ensure its stability. As far as practicable, the labeling of the sample containers should be performed before the sampling at a station starts. The labeling should be well documented. It is preferable to label the bottom of the containers. This is especially important if it is necessary to remove the lids before any further handling such as freeze drying.

It is usually an advantage to pre-weigh the containers as this will make it possible to determine the water content of the samples directly in the original containers.

Samples from one site/core or which are otherwise related should preferably be kept together, e.g. in a large, labeled plastic bag or box to minimize the risk of exchanging samples. During one cruise, thousands of unique valuable samples/subsamples can be collected and stored together until further work in the analytical laboratory and it is of crucial importance to the quality that the identification of the samples is not compromised.

There should also be one designated person who is responsible for the "sample bank" and the documentation of the handling ("chain of custody") procedures of the samples in a sample log-book.

The allowable storage time depends on the storage temperature, the purpose of the study, and the type of contaminant. Redox potential must be measured immediately. Refrigerator temperatures (+4 °C) are usually only suitable for short-term and temporary storage as biological activity in the sample is not arrested. Trace element samples can be kept at least one year deep-frozen without introducing appreciable changes. Mercury samples have been reported to be an exception [14]. The allowable storage period for organic contaminant samples may vary from a few days to several months, depending on the chemical and physical properties, e.g. the vapor pressure, of the compound of interest. Therefore, no general recommendations can be given, but the effect of storage conditions on the measured concentrations should be tested in advance. In many cases, freezing of the sample is appropriate. However, samples meant for grain size analyses and speciation studies must not be deep-frozen. Also samples for toxicity testing require special handling and storage procedures. Malueg et al. [17] warn against the use of freezing as a storage procedure prior to copper toxicity testing, and Schuytema et al. [18] have reported a decrease in the toxicity of sediments contaminated with DDT and endrin during freezing.

The recommendations given above are summarized in Table 5-1 for the determinants most often studied in sediments:

Table 5-1: Summary of recommendations for sediment storage

Parameter	Sample container	Storage temperature	Storage time
Nutrients	PE, PS	-20 °C	1 month
Trace elements	PE, PS	-20 °C	1 year
Trace organics	Glass	-20 °C	1-6 months
^{210}Pb	PE, PS	-20 °C	1 year
Redox	PE, PS	Immediate analysis	---
^{137}Cs	PE, PS	-20 °C	6 months
Organic carbon	PE, PS	-20 °C	6 months
Grain size	PE, PS	+4 °C	---

PE = polyethylene
PS = polystyrene

For storage for a longer period, e.g. sample banking special storage conditions are needed.

5.5 Internal quality control procedures

5.5.1 Split samples

Split samples can be used to assess the uncertainty (variance) in the results. Uncertainties in all sample handling and subsampling procedures contribute to differing extents to the overall variance. The splitting can take part at any stage of sediment handling, depending on the sources of uncertainty to be studied. If the analytical variance is to be studied, splitting should be carried out after any pretreatment such as drying, sieving, and homogenization of the sample.

5.5.2 Blanks

Whatever the precautions taken, there is always a risk of samples being contaminated under field conditions by sampling, subsampling, pretreatment equipment, storage containers, and environmental conditions such as dust in air. Blank sample data can be used to estimate the effect of the contamination sources on the variance of the results.

Different types of blank can be used to test the effect of contamination at different steps during sampling and (analytical) processing (see, e.g. [2]). The most important are:

- field blank;
- sample preparation blank;
- matrix blank;
- reagent blank.

It is much easier to collect many blank samples than to analyze them all, as this could be rather expensive. It has, therefore, been suggested that only the field blank sample belonging to a series of samples should be analyzed in the first place; if problems arise the individual blanks belonging to the same series can then be analyzed to determine the source of the contamination [2].

The field blank is a blank sample which has been prepared in the field and taken through all the different handling procedures, along with the real samples from the field, to the final analytical work in the laboratory. It can, e.g., consist of demineralized water or a sediment reference material. The effects of the sampling equipment and the sampling procedure itself are, however, difficult to assess.

The variance arising from a certain step in the procedure (e.g. transport) can also be tested by introducing the test sample (blank, reference material, or split samples) several times to this specific step.

Analyzing split samples gives valuable information concerning variation which to a high degree can be a result of contamination.

5.5.3 Documentation

Where and how to document and present the QA sampling data are always questions of concern. It is not possible to give any strict guidelines as this depends on the objectives of the sampling and the further use of the data. QA documentation should, however, always include enough information to be able to guarantee the desired quality of the data, e.g.

- provide the basis for the audit system set up to ensure that the protocols have been followed; and
- provide and verify the information about all types of variability of the data.

Throughout the sample history, from selection of the sampling site to the numeric result, accurate bookkeeping must, therefore, be maintained and the process must be followed by the responsible scientist/QA coordinator, who should record any deviations from the protocols.

The possibility of using techniques such as video uptake and photographic pictures for documentation of certain procedures, e.g. sampling, could sometimes be very efficient as a supplement to a more conventional log-book system.

The results from replicate-split and blank data and other forms of quality control data giving information about variability should be stored and presented in an easily assessable way, e.g. using different types of control card to ensure that future data users have the necessary information about all possible variability in the data set.

A list of recommended background information to be recorded is given in Appendix A; Appendix B gives a list of items to include in a documentation protocol for environmental sediment sampling (modified from that used during the 1993 ICES/HELCOM Baltic Sea Sediment Baseline Study).

It is not always necessary or possible to retrieve all the information suggested in the protocol and also discussed above. But it should as a minimum be discussed and if it is not included the consequences on data quality should be evaluated.

5.6 References

[1] D.S. Barth and T.H. Starks, *Sediment Sampling Quality Assurance User's Guide*, Cooperative Agreement CR 810550-01. Environmental Systems Laboratory, Las Vegas. Reproduced by National Technical Information Service (NTIS). U.S Department of Commerce, Springfield, VA 22161, USA (1985)

[2] L.H. Keith Ed., *Principles of Environmental Sampling*, American Chemical Society, Professional Reference Book (1988)

[3] L.H. Keith, *Environ. Sci.Technol.*, **24**, 610 (1990)

[4] WELAC, Eurachem Guidance Document No. 1, Welac Guidance Document No. WGD 2, *Accreditation for Chemical Laboratories* (1993)

[5] R. Mead, *The design of experiments. Statistical principles for practical application*, Cambridge University Press, Cambridge, UK (1988)

[6] A. Murdoch and S.D. MacKnight, in: *CRC Press Handbook of Techniques for Aquatic Sediment Sampling*, A. Murdoch and S.D. MacKnight Eds., N.V. Boca Raton. Florida, USA (1991)

[7] H. Lee II and R.C. Swartz, in: *Contaminants and Sediments*, Volume 2. Analysis, Chemistry, Biology. R.A. Baker Ed., Ann Arbor Science Publishers Inc. Ann Arbor, Michigan, USA (1980)

[8] B. Larsen and A. Jensen, in: *Review of Contaminants in Baltic Sediments*, ICES Cooperative Research Report 180, Compiled by M. Perttilä and L. Brügmann (1992)

[9] L. Brügmann, in: *Review of Contaminants in Baltic Sediments*, ICES Cooperative Research Report 180, Compiled by M. Perttilä and L. Brügmann (1992)

[10] Z. Kajak, K. Kacprzak and R. Polkowski, *Ekol. Pol.*, **B11**, 159 (1965)

[11] S. Blomqvist, *Sedimentology*, **32**, 605 (1985)

[12] H. Nies, H. Albrecht, V. Rechenberg, J. Goroncy, H. Dahlgaard, D. Weiss and L. Brügmann, *Dt. hydrogr. Z.*, **43**, 27 (1990)

[13] D.H. Loring and R.T.T Rantala, *Earth-Science Reviews*, **32**, 235 (1992)

[14] A. Murdoch and R.A. Bourbonniere, in: *CRC Press Handbook of Techniques for Aquatic Sediment Sampling*, A. Murdoch and S.D. MacKnight Eds., N.V. Boca Raton. Florida, USA (1991)

[15] A. Barbanti and M.H. Bothner, *Environ. Geol.*, **21**, 3 (1993)

[16] R Metz, *Sedimentology*, **32**, 613 (1985)

[17] K.W. Malueg, G.S. Schuytema and D.F. Krawczyk, *Environ. Toxicol. Chem.*, **5**, 245 (1986)

[18] G.S. Schuytema, A.V. Nebeker, W.L. Griffis and C.E. Miller, *Environ. Toxicol. Chem.*, **8**, 883 (1989)

Appendix A

Background information

Sampling site information

Date
Position and positioning equipment
Depth
Weather conditions
Echogram
Water salinity profile
Water nutrient profile
Sea bed water oxygen/hydrogen sulphide content
Sea bed topography
Echogram
Side scan sonar recording
Sea bed video
Rate of sedimentation (Pb210 method)

Sample/core information

Date
GPS coordinates
Depth
Nutrient (P and N) and carbon content of the sample
Water content of the sample
Visual inspection of the sample
Microscopic inspection of the sample
X-ray inspection of the sample
Sediment sample photography
Grain size distribution

Appendix B

Protocol

1. Station information

Station number
Station name
Purpose of sampling
Sampling date (dd.mm)
Sampling time (hh.mm)
Latitude (degrees, minutes, fractions of minutes)
Longitude (degrees, minutes, fractions of minutes)
Water depth (m)
Total number of cores taken
CTD data indicator (Yes/No)
Salinometer data indicator (Yes/No)
O_2 data indicator (Yes/No)
H_2S data indicator (Yes/No)
Sea bed video (Yes/No)
Remarks

2. Sample information

Station number
Sampling device
Sample/core/purpose number (list should be provided, see below)
Winch responsible
Responsible for slicing and storage
Responsible for redox measurement
Remarks on sampling

Sample/core number and purpose
(Example: Sediment baseline study)
1   X-ray
2   Redox potential
3   Age determination (^{137}Cs)
4   Trace elements
5   Grain size inspection
6   Nutrients
7   Age determination (^{210}Pb)
8   Backup
9   Organics
10   Age determination (laminae counting by photography)

Appendix B (continued)

3. General core information

Station number
Visual description of the core
Microscopic description of the core

4. Subsample information

Station number
Core purpose
Top of the sediment subsample (cm)
Bottom of the sediment subsample (cm)
Wet weight (with box and cover)
Dry weight (with box and cover)
Identifier of contaminant to be determined
Identifier of method of analysis
Measured concentration
Remarks

6.

Quality assurance of analysis of organic compounds in marine matrices: Application to analysis of chlorobiphenyls and polycyclic aromatic hydrocarbons

R.J. Law[1] and J. de Boer[2]
[1] Ministry of Agriculture, Fisheries and Food, Fisheries Laboratory, Burnham on Crouch, Essex CM0 8HA, United Kingdom
[2] DLO - Netherlands Institute for Fisheries Research, P.O. Box 68, NL-1970 AB Ijmuiden, The Netherlands

6.1   Sampling

Variability is an intrinsic property of the environment and nature, and is unavoidable. In addition to the spatial and temporal variability in concentration of contaminants seen as a result of environmental processes, analytical data also incorporate variability introduced as a result of the sampling and analytical processes themselves. The overall variability can be expressed as [1]:

$$\sigma^2 = \sigma^2 \text{ spatial} + \sigma^2 \text{ temporal} + \sigma^2 \text{ sampling} + \sigma^2 \text{ analysis} \qquad (1)$$

In order to obtain interpretable results from the environment it is essential that the technical component (sampling and analysis, the latter including pre-treatment and storage) does not dominate the overall variability of the data. It is important, therefore, to be able to estimate the variances associated with sampling and analysis; in an ideal world these would be zero and the variance of the data would be a perfect representation of the true distribution of the analytes in space and time. Realistically this is not possible, and it has been suggested that the technical variance should not contribute more than 20 % to the total variance, and that the analytical variance should not exceed 20 % of the technical variance [2]. Youden [3] has also suggested that once the analytical uncertainty has been reduced to one-third or less of the sampling uncertainty, then further reduction of the analytical uncertainty is of little importance. The analytical variance can be established for each of the steps of a specific procedure by the use of appropriate blank and control samples [4], and controlled by the operation of a suitable quality assurance program. Sampling errors cannot be controlled by the use of blanks, standards and reference samples, and so are best treated independently [5]. Sampling variance may contain systematic and random components arising from population and sampling considerations [6]. The representative sampling of populations requires the design of statistically sound sampling strategies, but is beyond the scope of this article. Sampling operations can also contribute both random and systematic components to the sampling uncertainty, both as a result of inadequate sampling strategy and the use of inappropriate sampling methodology. The selection of appropriate sampling devices for analysis of chlorobiphenyls (CBs) and polycyclic aromatic hydrocarbons (PAHs) will be considered in more detail below.

Aspects of quality assurance can be applied to sampling operations, and have been carried through to accreditation for the collection of water samples within at least one laboratory in the United Kingdom. Sampling devices must be selected and sampling procedures developed for each specific set of determinants, and in studies involving the determination of more than one class of determinants this can lead to the collection of multiple samples at each site, each following a different defined protocol. Once a sampling method has been selected then it should be fully tested and documented in full, step-by-step detail [7] and must be adhered to on every occasion samples are collected. Adequate training must be provided for all staff collecting samples, and their competence recorded. Sample log-sheets to be completed in the field should list the sampling sites, the samples to be taken, the containers to be used for each sample and the procedure to be followed. The person sampling should then record the date and time of sampling at each location, whether there was any deviation from the procedure laid down, and any unusual observations made at the time. Samples should also be provided with an indelible and unique identifier in a manner appropriate to the type of sample container used in each case. Samples taken for evidential purposes must (in many countries) also have a chain of custody record [8]. In fact this is probably a good idea for all samples whether or not litigation is likely to result, as it both confirms that the correct storage procedures were followed post-sampling, and ensures proper traceability from sampling to data reporting [9].

6.1.1 Sea water

Many organic compounds are hydrophobic and have very low solubility in sea water, being preferentially adsorbed by particulate materials such as plankton or suspended sedimentary particles in the water column. This means that relatively large water samples must often be collected and processed when dissolved concentrations are to be determined, and this usually involves operational difficulties [10]. The concentrations of compounds such as chlorobiphenyls in the dissolved phase are often so low as to render analysis very difficult, so that sea water is not the preferred matrix in which to monitor these compounds. The same is true of the larger PAH molecules (<3 rings); the lower PAH of 2 and 3 rings have appreciable solubility. Concentrations of the more hydrophobic PAHs and CBs are generally higher in fine sediments and some biological tissues than in sea water, and monitoring in these matrices may be more appropriate in many instances.

Devices for collecting sea water samples for organic trace analysis must be constructed of materials which will not leak interfering compounds into samples, and which can readily be cleaned with organic solvents (e.g. [11]). Because compounds such as CBs and PAH are hydrophobic they will tend to partition to the inside walls of samplers, and the proportion adhering within the sampler must be recovered with rinsing solvent and addition to the sample extract if negative biases are to be avoided. For this reason it is also not desirable to divide samples collected for the analysis of trace organics - the person sampling should collect a sample of the desired volume, and replication must be achieved by deploying samplers either multiply or repetitively. For air sampling the adsorptive capacity of the internal surfaces of stainless steel sampling canisters to trace organic compounds has been reduced by coating with a pure chrome-nickel oxide layer, but this technique has not yet been tested in water samplers [12]. All plastics probably contain some compounds added to alter their physical properties (plasticizers such as phthalate esters, UV absorbers, etc.) which could be leached by solvents [13]; plastics can, in addition, adsorb trace organics from water [14,15], and are therefore unsuitable. Leached phthalate esters may also interfere in the analytical procedures, such as the analysis of CBs by gas chromatography with electron-capture detection (GC/ECD). Preferred materials are glass, stainless steel, and polytetrafluoroethylene (PTFE). The type of working platform(s) to be used and the desired sampling depth are also important factors in sampler design. Two examples of successful devices for the collection of discrete sea water samples for analysis of trace organics have been described by Law et. al. [16] and Stadler and Schomaker [17]. The former is a small device which can be deployed by hand from quays, jetties, river banks, etc., as well as used from research vessels. It collects a sample of 2.7 L volume in a Winchester solvent bottle which is fitted within a weighted stainless steel frame. The frame is attached to a nylon line by which it is deployed, and a second line opens and closes a spring-loaded PTFE stopper. The bottle may therefore pass unopened through the organic-rich surface layer of the sea, thereby obviating contamination from that source. The sampler has been in routine use for approximately ten years, and has been used to collect samples for the analysis of a wide range of trace organics, including PAHs and organochlorine determinants, to a sampling depth of *ca.* 50 m [18]. The latter sampler is a larger device, collecting a sample of 10 L within a glass sphere. The sphere is mounted within a stainless steel frame, and is deployed from the hydrographic winch of a research vessel. It can be used for sampling at greater depths, and also has the advantage that extractions are conducted within the sampling vessel itself. Once again, this device has been proven over a considerable time in routine use, both for PAHs and for other determinants. Contamination from lubricants used on traditional steel hydrowires can be avoided by using Kevlar rope instead, as is now done for trace metal sampling. A depth-integrating bag sampler constructed of PTFE has also been described, and deployed in studies of PAHs in the St. Lawrence River [19]; a stainless steel system validated for the determination of organochlorine compounds (including CBs) in 28 L sea water samples has also recently been described [20,21]. A further method which is the subject of study is the continuous pumping of water from depth. This is now routinely employed for trace metal analysis, water being pumped from shallow depths through perfluoroalkoxy PTFE (PFA) tubing by means of an air-driven PTFE bellows pump situated in a clean laboratory on board ship [22]. For trace organics sampling both the tubing and the internal surfaces of the pump (rotors, etc.) contacted by the water must be constructed of PTFE (e.g. [23]) as this material exhibits the lowest adsorption and leaching problems [24]. For pumping from greater depths, a submersible pump at the sampling depth would be preferred to a pump at the surface, and pneumatically or electrically driven pumps of this type can be obtained. In some cases it may be necessary to determine the dissolved and particulate concentrations of trace organic contaminants separately. Particulate material is best collected by a technique such as continuous-flow centrifugation [25] rather than by filtering as for trace metals, as filtering is likely to alter the equilibrium within the sample and influence the distribution of the contaminants. Filtering is often used, however, particularly in large-volume continuous pumping samplers such as the Seastar [26,27], although it is unlikely, therefore, to yield more than a rough estimate of the partition. A recent comparative study of filtering and continuous-flow centrifugation for the determination of CBs in surface waters has showed that filter clogging and adsorption affect the accuracy of the filtering technique, but that even centrifugation is inadequate, because freely dissolved CBs and those bound to dissolved organic matter cannot be distinguished [28]. Some commercial oceanographic sampling bottles (e.g. Go-Flo samplers manufactured by General Oceanics) are also available with internal PTFE coatings, and after substitution of a number of components (seals, etc.) these may also be suitable for the collection of discrete samples. The needs of a specific program may dictate the use of one or other of these samplers, or require the development of a novel device applying the principles outlined here. Two alternative approaches which have been applied to the analysis of PAHs and CBs in waters are solid-phase extraction (SPE) and solid-phase microextraction (SPME). For preconcentration of PAH from drinking water samples the best results were obtained with combined octadecylsilane/ammonia (C_{18}/NH_2) solid-phase cartridges [29]. In SPME analytes are concentrated in a polymeric phase immobilized on to a fused-silica fiber, and later desorbed directly into a GC/MS system for analysis [30]. Detection limits for a range of target compounds were between 1 and 20 ng μl^{-1} with a sampling time of only 10 min, and with no use of organic solvents.

As some trace organics (including PAHs) are enriched in the surface microlayer of the water column, this area has also been sampled selectively [31]. The thickness of the microlayer is less than 1 mm, and the depth sampled is dependent on the device used. The Garrett screen is one of the simpler devices to construct and use [32].

6.1.2   Sediments

No overall review of sediment sampling methods has been prepared, although ICES (the International Council for the Exploration of the Sea) has prepared guidelines on the use of sediments in environmental monitoring programmes, such as that coordinated by the Joint Monitoring Group of the Oslo and Paris Commissions [33]. These guidelines incorporate general guidance on the sampling and analysis of sediments and the interpretation of data.

Surface sediments can be collected intertidally, or by means of a variety of grab samplers from a vessel. The choice of grab samplers is often determined by the substrate to be sampled, with bottom closing samplers deployed over muds and sands and side-closing grabs deployed over gravels. Whichever grab is used the bucket(s) in which the sample is collected should be constructed of stainless steel, to minimize contamination and so that the buckets can be rinsed with solvent prior to use. Such samplers are suitable for collecting samples of the upper few centimeters of consolidated sediment, but will disperse the unconsolidated floc at the sediment-water interface during descent. If this floc is to be sampled then multiple damped-array shallow coring devices are to be preferred. In addition, passive sediment traps can be used to collect samples of material sedimenting through the water column or resuspended from the seabed [34], although the design and deployment of these devices is still far from straightforward [35,36]. In depositional areas, the changing fluxes of contaminants to the seabed over time can be reconstructed by analysis of dated sediment cores and the "background" concentrations established [37-40]. For this purpose cores of a meter or more in length may be required; these may be collected by means of box corers, vibrocorers, or other specialized equipment. It is essential either that the tube in which the core is collected is constructed of noncontaminating material, or that sediment in contact with the core tube is not taken for analysis of trace organics.

6.1.3   Biota

Concentrations of most trace organic contaminants in biological tissues are very variable, and great care must be taken to make the sample representative of either the population as a whole or that part of the population for which information is needed [41]. The variability can be related to geographical location, tissue type, species, age, sex, diet, breeding status, and season, amongst other factors. Also, the aim of the study may dictate some aspects of the analyses to be conducted - for instance, it the aim is to assess the risk to human consumers then only edible tissues of commercial species of exploitable size may be considered. CBs are metabolized relatively slowly, and tend to be accumulated in tissues rich in neutral lipids. The highest concentrations are found in the blubber of marine mammals, which are top predators and also have large reservoirs of lipids deposited in their blubber. PAHs, by contrast, are readily metabolized in fish, polar metabolites being excreted via the bile [42]. Exposure of fish to PAHs is, therefore, often assessed by measurement of these polar metabolites in bile rather than by the analysis of tissues [43,44]. As PAH concentrations are generally so low, analysis of fish muscle should be contemplated for food assurance purposes only. In molluscs and crustaceans the metabolic processes are less efficient and accumulation of PAHs in tissues can occur, and as molluscs such as mussels are also sessile they can be used for studies of the spatial distribution of PAHs. The exposure of filter-feeding bivalves such as mussels (e.g. *Elliptio complanta*) is primarily *via* water, and therefore reflects ongoing rather than historical inputs [45]. Samples can be collected in a variety of ways: benthic animals can be taken intertidally from rocks or beaches, or sampled by means of a grab dredge or trawl in deeper water. Fish can be taken by rod and line, or in a variety of trawls and other nets depending on the habitat and habits of the species under investigation. The guidelines issued to participants in the Joint Monitoring Programme of the Oslo and Paris Commissions provide an instructive example of the complexity associated with the collection of representative samples of fish and shellfish populations in an international collaborative program [46].

6.2   Sample pre-treatment and storage

Environmental samples must often be stored for considerable periods of time prior to analysis, and during this period changes may occur which compromise the accuracy of the data produced. Unfortunately, no definitive studies have been conducted to establish, for each environmental matrix, the period of time for which samples can be stored without such problems [47]. Some experience can, however, be gained from laboratories involved in ongoing monitoring programs.

The major problems result from processes which lead to higher concentrations of analytes (principally contamination from storage containers, etc.) or losses of analytes (chemical or biological degradation, photodegradation if kept in the light, volatilization, etc.). Other factors such as the concentration and speciation of the analytes may also be important in certain instances.

6.2.1   Sea water

The storage of unextracted water samples for analysis of trace organic contaminants cannot be recommended, and they should generally be extracted as soon as possible after sample collection. Solvent extracts can be stored for long periods (months) in a freezer at -20 °C, if sealed in pre-cleaned glass containers sealed with PTFE-faced caps. Such containers should always be evaluated prior to their use for storage of valuable sample extracts. Losses and contamination during storage can be quantified by the analysis of standard solutions and clean solvents sealed in the containers and kept for a period similar to that expected for the real extracts. As a compromise measure, if extraction cannot be conducted promptly for some reason, an aliquot of the extracting solvent can be added to the sample in the sampling-bottle prior to short-term storage. The compounds of interest will partition into the solvent, and thus will not undergo degradation in the water phase. Samples for which such treatment is not possible (i.e., those in which volatile organics are to be determined) have been the subject of storage trials [47]. Standard EPA vials of 40 mL volume were filled with spiked water samples, sealed with a PTFE-faced septum and no headspace, and stored for 56 days at 4 °C. No volatilization and loss of the very volatile compound chloromethane (boiling point -24.2 °C) was observed over the storage period, and so no such losses could be expected for even the most volatile CBs or PAH. In real samples microbial degradation would probably be the most important process leading to lower concentrations of determinants; adsorptive effects would need to be obviated by rinsing the container with solvent and adding the rinsings to the sample extract, as outlined above.

6.2.2   Sediments

Subsamples of sediment are transferred to solvent-cleaned, wide-mouthed glass jars for storage in a freezer at -20 °C prior to analysis. The jars should not be overfilled, particularly when the sediment is muddy, or the jar can break as the water held within the sediment sample expands on freezing. These jars often have plastic lids with waxed inserts, and as an additional safeguard against contamination it is desirable to place solvent-rinsed aluminum foil over the mouth of the jar before screwing on the lid [16]. Whilst stored in a freezer aboard ship it is advisable to store sediment jars in plastic crates, and to secure these to racks or shelving within the freezer; bad weather might otherwise lead to a drastic reduction in the number of usable samples!

Storage, homogenization, and extraction of sediments are much easier when the samples are dry, and many analysts dry sediments as a matter of routine. A number of procedures are commonly used, but each can carry risks for the integrity of the samples [48]. Chemical drying can be performed by grinding with Na_2SO_4 or $MgSO_4$, but this can cause irreversible adsorption of some analytes. This problem may be avoided by drying with water-adsorbing materials such as alumina or silica, but water may then be released during extraction if polar solvents are used [49]. Oven-drying at relatively low temperatures (ca. 40 °C) can result in losses of analytes by evaporation, particularly in samples with a low organic carbon content (e.g. most sandy samples) [48]. Neither of these last two procedures can be generally recommended for the reasons outlined. In addition, air-drying at ambient temperature and freeze drying are both commonly used. Both losses of and contamination with target compounds can occur during these procedures [50], and analysts are urged to validate their own procedures thoroughly **before** applying them routinely. Contamination can be checked either by submitting a low-level reference material to the process, or by analyzing a check sample such as C_{18}-bonded silica which would adsorb contaminants encountered during the drying process [48]. Similarly losses can be checked either by submitting a high-level reference material or a spiked matrix to the process prior to analysis. It is further recommended that drying procedures be verified by application of a second technique.

6.2.3 Biota

Animals should be stored in a freezer at -20 °C prior to analysis. If possible they should be stored whole and any tissues required for study removed under appropriate conditions of cleanliness in the laboratory just prior to analysis. If biomarker studies (e.g. ethoxyresorufin-O-deethylase (EROD) determinations [51]) or fish disease studies are to be conducted on the same animals then the liver tissue will normally be removed at sea and appropriate precautions must be taken there. Ideally individual fish should be wrapped in solvent-rinsed aluminum foil, placed in plastic bags, and frozen quickly. If fish are frozen in bulk it can take a considerable time (periods of hours to days) for the whole sample to be frozen through to the center, even in a trawler blast-freezer at -50 °C. If these fish were warm prior to freezing (following a period of time on the deck of a ship in summer, for instance) then vulnerable tissues such as liver can undergo considerable decomposition before it is arrested by the falling temperature. Freeze drying is also commonly applied to biological tissues prior to extraction, and the same cautionary note applies as to sediments (above). Whilst discussing the results of an international intercomparison exercise Farrington et al. [52] reported that freeze drying of mussel tissue resulted in losses of 2- and 3-ring aromatic hydrocarbons (naphthalenes and phenanthrenes), and led to changes in the tissue matrix which reduced both overall extraction efficiency, and the recovery of deuterated PAH standards added for the purposes of quantification.

There is currently no information available from rigorously conducted studies of the effects of storage on the results obtained from the analysis of either biota or sediments for CBs or PAHs. One such study is in progress under the QUASIMEME (Quality Assurance of Information for Marine Environmental Monitoring in Europe, see chapter 10) programme [53,54], in which the effects of storage conditions (temperature, freeze drying) for biota on the determined concentrations of CBs, organochlorine pesticides, and trace metals will be investigated. Apart from variations in the concentrations of the analytes themselves, other factors such as the oxidation of lipids and the dry weight of the tissue will be monitored, as changes in the lipid composition and moisture content may influence the concentrations and the ease of extraction of the analytes. A preliminary report on the results of this study should be available early in 1996.

6.3 Analytical aspects

When selecting or developing a method for the determination of CBs or PAHs in environmental samples, it is essential that both the method as a whole, and all individual steps of the method, are fully validated, and that the performance of the method is regularly checked. Before the final determination can be made by a chromatographic method, rather complex and laborious extraction and clean-up is required. This extraction and clean-up, and the HPLC or GC analysis, can all be sources of error yielding concentrations either too low as a result of losses (evaporation, adsorption, etc.) or too high (blank values, co-elution, etc.). There is no single method which can be recommended as the best for the determination of CBs or PAHs. Method validation and QA are therefore extremely important. The validation of a method should include: (i) validation of each individual step: calibration (preparation of calibration solution, optimization of GC or HPLC), extraction (recovery, blank values), clean-up (recovery, blank values), GC (or HPLC) analysis separation/co-elution, use of internal standards, calibration curves, integration, etc.), (ii) validation of the method as a whole by analyzing certified reference materials; and (iii) a regular check of the performance of the method by the analysis of internal laboratory reference materials (LRMs) together with each batch of samples and the recording of the results on a quality control chart (cf. Fig.1), and regular participation in national or international interlaboratory studies. Additionally it is of utmost importance that the methods used are well documented, and that fully detailed protocols are readily available to the analysts. Because of the complexity of these methods, they are undeniably subject to personal influence, and the results of interlaboratory studies have shown that laboratories' data differ when different individuals conduct the analyses [55]. Protocols should be designed in such a way as to minimize the scope for personal interpretation, and the quality control data should be studied for each analyst so as to identify and encourage those whose results are least reproducible.

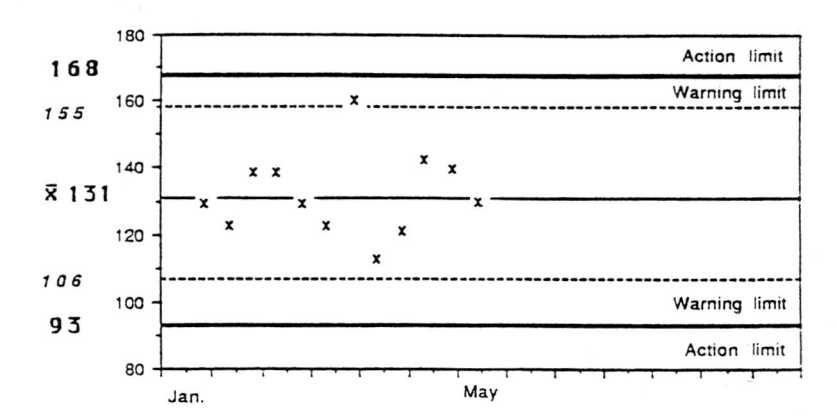

Figure 6-1: Quality control chart for the determination of CB105 in cod liver; the concentrations are expressed in μg/kg wet weight.

6.3.1 Chlorobiphenyls

The analysis of CBs in environmental samples normally comprises three stages: extraction, clean-up, and gas chromatographic (GC) determination. These steps will be discussed with special attention to quality assurance and quality control (QA/QC). Calibration is dealt with under GC analysis and further attention is given to interlaboratory studies and the use of certified reference materials (CRMs).

6.3.1.1 Extraction

As described above the extraction of CBs from sea water should ideally take place at the sampling site, because any time left between sampling and extraction may lead to adsorption of CBs on to the walls of sampling devices, tubing, and containers. Because of the extremely low levels of CBs in sea water and the related problems of obtaining sufficiently low blank values, only a few specialized laboratories in the world are able to produce reliable data on CB concentrations in sea water. Methods on the determination of CBs in sea water were reported by Tanabe et al. [56] and Schulz-Bull et al. [57], who both used extraction by Amberlite XAD-2 resin columns, and Hermans et al. [28], who used extraction with *n*-pentane in a continuous batch extractor. Glassware and tubing were cleaned by either pumping of seawater for 3h at a pumping rate of 1.2 m^3 h^{-1} through the system used [57] or rinsing with acetone, washing with a detergent solution, washing with demineralized water and kiln-firing at 250 °C [28]. Insufficient attention to the extensive pre-cleaning of glassware and tubing used may in the past have suggested a decreasing trend of CB concentrations in sea water, whereas in reality this was probably caused by differences in the blank values obtained [58]. Recoveries should also be checked by extraction and analysis of sea water which has passed the XAD-2 resin columns or the batch extractor. The reproducibility of the extraction should also be checked, e.g. by extraction of five aliquots of a single sample. Other solid sorbents that have been used are Tenax [59] and polyurethane foam [60]. Hexane [61,62] or dichloromethane were usually used for extraction of water with organic solvents [63-65]. Water volumes taken for analysis usually vary between one and several hundred litres [61,64,66-68].

Overviews of extraction methods used for biota and sediments have recently been published [69,70]. Extraction of CBs from biota and sediments can successfully be carried out by Soxhlet extraction [71], saponification [71,72], ultrasonic extraction [73,74], liquid-liquid extraction [75], and supercritical fluid extraction (SFE) [76-78]. Soxhlet extraction is the most frequently used technique [55]. It has been demonstrated, however, that Soxhlet extraction does not always result in the highest recoveries [71]; one advantage is that the more labile compounds (such as chlorinated pesticides) are not degraded during such extraction. As these pesticides are often determined simultaneously with CBs, Soxhlet extraction is often selected as the most appropriate method for the combined determination. More aggressive methods such as saponification can, however, yield slightly better recoveries. The conditions for saponification need to be selected carefully, because Van der Valk and Dao [79] have demonstrated that saponification of sewage sludge may lead to dechlorination of higher chlorinated CBs, metal particles present in the sludge acting as a catalyst for this reaction. A maximum temperature of 70 °C and a maximum saponification time of one hour were recommended by these authors.

The conditions and solvents needed for extraction of sediments may differ, depending on the type of sediment sampled. In sediments with a low organic carbon content, extraction with a medium-polarity solvent such as acetone or dichloromethane may be adequate. If more organic material is present, a combination of a polar and a non-polar solvent is required for extraction. If graphitic carbon-like materials (e.g. coals) are present, toluene may be preferred; these materials are, however, uncommon in marine sediments [48].

Methods utilizing SFE for the extraction of CBs from biota and sediments are being developed. The use of SFE for extraction from sediments is promising, as the lipid content of sediments is generally lower than in biota and co-extraction of lipids does not play a role. By using SFE the total analysis time is considerably reduced, which makes this method very attractive. SFE instrumentation is, however, relatively expensive, and the benefits of this technique must be balanced against the investment before a decision on its use can be made. Bøwadt et al. [80] reported the application of SFE to the extraction of CBs from fish tissues. Although there are various advantages, it must be considered that (i) the sample should be freeze dried; (ii) additional clean-up may be needed for extracts from lipid-rich tissue because of the co-extracted material; and (iii) the maximum sample intake is relatively small (*ca.* 2 g) which makes the detection of compounds present at the ultra trace level, such as planar CBs very difficult [81,82].

Whichever technique is used, it is essential that the recoveries of the analytes are checked. This can be achieved by spiking (although there is always an uncertainty to which extent the spike will represent real-life conditions), by prolonged extraction with different solvents [80], or by comparison with other methods such as saponification. It is recommended that a surrogate (internal) standard should be added to all solutions analyzed as a check on whether losses have occurred. A standard solution should be taken through the whole procedure in order to establish the recovery for each determinant. This recovery standard should not be used to correct any data, but as a control on the whole procedure. If major losses have occurred then data should not be reported. CB29 (2,4,5-triCB) can be recommended as a recovery standard. Because of its high volatility, losses resulting from evaporation are easily detected, and, because it elutes relatively late from alumina and silica columns, losses during clean-up may easily be detected by use of this compound. Small peaks which may be present at the retention time of CB29 in the chromatogram do not hinder the use of CB29 for this purpose, as the recovery standard is only used for the control of major errors.

Further, blank values should be sufficiently low - clean glassware, nanograde quality solvent - and the reproducibility should be tested. If difficulties are encountered in reducing blank values, CB concentrations in laboratory air should be measured. Typical CB concentrations in air are 0.1 ng m^{-3} for less volatile CBs and 2 ng m^{-3} for more volatile CBs [48]. Laboratories built between 1960 and 1970 should, however, be suspected of having much higher CB concentrations in air as a result of the possible use of CB-containing paints and rubbers [83]. A check for CB contamination of laboratory air can be carried out by placing a Petri dish containing 2 g of C18-bonded silica in the laboratory for two weeks. Elution of the material in a column with 10 ml 10% diethyl ether/hexane and determination by GC/ECD should result in CB concentrations <1 ng per CB congener if there is no significant contamination of the laboratory air [48].

6.3.1.2 Clean-up and fractionation

Overviews on clean-up methods for CBs have been described recently [69,70]. Clean-up of extracts of biota and sediments is needed because extraction methods used are normally not sufficiently selective. This means that the final determination can be disturbed by: (i) lipids, which can cause a deterioration of the GC column and contamination of injector and detector; (ii) co-elution with other contaminants such as organochlorine pesticides, polychlorinated camphenes (toxaphene), polychlorinated naphthalenes and similar compounds; and (iii) compounds, such as oils, which cannot be detected by a specific detector such as the electron capture detector (ECD), but create negative peaks or an erratic response to the determinant [70]. Furthermore, it may be necessary to make a pre-separation of the CBs themselves, particularly if CBs, present at a ultra-trace level (such as non-*ortho* substituted CBs) are to be determined [72].

Frequently used techniques for fat separation are column chromatography over alumina or Florisil columns, gel permeation chromatography (GPC), and sulfuric acid treatment. The last technique normally also removes most compounds which yield negative peaks by GC/ECD. Less stable compounds (such as some organochlorine pesticides) are degraded by the action of sulfuric acid. The use of saponification, which is also relatively aggressive, also disrupts the lipids. Silica gel or Florisil columns are mostly used for the fractionation of CBs and other compounds such as organochlorine pesticides. Other complex mixtures such as polychlorinated terphenyls (PCTs) or polybrominated biphenyls (PBBs) are normally not separated from CBs on silica gel columns. These compounds can, however, be separated by GC because they elute at higher temperatures than the CBs. Pre-separation of CBs prior to the determination of non- and mono-*ortho*-substituted (planar) CBs is normally carried out by HPLC using porous graphitic carbon (PGC) columns [77,81] or 2-(1-pyrenyl) ethyldimethylsilylated silica (PYE) columns [84]. The use of charcoal columns for this separation is discouraged, mainly because of batch-to-batch variations of the charcoal and less precise separations [85]. Overviews of methods for the determination of non- and mono-*ortho* substituted CBs are given by Wells [70] and Hess et al. [82]. Sediments generally require an extra treatment to remove the sulfur [48], as sulfur can disturb the chromatogram to a large extent (Fig. 6-2). These extra treatments include shaking with tetrabutylammoniumsulfide [77], or treatment with copper [86]. Application of saponification or GPC [87] makes an extra sulfur removal step redundant, as sulfur is already removed by the use of these techniques [48].

It is essential to test the different clean-up and fractionation steps used. The reproducibility of the elution patterns should be verified. Shifts in elution patterns may occur between standard solutions and samples, because lipids may affect the elution profile. Losses resulting from adsorption should be checked by recovery experiments. Losses due to evaporation should also be checked, especially when extracts are concentrated between different clean-up steps. When a rotary evaporator is used, it is recommended to use a few milliliters of a high boiling solvent, e.g. *iso*-octane, as a keeper, to prevent evaporation to dryness and the occurrence of considerable losses. The rotary evaporator is a source of contamination and should, therefore, regularly be thoroughly rinsed with pentane or hexane. The use of a Kuderna-Danish evaporator may be an alternative. Materials such as silica and Florisil are very sensitive to moisture and should be kept in a desiccator. A standard treatment is recommended (heating at a fixed temperature for a standard time, adding a fixed percentage of water) prior to their storage or use. Solvent blanks, glassware, and chromatographic material should be analyzed. It is well known that commercial silica gel may contain PCBs [88].

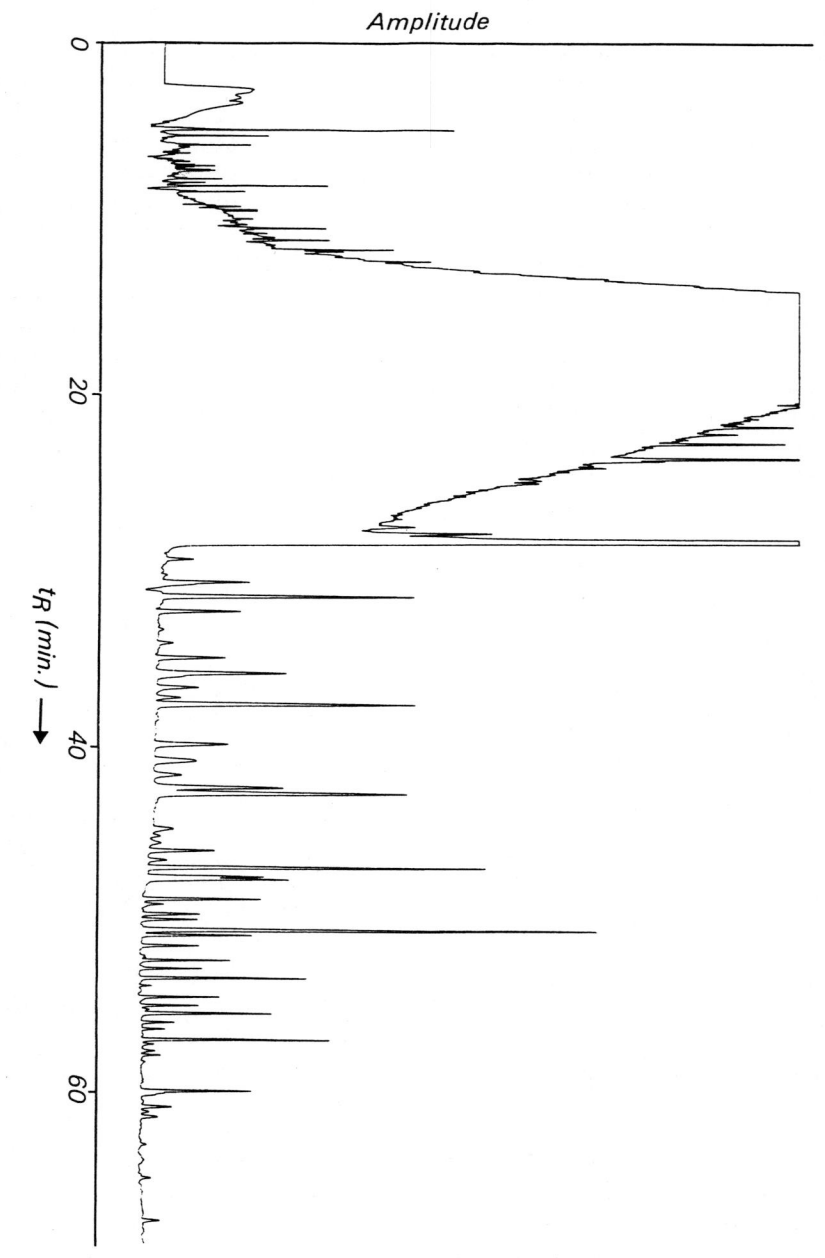

Figure 6-2: GC/ECD chromatogram of a sediment sample showing the disturbance caused by the presence of sulfur in the extract injected.

6.3.1.3 Calibration and analysis

With regard to the GC analysis there is a broad choice of methods and of conditions within the methods. These will not be discussed in detail; apart from the inclusion of some references we will confine ourselves to the discussion of quality control of the calibration and analysis.

For complex mixtures such as CBs a high degree of GC resolution is essential. The results of several intercomparison exercises have shown that capillary columns with a minimum length of 50 m and a maximum internal diameter of 0.25 mm are required for an acceptable separation of CBs [55,89,90]. The use of two different columns is also strongly recommended to enable an unambiguous determination of all relevant CBs. A further reduction of the internal diameter to 0.15 mm will improve the separation, but the pressure regulators of most GCs will need to be changed as pressures up to 400 kPa are required [92]. As a result of these higher gas pressures, leakages may occur more easily. The use of hydrogen is recommended as a carrier gas because it offers a better resolution at a higher gas velocity. Information on the retention times of CBs in single-column GC can be obtained from elsewhere [80,91-95]. Mathematical modeling by means of a single ring indix system for CB retention times was reported [96]. More complex systems have also been used for the determination of CBs. Larsen et al. [93] reported the use of serial GC, Storr-Hansen [97], Ojala [98] and Bøwadt et al. [78] reported the use of parallel GC, and the use of multi-dimensional GC was also reported [57,86,99-104].

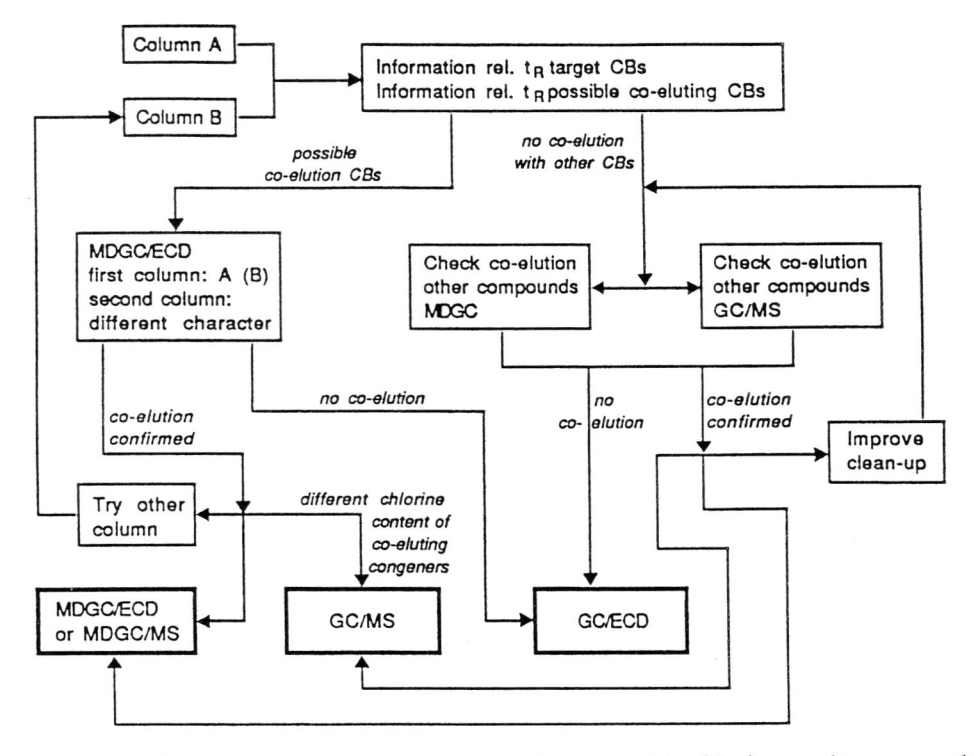

Figure 6-3: Protocol for the selection of an analytical method for CBs in an unknown sample.

The choice of single-column GC is determined by a number of requirements, such as the number of CBs to be determined, which CBs they are, whether the determination is to be applied on a routine or research basis, etc. Whichever system is selected, it is essential that absence of co-elution of contaminants or other CBs with the target CBs is guaranteed. The scheme in Fig. 6-3 shows how co-elution can be checked. Multidimensional GC (MDGC) plays an important role in this scheme, because with this technique unambiguous determination of CBs can always be achieved. The use of MDGC in certification procedures is, therefore essential [101,102, 105,106]. If MDGC is not available, it may be necessary to try a number of different columns before the absence of co-elution can be guaranteed. Instead of using ECD for detection, mass spectrometric (MS) detection may furnish additional information. For the determination of CBs the use of MS is, however, limited, because for CBs with an identical number of chlorine atoms a distinction can not usually be made. The use of GC/MS for the determination of planar CBs has been reported [72,81,107-109]. An advantage in this respect is the possibility of using ^{13}C-labeled CBs as internal standards.

Apart from the verification of absence of co-elution a number of other conditions should be controlled to guarantee high quality analysis. A number of sources of error that can occur in CB analysis are given by Wells [70]. Here we will discuss the quality control of injection, detection, calibration, and long-term stability.

6.3.1.4 Injection

As for the other parameters, optimization is essential prior to use of any injection technique. The two major systems used are splitless and on-column injection. Split injection should not be used as it causes severe discrimination effects.

In splitless injection the volume of the liner should be large enough to contain the total gas volume after evaporation of the quantity of solvent injected. For example, 1 µL *iso*-octane requires a 1 mL liner volume. Memory effects, as a result of contamination of the gas tubing attached to the injector, can also occur. This can be investigated by injection of *iso*-octane after an analysis of a CB standard. Small injection volumes are preferred, but this of course increases the detection limit. The use of an autosampler is a prerequisite for obtaining an acceptable reproducibility. Larger liner volumes can cause a poor transfer of early eluting components, so that peaks of those analytes will be reduced or even disappear altogether. The use of loose packing of (silylated) glass wool in the liner improves the response and reproducibility of the injection [110] although some organochlorine pesticides, e.g. DDT, may easily degrade when this technique is applied. Positioning of the column in the injector is important and should be done according to the instructions of the GC manufacturer. The splitless time should be optimized to avoid discrimination. This can be done by injecting a solution containing an early eluting and a late eluting CB, using different splitless times. Starting with a splitless time of e.g. 0.5 min, the peak height of the late eluting compound will presumably increase relative to that of the first compound as the splitless time is increased. The optimum is found at the time where the increase does not continue any further. The split ratio is normally set at 1:25 and is not really critical. The septum purge, normally 2 mL/min, should be stopped during injection until the splitter is opened. The temperature of the injector should preferably be about 10 °C above the maximum temperature of the oven program, but should not exceed the maximum temperatures of the column and septum used. Regular replacement of the septum is recommended, in order to prevent small particles of the septum entering the liner. For on-column injection, the optimum oven temperature program and the optimum initial temperature should first be selected. Because of the variety in on-column injectors, a detailed optimization procedure cannot be given; considering that the number of important operating conditions is greater than for splitless injection, a simplex procedure is recommended. More information on optimization of on-column injection may be obtained from Snell et al. [111].

6.3.1.5 Detection

The most frequently used detector for CB analysis is the electron capture detector (ECD). Modern ECDs generally contain a piece of radioactive ^{63}Ni foil with an activity of 370 MBq. The advantage of this detector is its sensitivity for halogenated compounds. Detection limits of 0.1 pg may be obtained for higher chlorinated CBs. ECD response factors for CBs are given by Mullin et al. [92]. Injection of chlorinated or oxygen containing solvents should be avoided. The temperature should be optimized. A range is normally given by the manufacturer; depending on the type of detector, optimum temperatures vary between 280 and 350 °C. A higher temperature, especially in combination with hydrogen as a carrier gas, is preferred because this may keep the nickel foil clean. When cleaning of the detector is necessary (this can be seen from the chromatogram by negative peaks occurring immediately after positive peaks and an higher signal together with a lower signal to noise ratio). This should be carried out by the manufacturer because of risk of radioactive contamination.

A drawback of the ECD is its poor linearity, especially in the lower range (0-50 pg) (Fig. 6-4). Therefore, calibration is of utmost importance. Injection of different dilutions of a CB standard solution during a series of samples (multilevel calibration) is strongly recommended. Alternatively the linear range of the ECD should be determined [112, 113]. Samples should then be diluted or concentrated to fit in this range. As the different CB congeners are present in the sample at divergent concentrations, more than one injection per sample may be required. If CB concentrations in the extracts are very low and the extracts are already maximally concentrated, multi-level calibration is preferred.

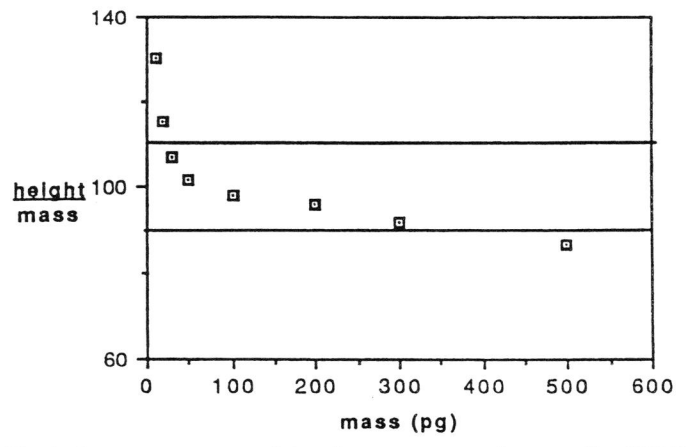

Figure 6-4: Typical response curve of the electron capture detector for CB153, which shows the nonlinear relationship between the ratio of the detector response/mass to the mass of CB injected.

When using GC/ECD, the system should be equilibrated by injecting at least one sample prior to a series of samples and standards. The data from this injection(s) should be ignored as a lower than expected response will normally be obtained. The injector, column, and detector take some time to stabilize.

A sample batch should normally consist of:

- a procedure blank;
- an internal reference material;
- 4 or 5 standards to enable proper multilevel calibration;
- one standard that has been treated similarly to the samples (recovery determination); and
- the samples, of which at least one should be in duplicate.

The standards used for the multilevel calibration should be regularly distributed through the batch. The use of mass spectrometry (MS) as detection technique in CB analysis is becoming more popular, mainly because of the availability of simpler and more sensitive mass spectrometers such as ion-trap instruments. Negative chemical ionization (NCI), also called electron capture negative ion (ECNI) mass spectrometry, is extremely sensitive, *ca.* 10-fold better than ECD for penta-deca CBs, for example. Unfortunately lower CBs (except some tetra CBs cannot be measured by this technique. Electron impact (EI) may be used as alternative ionization method, but for most CBs the sensitivity of this method is 10-fold lower than for ECD.

6.3.1.6 Calibration

Although weighing and diluting seem to be simple manipulations which should not cause major errors, results from recent intercomparison exercises have shown that at many laboratories these procedures are not under control [55,89,90]. Detailed information on the preparation of calibration solutions is given by Wells et al. [112]. Calibration solutions for CBs should preferably be made from crystalline CBs. A number of certified CB standards is available from the EC Measurements and Testing Programme, BCR, Brussels. At least 150 of the 209 CB congeners are now commercially available from different suppliers, most in minimum purities of 98% or 99%. The cost of the preparation of a CB calibration solution is relatively high. Most congeners cost at least US $100 for 5 mg, this is the minimum quantity advised for weighing. Two independent solutions should be made simultaneously to enable a check of the concentrations [112]. It is, therefore, a waste of money if a substantial part of these solutions must be discarded after some time, if doubt has arisen as to the exact concentrations present in the solutions, owing to evaporation during storage. Most calibration solutions are stored in graduated flasks in the refrigerator. The use of screw-cap bottles in the refrigerator can cause even more errors, as caps can loosen in the refrigerator, as a result of temperature differences, leading to evaporation, and tighten again when brought to room temperature [90]. Corrections may of course be applied to the concentrations, but sooner or later errors will result. The use of ampoules for storage of CB calibration solutions is, therefore, strongly recommended. Depending of the number of routine analyses carried out in a laboratory a few hundred, or thousands, of ampoules can be made with the required CBs in e.g. 5 mL *iso*-octane at a concentration *ca.* 10-fold higher than the normal working range, of which one is used for each series of samples, several dilutions being made for the purpose of constructing a calibration curve. Distinct advantages are: all ampoules of both solutions can be used; there are no errors as a result of evaporation; and exchange of solutions with colleagues in other laboratories is easy.

One or more internal standards, added in a fixed volume or weight to all standards and samples, should always be used to control the final concentration step and the GC analysis. Preferably a CB should be used, for example CBs 29, 112 (2,3,3',5,6-penta CB), 155 (2,2',4,4',6,6'-hexa CB) and 198 (2,2',3,3', 4,5,5',6-octa CB). Alternatively 1,2,3,4-tetrachloronaphthalene (TCN) or homologs of dichloroalkyl benzyl ether can be used [114].

6.3.1.7 Long-term stability

One internal (laboratory) reference sample should be included with each batch of samples. This sample should be taken from a large, homogeneous batch of sample that can serve as an internal reference material over a long period. A quality control chart (Fig. 6-1) should be prepared using the data for this material [81]. If the warning limits are exceeded, routine analysis should be suspended whilst the method is checked for errors, and the data from the last batch should not be reported. At least once per year a certified reference material should be analyzed, and each laboratory should regularly take part in interlaboratory studies on the determination of CBs.

6.3.1.8 Interlaboratory studies

Since the 1970s when the detrimental effects of PCBs in the environment were recognized and more laboratories became involved in PCB analysis, interlaboratory studies have been conducted on the determination of total PCBs and CBs. Until the mid 1980s little progress had been made. Most intercomparison exercises resulted in coefficients of variation (CVs) of 50-100% [115,116]. In a stepwise designed interlaboratory study, organized for a relatively small group of laboratories (*ca.* 10) prior to the certification of materials for CB analysis, it was shown that CVs could be reduced by identification and elimination of sources of error [112,117]. This idea of a step-by-step approach appeared to be essential for improvement in comparability of results from the analysis of such a complex mixture. For a large group of laboratories (ca. 60), all involved in international marine monitoring programs, a significant reduction in CVs was obtained by such an approach [55,72,89]. An impression of the reduction in CVs obtained is given in Fig. 6-5. For the major CBs, 118, 138, 153, and 180, CVs of around 15% can now be obtained for the matrices seal blubber oil and dried sediment. This means that differences between two CB concentrations in a sample of about 50% can now be recognized by such a group of laboratories, disregarding natural variation. Although improvements have been made, it is clear that further improvements and consequent reduction of CVs are needed. Areas in which improvement can still be made are: further avoidance of co-elution by improved chromatography and better selection of GC columns; better calibration and storage of calibration solutions in ampoules; control of long-term precision by the analysis of certified and internal laboratory reference materials; the recording of quality control charts and detailed documentation of procedures; and improvement of extraction conditions, especially for lean fish tissues.

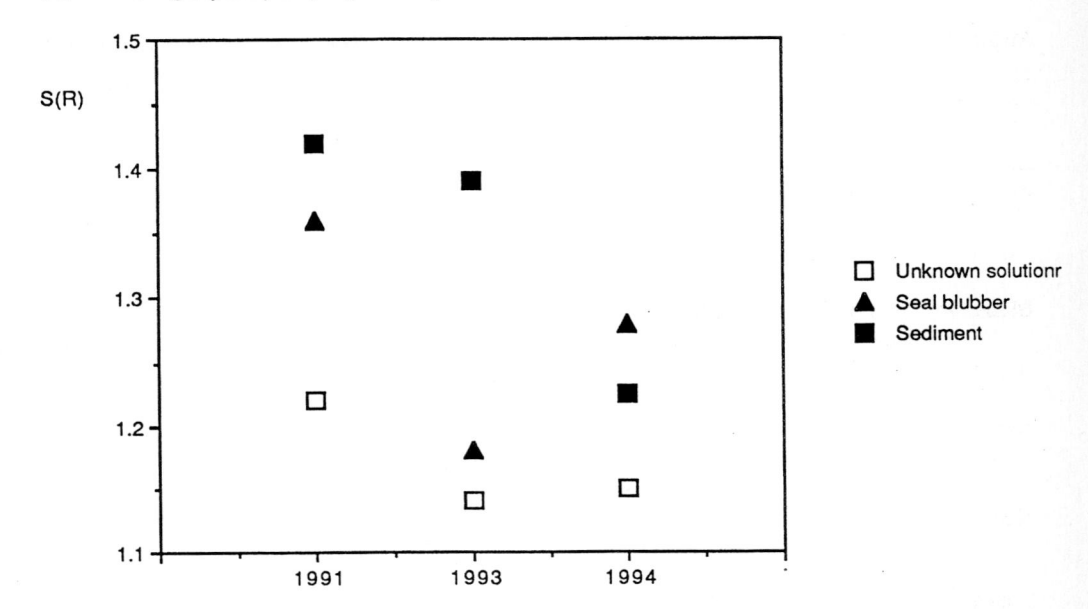

Figure 6-5: Reduction of the relative standard deviations for the reproducibility (S(R)) during the ICES/IOC/OSPARCOM CB intercomparison exercise on the determination of CBs in marine media; S(R) expressed as a relative size: 1.1 ≈ 10%, 1.2 ≈ 20%

6.3.1.9 Reference materials

Over the last ten years a range of certified reference materials (CRMs) has become available. Table 6-1 gives an overview of CRMs certified for CBs which are available at the moment from different organizations. When selecting a specific CRM, one should realize that there may be considerable differences in precision of CRMs. This is caused by the procedures used for certification. The uncertainties of the various CRMs as far as known are given in Table 6-1. Single stage intercomparisons, as used by the IAEA [118], may not be sufficient to furnish CRMs with sufficiently small uncertainties. Such CRMs can even counterproductive because laboratories may claim that their analyses are quality assured as they have used CRMs in their AQC program, whereas the large uncertainties associated with the certified values mean that very variable data can be produced and yet found acceptable.

Table 6-1: Reference materials certified for CBs and their suppliers.

Source	Code	Type	No. of CBs	Mean uncertainty (%)
BCR[a]	349	Cod liver oil	7	15
	350	Mackerel oil	7	25
	382	Dried sludge	6	19
IAEA[b]	MA-B-3/0C	Garpike tissue	2	[e]
	351	Lyophilized tuna	1	[e]
	357	Coastal sediment	10	[e]
NIST[c]	1588	Cod liver oil	5	18
	1939	River sediment	3	[e]
NRC[d]	HS-1	Harbour sediment	10	16
	HS-2	Harbour sediment	10	19

[a] BCR: Community Bureau of Reference, Brussels, Belgium.
[b] IAEA: International Atomic Energy Agency, Vienna, Austria.
[c] NIST: National Institute of Standards and Technology, Gaithersburg, USA.
[d] NRC: National Research Council, Halifax, Canada.
[e] not known.

6.3.2 Polycyclic aromatic hydrocarbons

Many of the principles of QA outlined above for the analysis of CBs can also be applied to PAH analysis. The quality of solvents and other materials and reagents must be assessed, and method blanks and recoveries established. All steps of the procedure should be validated and documented, and the use of internal (surrogate) standards (such as deuterated PAH compounds) is recommended for purposes of quantification.

6.3.2.1 Extraction

Traditional extraction methods have utilized solvent extraction (direct liquid-liquid extraction for waters; Soxhlet extraction for biota and sediments, or ultrasonication for sediments) or alkaline saponification (biota and sediments) [127]. A new technique which is being widely applied to the analysis of both CBs and PAHs in sediments is SFE (supercritical fluid extraction), often because of the relative ease with which it can be coupled to chromatographic techniques such as GC and SFC (supercritical fluid chromatography) [29, 119,120].

6.3.2.2 Clean-up

Sample extracts need to be purified prior to chromatographic analysis in order to remove interfering co-extractants, such as lipid materials, as described above for CBs. Sediment samples may also require an additional clean-up step to remove elemental sulfur [121]. Traditionally clean-up has been accomplished either by chromatography on alumina and silica, in some cases combined with solvent partitioning of aromatic from aliphatic hydrocarbons. The adsorbents can be obtained commercially in the form of SPE cartridges, which are disposable, quick and easy to use, and are uniformly packed [122]. All of these procedures are time-consuming, and must be carefully monitored if consistently good results are to be obtained. Variations in the effectiveness of extract clean-up can lead to difficulties in quantification, and in confirming the identity of analytes by GC/MS techniques [122]. Clean-up utilizing HPLC techniques is now becoming more common [29], as the procedure is both more reproducible and easier to automate. One such technique involves chromatography on silica chemically bonded to 2,4-dinitroaniline, and isolation of fractions containing 3-5-ring PAHs [123]. The use of GPC (gel-permeation chromatography) eliminates high molecular weight matrix components which can interfere with GC/MS determinations of PAHs [122]. A third example utilizes an automated solid-phase extraction unit mounted in-line between an HPLC autosampler and the analytical column, enabling the direct injection and analysis of uncleaned water extracts [124]. Such methodology will become increasingly common in the future as they minimise the use of solvents.

6.3.2.3   Analysis

Two major approaches are generally followed in the analysis of PAHs; these utilize either gas or liquid chromatography. Fluorescence spectrometry is also commonly employed as a rapid screening technique, and can enable determination of individual PAHs if the method is applied at low temperatures [125,126]; the instrumentation required is not, however, generally available. The greatest sensitivity and selectivity are achieved by combining HPLC with fluorescence detection, and capillary gas chromatography with mass spectrometry (GC/MS); in some studies these techniques have been employed in tandem [43]. Inter-comparison exercises have demonstrated that good comparability can be obtained for total hydrocarbon determinations using techniques such as fluorescence spectrometry, but that there is a serious lack of comparability between specific hydrocarbon concentrations measured in different laboratories and using both of these analytical approaches [126]. The reasons for this lack of comparability have not been rigorously investigated to date. Investigations within a single laboratory have shown that comparable results can be obtained from the two methods when applied in parallel, and that HPLC separation of some isomers of parent PAH is superior (Figure 6-6) [127]. A systematic stepwise intercomparison programme to investigate the reasons for the poor comparability of these techniques is being conducted under the QUASIMEME program, which began in 1994 [128,129].

Environmental PAHs arise from two main anthropogenic sources, the combustion of fossil fuels, and the direct release of oil and oil products. Certain localized areas have also received significant inputs of PAHs generated by industrial processes (such as the smelting of metals [130] and these tend to be similar in composition to those arising from combustion. PAH mixtures arising from combustion sources consist predominantly of parent, unalkylated PAH, whereas in oils a major fraction of the PAHs can be alkylated. In consequence, chromatograms of the aromatic fractions of samples contaminated by oil and oil products are usually more complex than those for samples where combustion is the major source. This has analytical implications, as low-resolution methods may be incapable of resolving such mixtures and many compounds may co-elute.

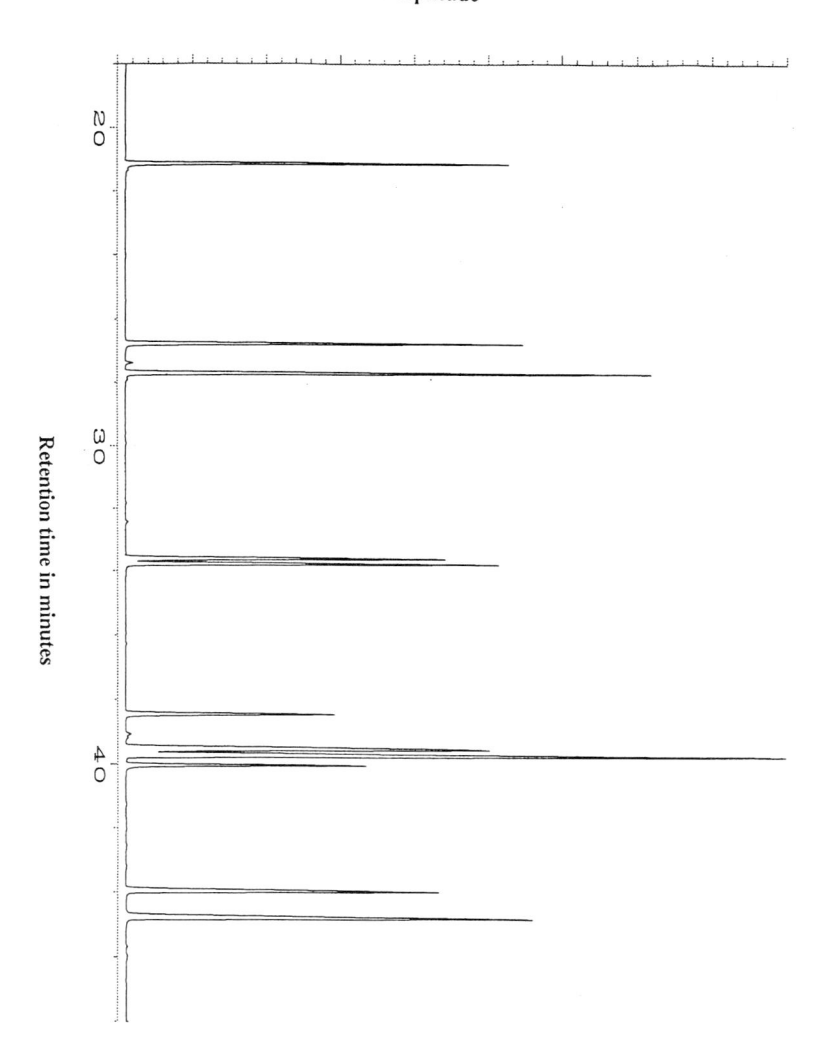

Figure 6-6a: Comparative PAH chromatograms obtained by GC and HPLC techniques, illustrating the lower resolution of the HPLC technique. (a) an 11-component PAH standard analyzed by gas chromatography with flame-ionization detection (analogous to the total ion current obtained from GC/MS in the electron-impact mode). In elution order the peaks are: phenanthrene, fluoranthene, pyrene, benz[*a*]anthracene, chrysene, benzo[*b*]fluoranthene, benzo[*e*]pyrene, benzo[*a*]pyrene, perylene, indeno[1,2,3-*cd*]pyrene and benzo[*ghi*]perylene.

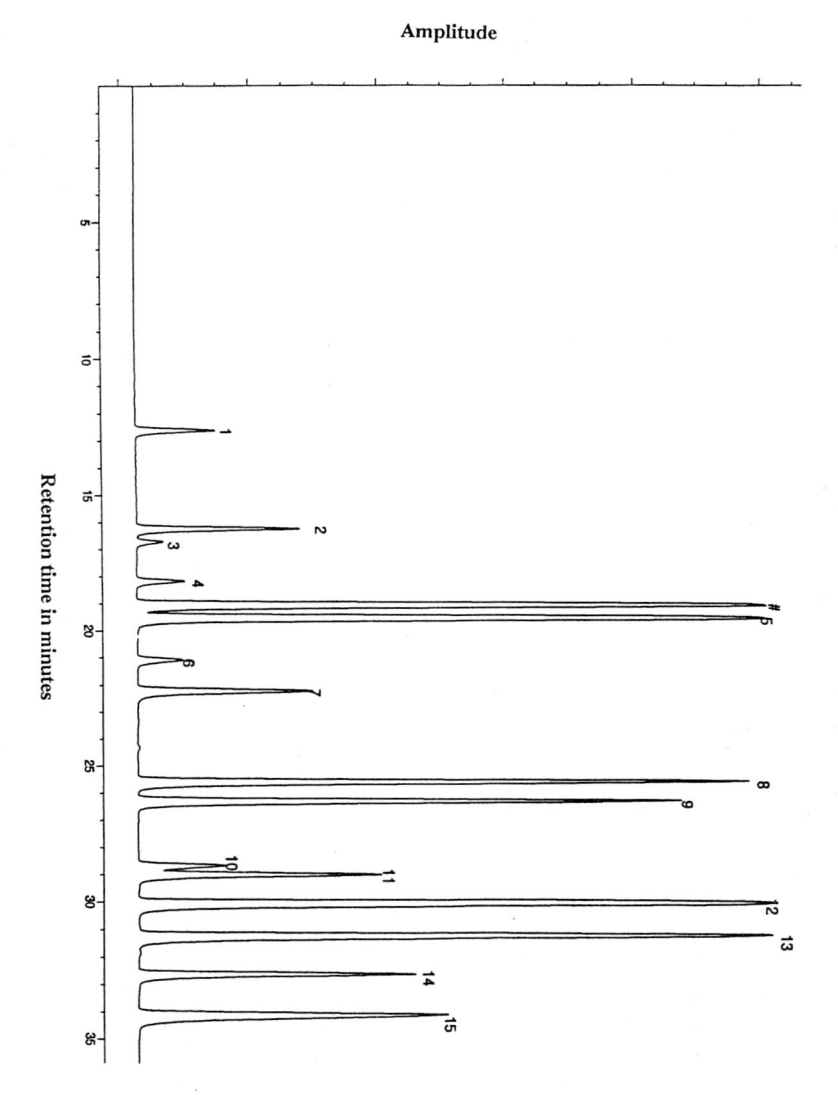

Figure 6-6b: Comparative PAH chromatograms obtained by GC and HPLC techniques, illustrating the lower resolution of the HPLC technique. (b) an HPLC chromatogram (fluorescence detection) of a 16-component PAH standard. The peaks numbered 1 to 15 are: naphthalene, acenaphthene, fluorene, phenanthrene, anthracene, fluoranthene, pyrene, benzo[*a*]anthracene, chrysene, benzo[*e*]pyrene, benzo[*b*]fluoranthene, benzo[*a*]pyrene, dibenzo[*a,h*]anthracene and benzo[*ghi*]perylene. The peak marked # is anthracene-d_{10} (surrogate standard).

Neither of the main analytical approaches described has been studied in the same detail as the CB methods described above. Also, full internal validation of methodology is limited both by the availability of certified reference materials, and also because the confidence intervals around the certified values are often rather wide. There is a need for more CRMs to be made available in order to facilitate matrix matching, and for low level research materials with indicative "less than" values for use as matrix blanks and for spiking [131].

A list of CRMs currently available (comprising 1 biological tissue sample and 7 sediments) is given in Table 6-2, with their suppliers and an indication of the number of PAHs certified in each case.

Table 6-2: Reference materials certified for PAHs, and their suppliers.

Source	Code	Type	No. of PAHs	Mean uncertainty (%)
BCR[a]	088	Sewage sludge (e)	8	17
IAEA[b]	357	Coastal sediment	9	23
NIST[c]	1941a	Marine sediment	11	11
	1974	Mussel tissues	9	(f)
NRC[d]	HS-3	Harbour sediment	16	27
	HS-4	Harbour sediment	16	22
	HS-5	Harbour sediment	16	37
	HS-6	Harbour sediment	16	24
	SES-1	Estuarine sediment	15	(g)

[a] BCR: Community Bureau of Reference, European Commission, Brussels, Belgium.
[b] IAEA: International Atomic Energy Agency, Vienna, Austria.
[c] NIST: National Institute of Standards and Technology, Gaithersburg, USA.
[d] NRC: National Research Council, Halifax, Canada.
(e) A freshwater sediment certified for PAH is expected to be available in 1996.
(f) Not known - replacement CRM 1974a due for release in 1995.
(g) Spiked matrix - determined concentrations vary by analytical method.

Acknowledgements

The authors would like to thank Dr. E.L. Donaldson of the Industrial Science Centre, Lisburn, Northern Ireland, for provision of information on accredited sampling procedures undertaken within his laboratory; and Vanessa Dawes for the HPLC chromatogram reproduced as Figure 6-6b.

6.4 References

[1] ICES, *ICES Coop. Res. Rep.*, **190**, 49 (1992)
[2] M.H. Ramsey, *Anal. Proc.*, **30**, 110 (1993)
[3] W.J. Youden, *J. Assoc. Off. Anal. Chem.*, **50**, 1007 (1967)
[4] S.C. Black, in: L.H. Keith (Ed.), *Principles of Environmental Sampling*, American Chemical Society, 109 (1988)
[5] B. Kratochvil, D. Wallace and J.K. Taylor, *Anal. Chem.*, **56**, 113R (1984)
[6] J.K. Taylor, in: L.H. Keith (Ed.), *Principles of Environmental Sampling*, American Chemical Society, 101 (1988)
[7] M.J. Barcelona, in: L.H. Keith (Ed.), *Principles of Environmental Sampling*, American Chemical Society, 3 (1988)
[8] B. Kratochvil and J.K. Taylor, *Anal. Chem.*, **53**, 924A (1981)
[9] W.P. Cofino, in: D. Barcelo (Ed.), *Environmental Analysis: Techniques, Applications and Quality Assurance*, Elsevier Science Publishers B.V., 359 (1993)
[10] I. Bouloubassi and A. Saliot, *Mar. Pollut. Bull.*, **22**, 588 (1991)
[11] M.B. Yunker, R.W. MacDonald, W.J. Cretney, B.R. Fowler and F.A. MacLaughlin, *Geochim. Cosmochim. Acta*, **57**, 3041 (1993)
[12] J.-P. Hsu, *VB Monographs in Mass Spectrometry*, **6**, ISSN 0965-6758 (1994)
[13] U.M. Cowgill, in: L.H. Keith (Ed.), *Principles of Environmental Sampling*, American Chemical Society, 171 (1988)
[14] M.J. Barcelona, J.A. Helfrich and E.E. Garske, *Anal. Chem.*, **57**, 460 (1985)
[15] M.J. Barcelona, J.A. Helfrich and E.E. Garske, *Anal. Chem.*, **57**, 2752 (1985)
[16] R.J. Law and T.W. Fileman and J.E. Portmann, *Aquat. Environ. Prot.: Analyt. Meth.*, MAFF Direct. Fish Res., Lowestoft, **2**, 25 (1988)
[17] D. Stadler and K. Schomaker, *Dt. Hydrogr. Z.*, **30**, 20 (1977)
[18] R.J. Law and J.A. Whinnett, *Oceanologica Acta*, **16**, 593 (1993)
[19] R.H. Meade and H.H. Stevens jr. *Sci. Total Environ.*, **97/98**, 125 (1990)
[20] I. Cruz, D.E. Wells, and I.L. Marr, *Anal. Chim. Acta*, **283**, 280 (1993)
[21] A.G. Kelly, I. Cruz and D.E. Wells, *Anal. Chim. Acta*, **276**, 3 (1993)
[22] D.J. Harper, *Mar. Chem.* **21**, 183 (1987)
[23] J. Albaigés, J. Grimalt, J.M. Bayona, R. Risebrough, B. de Lappe and W. Walker II, *Org. Geochem.*, **6**, 237 (1984)
[24] M.J. Barcelona, J.A. Helfrich, E.E. Garske and J.P. Gibb, *Ground Water Monit. Rev.*, **4**, 32 (1984)
[25] J.L. Domagalski and K.M. Kuivila, *Estuaries*, **16**, 416 (1993)
[26] M.G. Ehrhardt and K.A. Burns, *J. Exp. Mar. Biol. Ecol.*, **138**, 35 (1990)
[27] A. Germain and C. Langlois, *Water Poll. Res. J. Canada*, **23**, 602 (1988)
[28] J.H. Hermans, F. Smedes, J.W. Hofstraat and W.P. Cofino, *Environ. Sci. Technol.* **26**, 2028 (1992)
[29] K.G. Furton, E. Jolly and G. Pentzke, *J. Chromatogr,* **642**, 33 (1993)
[30] D.W. Potter and J. Pawliszyn, *Environ. Sci. Technol.*, **28**, 298 (1994)
[31] J.R. Kucklick and T.F. Bidleman, *Mar. Environ. Res.*, **37**, 63 (1994)
[32] W.D. Garrett, *Limnol. Oceanogr.*, **10**, 602 (1965)

6.4 References

[33] ICES, *ICES Coop. Res. Rep.*, **198**, 45 (1994)
[34] E. Lipiatou, J.-C. Marty and A. Saliot, *Mar. Chem.*, **44**, 43 (1993)
[35] W.D. Gardner, *J. Mar. Res.*, **38**, 17 (1980)
[36] G. Gust, A.F. Michaels, R. Johnson, W.G. Deuser and W. Bowles, *Deep Sea Res. I*, **41**, 831 (1994)
[37] H. Compaan, T. Smit, R. Law, A. Abarnou, E. Evers, H. Hünnerfuss, E. Hoekstra, J. Klungsøyr, L. Holsbeek and C. Joiris, Rijkswaterstaat, Tidal Waters Division, Report DGW-92.033 (1992)
[38] B. Pavoni, A. Sfriso and A. Marcomini, *Mar. Chem.*, **21**, 25 (1987)
[39] G. Sanders, K.C. Jones and J. Hamilton-Tayler, *Environ. Toxicol. Chem.*, **12**, 1567 (1993)
[40] R. van Zoest and G.T.M. van Eck, *Mar. Chem.*, **44**, 95 (1993)
[41] R. Albert and W. Horwitz, in: L.H. Keith Ed., *Principles of Environmental Sampling*, American Chemical Society, 337 (1988)
[42] A.A. Elskus and J.J. Stegeman, *Mar. Environ. Res.*, **27**, 31 (1989)
[43] M.M. Krahn, G.M. Ylitalo, J. Buzitis, S.-L. Chan, U. Varanasi, T.L. Wade, T.J. Jackson, J.M. Brooks, D.A. Wolfe and C.-A. Manen, *Environ. Sci. Technol.*, **27**, 699 (1993)
[44] F. Ariese, S.J. Kok, M. Verkaik, C. Gooijer, N.H. Velthorst and J.W. Hofstraat, *Aquat. Toxicol.*, **26**, 273 (1993)
[45] P.B. Kauss and Y.S. Hamdy, *Hydrobiologia*, **219**, 37 (1991)
[46] Oslo and Paris Commissions, Manual on the Principles and Methodology of the Joint Monitoring Progamme (1994)
[47] M.P. Maskarinec and R.L. Moody, in: L.H. Keith Ed.; *Principles of Environmental Sampling*, American Chemical Society, 145 (1988)
[48] F. Smedes and J. de Boer, *Quím. Anal.*, **13**, S100 (1994)
[49] J. Japenga, W.J. Wagenaar, F. Smedes and W. Salomons, *Environ. Technol. Lett.*, **8**, 9 (1987)
[50] R.J. Law, J. Klungsøyr, P. Roose and W. de Waal, *Mar. Pollut. Bull.*, **29**, 217 (1994)
[51] F. Galgani and J.F. Payne, ICES, Techniques in Marine Environmental Sciences, 13 (1991)
[52] J.W. Farrington, A.C. David, J.B. Livramento, C.H. Clifford, N.M. Frew and A. Knap, *ICES Coom. Res. Rep.*, **141** (1986)
[53] D.E. Wells, W.P. Cofino, Ph. Quevauviller and G. Griepink, *Mar. Pollut. Bull.*, **26**, 368 (1993)
[54] D.E. Wells, *Mar. Pollut. Bull.*, **29**, 143 (1994)
[55] J. de Boer, J. van der Meer, L. Reutergårdh and J.A. Calder, *J. Assoc. Off. ANal. Chem.*, **77**, 1411 (1994)
[56] S. Tanabe, M. Kawano and R. Tatsukawa, *Transact. Tokyo Univers. Fish.*, **5**, 97 (1982)
[57] D.E. Schulz-Bull, G. Petrick and J.C. Duinker, *Mar. Chem.*, **36**, 365 (1991)
[58] G.R. Harvey, W.G. Steinhaver, H.P. Miklas, *Nature*, **252**, 387 (1974)
[59] V. Leoni, G. Puccetti, R.J. Colombo and A.M.D. Ovidio, *J. Chromatogr.*, **125**, 399 (1976)

[60] P. Larsson, L. Okla, S.O. Ryding and B. Westöö, *Can. J. Fish Aquat. Sci.*, **47**, 746 (1990)
[61] B.G. Oliver and K.D. Nicol, *Chromatographia*, **16**, 336 (1982)
[62] A. Opperhuizen and P.I. Voors, *Chemosphere*, **16**, 2379 (1987)
[63] D.L. Swackhammer, B.D. McVeety and R.A. Hites, *Environ. Sci. Technol.*, **22**, 664 (1988)
[64] J.C. Duinker and F. Bouchertall, *Environ. Sci. Technol.*, **23**, 57 (1989)
[65] D.J. Gregor and W.D. Gummer, *Environ Sci. Technol.*, **23**, 561 (1989)
[66] L. Morselli, S. Zappoli and S. Donati, *Ann. Chim.*, **79**, 677 (1989)
[67] M.Th.J. Hillebrand, J.M. Everaarts, H. Razak, D. Moelyadi Moelyo, L. Stolwijk and J.P. Boon, *Neth. J. Sea. Res.*, **23**, 369 (1989)
[68] M. Chevreuil, L. Garnier, A. Chresterikoff and R. Letolle, *Water Res.*, **24**, 1325 (1990)
[69] V. Lang, *J. Chromatogr.*, **595**, 1 (1992)
[70] D.E. Wells, in: Environmental Analysis Techniques, Applications and Quality Assurance: D. Barcelo,(Ed.), *Elsevier, Sci. Publ.*, Amsterdam, 79-109, (1993)
[71] J. de Boer, *Chemosphere*, **17**, 1803 (1988)
[72] J. de Boer, C.J.N. Stronck, F. van der Valk, P.G. Wester and M.J.M. Daudt, *Chemosphere*, **25**, 1277 (1992)
[73] T.A. Bellar, J.J. Lichtenberg and S.C. Lonneman, in: Recovery of Organic Compounds from Environmental Contaminated Bottom Material: R.A. Baker, (Ed.), Ann. Arbor (MI): *Ann Arbor Sci.*, (1980)
[74] R. Fuoco, M.P. Colombini and E. Samcova, *Chromatographia*, **36**, 65 (1993)
[75] D. Jensen, L. Renbergt and L. Reutergårdh, *Anal. Chem.*, **49**, 316 (1977)
[76] S.B. Hawthorne, D.J. Miller, J.J. Langenfeld and M.D. Burford, in: Proc. 15th Intern. Symp. Capill. Chromatogra., Riva del Garda, May 24-27: P. Sandra and G. Devos (Eds.), Heidelberg: Huething Verlag, 1701 (1993)
[77] J.J. Langenfeld, S.B. Hawthorne, D.J. Miller and J. Pawliszyn., *Anal. Chem.*, **65**, 338 (1993)
[78] S. Bøwadt, B. Johansson, P. Fruekilde, M. Hansen, D. Zilli, B. Larsen and J. de Boer, *J. Chromatogr.*, **675**, 189 (1994)
[79] F. van der Valk and Q.T. Dao, *Chemosphere*, 17 (1988)
[80] S. Bøwadt, H. Skejø-Andersen, L. Montanarella and B. Larsen, *Intern. J. Environ. Anal. Chem.*, (1994b, in press)
[81] J. de Boer, C.J.N. Stronck, W.A. Traag and J. van der Meer, *Chemosphere*, **26**, 1823 (1993)
[82] P. Hess, D. Wells, J. de Boer, W. Cofino and P. Leonards, *J. Chromatogr.*, **703**, 417 (1995)
[83] C. Benthe, B. Heinzow, H. Jessen, S. Mohr and W. Protard, *Chemosphere*, **25**, 1481 (1992)
[84] P. Haglund, L. Asplund, U. Järnberg and B. Jansson, *J. Chromatogr.*, **507**, 389 (1990)
[85] N. Kannan, G. Petrick, D. Schulz, J.C. Duinker, J. Boon, E. van Arnhem and S. Jansen, *Chemosphere*, **23**, 1055 (1991)
[86] L.M. Smith, D.L. Stalling and J.L. Johnson, *Anal. Chem.*, **56**, 1830 (1984)

[87] M. Schlabach, A. Biseth, H. Gundersen and M. Oehme, in: Proc. 13th Int. Symp. Dioxin '93, Vienna, Organohalogen Compounds, Vol.11, H. Fiedler, H. Franck, O. Hutzinger, W. Parzefall, A. Riss and S. Safe (Eds.), Federal Environ. Agency, Vienna, Austria (1993)
[88] Å. Bergman, L. Reutergårdh and M. Ahlman, *J. Chromatogr.,* **291**, 392 (1984)
[89] J. de Boer and J. van der Meer, Report on the results of the ICES/IOC/OSPARCOM intercomparison exercise on the determination of chlorobiphenyl congeners in marine media - step 4, *ICES Coop. Res. Rep.,* (in press)
[90] D.E. Wells and J. de Boer, *Mar. Pollut. Bull.,* **29**, 174 (1994)
[91] J. de Boer and Q.T. Dao, *J. High. Resolut. Chromatogr.,* **12**, 755 (1989)
[92] M.D. Mullin, C. Pochini, S. McGrundle, M. Rombes, S. Safe and H. Safe, *Environ. Sci. Technol.,* **18**, 468 (1984)
[93] B. Larsen, S. Bøwadt and R. Tilio, *Intern. J. Environ. Anal. Chem.,* **47**, 47 (1992)
[94] B. Larsen, S. Bøwadt, R. Tilio and S. Facchetti, *Chemosphere,* **25**, 1343 (1992)
[95] J. de Boer, Q.T. Dao and R. van Dortmund, *J. High Resolut. Chromatogr.,* **15**, 249 (1992)
[96] B.R. Larsen, S. Bøwadt, in: Proc. 15th Int. Symp. Capill. Chromatogr., Riva del Garda, May 24-27: P. Sandra and G. Devos (Eds.), Heidelberg: Huethig Verlag, 503-510 (1993)
[97] E. Storr-Hansen, *Intern. J. Environ. Anal. Chem.,* **66**, 67 (1994)
[98] M. Ojala, *J. High Resolut. Chromatogr.,* **16**, 679 (1993)
[99] J.C. Duinker, D.E. Schulz and G. Petrick, *Anal. Chem.,* **60**, 478 (1988)
[100] F.R. Guenther, S.N. Chessler and R.E. Rebbert, *J. High Resolut. Chromatogr.,* **12**, 821 (1989)
[101] J. de Boer and Q.T. Dao, *J. High Resolut. Chromatogr.,* **12**, 755 (1989)
[102] J. de Boer and Q.T. Dao, *Intern. J. Environ. Anal. Chem.,* **43**, 254 (1991)
[103] N. Kannan, D.E. Schulz-Bull, G. Petrick and J.C. Duinker, *Intern. J. Environ. Anal. Chem.,* **47**, 201 (1992)
[104] J. de Boer and U.A.Th. Brinkman, *Anal. Chim. Acta,* **289**, 261 (1994)
[105] E.A. Maier, J. Hinschberger, H. Schimmel, B. Griepink and D.E. Wells, *Rep. EUR 15254,* Commission of the European Communities, BCR, Brussels (1993)
[106] E.A. Maier, H. Schimmel, J. Hinschberger, B. Griepink and D.E. Wells, *Rap. EUR 15255,* Commission of the European Communities, BCR, Brussels (1994)
[107] L.G.M.Th. Tuinstra, J.A. van Rhijn, A.H. Roos, W.A. Traag, R.J. van Mazijk and P.J. W. Kolkman, *J. High Resolut. Chromatogr.,* **13**, 797 (1990)
[108] D.G. Patterson Jr., C.R. Lapeza Jr., E.R. Barnhart, D.F. Groce and V.W. Bruse, *Chemosphere,* **19**, 127 (1989)
[109] D.W. Kuehl, B.C. Butterworth, J. Libal and P. Marquis, *Chemosphere,* **22**, 849 (1991)
[110] K. Grob Jr. and H.P. Neukom, *Chromatographia,* **18**, 517 (1984)
[111] R.P. Snell, J.W. Danielson and G.S. Oxborow, *J. Chromatogr. Sci.,* **25**, 225 (1987)
[112] D.E. Wells, E.A. Maier and B. Griepink, *Int. J. Environ. Anal. Chem.,* **46**, 255 (1992)
[113] J. de Boer, J.C. Duinker, J.A. Calder and J. van der Meer, *J. Assoc. Off. Anal. Chem.,* **75**, 1054 (1992)

[114] D.E. Wells, M.J. Gillespie and A.E.A. Porter, *J. High Resolut. Chromatogr.*, **8**, 443 (1985)
[115] C.J. Musial, and J.F. Uthe, *J. Assoc. Off. Anal. Chem.*, **66**, 22 (1983)
[116] J.F. Uthe, C.J. Musial and R.K. Misra, *J. Assoc. Off. Anal. Chem.*, **71**, 369 (1988)
[117] L.G.M.Th. Tuinstra, A.H. Roos, B. Griepink and D.E. Wells, *J. High. Resolut. Chromatogr.*, **8**, 475 (1985)
[118] Anon., Standard and reference materials for marine sciences, revised edition, Manuals and Guides. Intern. Oceanographic Commission, Paris, **25** (1993)
[119] J.M. Levy, L.A. Dolata and R.M. Ravey, *J. Chromatogr. Sci.*, **31**, 349 (1993)
[120] A. Meyer, W. Kleiböhmer and K. Cammann, *J. High Res. Chromatogr.*, **16**, 491 (1993)
[121] M. Ehrhardt, J. Klungsøyr and R.J. Law, *ICES Techniques in Marine Environmental Sciences*, **12**, 47pp. (1994)
[122] P.J. Nyman, G.A. Perfetti, F.L. Joe jr. and G.W. Diachenko, *Food Additives and Contaminants*, **10**, 489 (1993)
[123] L. Nondek, M. Kuzilek and Š. Krupicka, *Chromatographia*, **37**, 381 (1993)
[124] V.J. Dawes and R.J. Law, *Proc. North Sea Symposium*, Ebeltoft, Denmark, 18-21 April 1994
[125] J.W. Hofstraat, H.J.M. Jansen, G.P. Hoornweg, C. Gooijer, N.H. Velthorst and W.P. Cofino, *Int. J. Environ. Anal. Chem.*, **21**, 299 (1985)
[126] R.J. Law and M.D. Nicholson, *ICES Coop. Res. Rep.*, **207**, 52 (1995)
[127] S.A. Wise, L.C. Sander and W.E. May, *J. Chromatogr*, **642**, 329 (1993)
[128] R.J. Law and J.L. Biscaya, *Mar. Pollut. Bull.*, **29**, 235 (1994)
[129] D.E. Wells, *Mar. Pollut. Bull.*, (in press)
[130] A. Bjorseth, J. Knutzen and J. Skei, *Sci. Total Environ.*, **13**, 71 (1979)
[131] D.E. Wells, *Fresenius Z. Anal. Chem.*, **332**, 583 (1988)

7.

Quality assurance of sampling and sample pretreatment for trace metal determination in soils

R. Rubio and M. Vidal

Departament de Química Analítica. Universitat de Barcelona, Av. Diagonal 647, 08028 Barcelona, Spain

In studies of soil pollution, the same importance should be given to sampling and sample pretreatment as is usually accorded to the subsequent analytical steps. Quality control procedures are required at each stage of the overall analysis.

A quality assurance (QA) program is a system of activities aimed at ensuring that information provided in environmental risk assessment meets the needs of the users of the data [1]. It is designed to give control of both field (including sampling) and laboratory operations. Within a QA program, a distinction can be made between quality control, which represents the activities aimed at controlling the quality of the data, and quality assessment, i.e. determination of the quality of the data produced. This chapter deals with QA in sampling and pretreatment of soils prior to the study of their trace metal content.

A critical aspect of the reliability of any analytical measurement when studying soil pollution is that of sample quality. Samples collected should be as representative as possible of the soil to be characterized and every precaution should be taken during sampling, pretreatment, and storage to ensure that samples remain unaltered. In environmental assessment studies this can be especially important since the results will be used to decide whether corrective actions should be taken [2].

To ensure safe storage and to prevent contamination, there must be adequate facilities for handling, shipping, receipt, and adequate storage of samples. Furthermore, it is necessary to write down standard operation procedures (SOPs) to be followed in performing any operation, including objectives, sampling plans, preparation of containers and equipment, maintenance, calibration and cleaning of field equipment, sample preservation, packaging and shipping, health and safety protocols, and chain-of-custody protocols [3]. As stressed in Chapter 1, the availability of reference materials and the design of interlaboratory exercises also contribute to quality assurance programs.

The presence and behavior of trace metals in soils are usually studied for two main purposes: availability to plants, since metals can be considered as nutrients, and the hazards connected with pollution. As a result of high levels of pollution in some areas there has been a significant increase in studies on the inputs of metals to soils and subsequent uptake by plants. In recent decades there has been increasing interest in the determination of the "forms" of metals (e.g. carbonate-bound, easily exchangeable) which can be present in soils and sediments by applying well defined extraction procedures. This is an operational concept to characterize metal "forms" and provides valuable information on their mobility and transport and their participation in biological processes with living organisms. Nevertheless, the determination of the total content of selected elements in soils enables, from the point of view of pollution, to identify hot spots, to design monitoring programs, and to establish remedial actions.

Soils are heterogeneous materials containing many organic and inorganic components; heavy metals are bound to these components by different mechanisms. Binding is influenced by changes of environmental conditions such as redox potential, pH, salinity, temperature, or microbial activity. Thus, adsorption-desorption, changes in buffering capacity, ionic exchange, solubilization, and precipitation determine the metal content of the soil particles. Particle size is an important property, because this has a significant influence on metal content; the highest percentage of metal is usually found in the finest sedimentary fractions. These factors have to be kept in mind not only to achieve representative sampling but in further steps, i.e. drying, sieving, milling, grinding, etc., and when selecting the particle size and the sample size for the final metal determination.

When studying the trace element content of soils, the soil solution must also be considered. The composition of the solution is important not only environmentally and because of its metal content, but also in other scientific areas, because the interstitial water is in close equilibrium with the solid phase, thus for environmental, agricultural, and ecological considerations the composition of this solution provides valuable information on the mobility and availability of metals [4].

Sampling and sample pretreatment of soils for trace metal determination is a problem which is far from solved and all the steps involved in these processes have to be revised. In addition, standardized procedures should be established. Aware of this situation, the BCR (now Measurement and Testing Programme of the EC) and the Department of Analytical Chemistry of the University of Barcelona organized a specific workshop devoted to quality assurance in sampling and sample pretreatment of soils and sediments [5]. These aspects were deeply discussed in the workshop by a group of European experts, and the final conclusions and papers presented have been published in a special issue of Química Analítica (volume 13, 1994). Some of the conclusions of the workshop and comments from papers are referred to in this chapter.

7.1 Steps in the analytical process

Analytical processes may be divided into five main steps: the model, in which the goals of the study are defined; planning (selection of analytical procedures, sampling sites, available personnel and apparatus, etc.); the sample (collecting and reducing it to test portions; analysis (preliminary operations, acquisition of data); and evaluation (selection of best values, estimation of reliability of values or assessment of the validity of the model...) [6]. Here, special attention is paid to "planning" and "sample" steps dealing with trace element determinations in soils. Therefore, sampling plan and strategies, pretreatment of samples (including drying, milling, crushing, grinding, sieving, or sterilization), subsampling and storage are studied in respect of their quality control and general aspects which must be considered in order to ensure the representativeness of the soil sample being analyzed.

7.2 Soil sampling in trace element analysis

7.2.1 Design of a sampling plan

Soil sampling characteristics depend on the type of soil, its location, and properties such as lime content, drainage status, or organic matter content. The importance of these properties depends on the particular trace metal to be studied [7].

The design of a sampling plan depends on the global objective of the work. Objectives can be either general (determination of soil quality, elaboration of soil maps, or study of anthropogenic changes) or specific (determination of physical, biological, or chemical soil characteristics or identification and quantification of products released by industrial processes). In studying trace metal pollution in soils, the goals of sampling may be to identify a pollution event, to monitor site conditions in order to determine whether remedial action is required, to design and implement remedial measures, to assess the total exposure and risks to the population, to determine appropriate treatment levels or monitoring criteria, or to predict future conditions [8]. Previous information of interest (such as pedological, geomorphological, or hydrological features) may be necessary to design a sampling plan. Preliminary sampling may often be useful before the final objectives can be defined, or instead, "on-site" testing for easy acquisition of information on the levels of substances of interest. For instance, field portable energy dispersive X-ray fluorescence (FP-EDXRF) spectrometry has been used to carry out a rapid monitoring in contaminated areas, the elements studied being Cr, Mn, Ni, Cu, Zn, As, Se, and Pb. Once a "hot spot" is located, a suitable sampling strategy can be applied [9].

The sampling program should include not only the goals of study, but also a proper statistical design (sample collection should always be guided by a statistical analysis), the instructions for sample collection, the types of quality control sample to be used, the safety conditions, and the form describing the sampling operation. In respect of sample collection, a complete sampling plan should include the analytes of interest, location of sampling area, number and size of samples to collect, choice of sampling points, sampling strategy, the use or not of composite samples, the instruments and containers to use in the collection, and the first protocols to preserve the samples.

Irrespective of the sampling approach followed, the sampling plan has to consider that the quantity of sample will be sufficient for replicate analyses, subsampling, homogenization and analysis. The collection of at least 500 g of fresh soil (single or composite samples) is recommended by the International Standards Organization (ISO) [10].

Prior to sampling, all the different steps should be listed in an SOP, so that all the steps can be performed correctly.

Technicians performing sampling should be trained and they could collaborate in writing down the SOPs for sampling [11].

7.2.2 Sampling strategies

To obtain representative samples from a population, strategies have been developed to reduce errors in sampling operations, considering that error in sampling is also related to the sampling approach. Therefore, objectives for data quality (DQOs) should take into account the confidence level required.

Two main sampling approaches can be distinguished: non-probability (judgmental sampling) and probability (random and systematic sampling) [6,12-14]:

7.2.2.1 Non-probability approach: judgmental sampling

Judgmental sampling is based upon previous knowledge of the sampling area and the sources of the pollutants. The judgmental decision prevails over statistical considerations. This approach can be used, for instance, to study the changes over time of a given process (by controlling the influence of variables which are not the aim of the study) or the evaluation of the degree of contamination, prior to a later systematic strategy. Because of this, this sampling strategy may lead to systematic errors (bias) and additional data or reference points may be necessary to evaluate the accuracy of the results in the environmental sample of interest. An advantage of this strategy is that it leads to the smallest number of samples to analyze.

7.2.2.2 Probability approach: random and systematic sampling

If information on the contaminant sources is not available and representative sampling is required in order to carry out an extensive pollution assessment, probability sampling is recommended; this can be divided into three categories: simple random, stratified random, and systematic.

The less known about the process being studied, the more random sampling is needed. Random sampling may be performed by considering the population of samples as a whole (simple random sampling) or by using previous information and performing the sampling in each subpopulation chosen according to the information and objectives of the study (stratified random sampling).

Simple random sampling is an unbiased method of analysis but it is difficult to perform. It must be pointed out that a sample selected haphazardly is not a random sample and that results may be biased as a result of the protocol selected for collecting the sample. In any case, the use of a table of random numbers to select the sampling units is recommended as an aid to sample collection. Selection criteria, based on random distances from a system of coordinates, and using the intersections as sampling points, can also be used.

Stratified random sampling reduces variability and bias arising from sampling error by means of the subdivision of the sampling area into areas (strata) with an equal number of subpopulations. Therefore, the percentage of samples taken in a stratum, relative to the total number, must be the same as the area of this given stratum as a percentage of the total sampling area. The strata may be defined according to environmental factors, such as texture or type of soil, or pollutant concentration levels. To improve precision in respect of simple random sampling, sampling units must be more homogeneous within strata than among strata. In general, an increase in stratification increases precision, but only up to a certain level, beyond which there is no a further gain. A comparison between stratified and simple random sampling in trace metal (especially Pb and Cu) monitoring in soils can be found in the literature [15]. Stratified random strategy led to a marked improvement in precision, quantified by means of the relative standard deviation, compared with simple random sampling (36 to 21% for Pb and 15 to 11% for Cu).

Systematic strategy, using either random or geometrical sampling grids, is often used for homogeneous soils, but grids should be avoided if clear variations (changes in slope or canopy, for instance) are noticed in the sampling area. Systematic strategy is more precise than random sampling if the correlation between plots decreases with the distance between them.

Different approaches can be followed when sampling with this strategy. The use of a geometrical sampling grid consists of placing the grid randomly and collecting the samples at specific positions (for instance, in the centre of each block or at the intersections). Random sampling grids, on the other hand, are fixed and sampling points are randomly selected in a block and, subsequently, samples are collected in all the blocks at the same positions. In any case, this sampling strategy is recommended for better determination of macrovariation in soil composition.

The choice of the kind of grid depends on the extent and orientation of the contamination and on the size of area under study. Circular grids may be used to delineate local pollution and to assess the influence of industrial plants, and regular grids to assess the degree of contamination, to locate the sources of pollution, and to establish the extent of contaminated areas [10]. In any case, a minimum area of 50 square meters is advised.

7.2.2.3 Geostatistics and data analysis techniques applied to soil sampling

It is sometimes important to study the spatial dependence of the variables (e.g. trace metal concentration) to be measured, namely the kriging factors after the application of the theory of regionalized variables [16]. Kriging is a linear interpolation technique based on statistical approximation of the spatial distribution of the variable studied. Usually the random residuals of the mean are distributed around zero and they are supposed to be uncorrelated. But if they are correlated and this fact is not considered, there may be an overestimation of the variance since residues may be more strongly related as the distance between points decreases, which may lead to an increase in the number of samplings [17]. By means of the construction of semivariograms, it is possible to characterize the structure of the spatial dependence, considering that the spatial variable has a constant mean and that the covariance function depends only on the distance between points. Geostatistics are of no advantage over classical methods if the spatial dependence is much less than the purely random contribution, but probability kriging is an essential tool in decision making processes [18].

A progressive sampling approach is a field sampling strategy based on the establishment of contamination maps using a given number of parameters as indicators, thus avoiding problems arising as a result of limited financial resources. It has been applied to provide fast identification of the most relevant data, studying the variability of the materials concerned. By carrying out a subsequent sampling it is possible to study further the most interesting zones [19,20]. Two approaches can be followed, namely random sampling or an ordered spatial structure, applying geostatistics and data analysis, which is necessary when there is an ordered spatial structure of site contamination and a relationship between the presence of a trace element and a contamination process.

Other sampling approaches, which apply statistical criteria such as cluster sampling or probability proportional to size sampling, are also found in the literature. Among them, the paper by Eberhardt and Thomas should be emphasized [21].

7.2.3   Soil solution sampling

Special devices and precautions have to be considered for soil solution sampling. Many different soil solution samplers have been described in the literature, and the common principle is to preserve as much as possible the integrity of the sample. Avoiding contamination of the sample from the main components of the materials used in the construction of the samplers is the most serious problem. Porous ceramic, fritted glass [22], and new porous plastic materials [23] are mainly used to avoid contamination. A wide variety of soil solution samplers is reported in a review by Litaor [24]. The majority of samplers operate by suction, and modifications, automation, and different installation systems in the field are described in this report. The major problem arises not only from the alterability of the original interstitial water caused by the disturbance of the soil during sampling and changes in redox potential and/or in pH, but is also related to the protection of samplers installed in the field from human or animal activity.

Field sampling of soil solution is difficult and for the time being no standardized methods exist. Further research is needed in this respect to guarantee the representativeness of the samples. The overall recommendation is that the sampling program, number of samples and replicates, as well as sample design, should be evaluated according to each specific problem.

7.2.4   Number of samples

The number of samples required to carry out a study is related to the variability acceptable in the final results. When using simple random strategy and considering a Gaussian distribution, the number of samples needed to obtain a mean value with a certain error and significance level is given by the following expression [25]:

$$n = \frac{t_\alpha^2 \cdot S^2}{(R \cdot x)^2}$$

where n is the number of samples, t_α is Student's t for a given significance level (usually 95%), with (n-1) degrees of freedom, S^2 is the estimated variance, and $R \cdot x$ is the accepted variability in mean estimation, usually lower than 10%.

S^2 and x are usually estimated in previous studies, with a (low) controlled number of samples.

7.2 Soil sampling in trace element analysis

It is not possible to reduce the natural variability of a soil by taking more samples, but the error of the mean may be reduced. Nevertheless, the range becomes wider with an increasing number of samples.

The number of samples resulting from this calculation is often too high for field campaigns, mainly because of the associated cost. It is clear that if the number of samples has to be reduced either the probability of error is reduced or the range of precision is increased. Therefore, a subsequent option is to choose an alternative number of samples in which variability stabilizes. This minimum value may be obtained by applying classical methods using relative precision curves or using geostatical methods working by kriging. As can be found in the literature, this figure may fall within the range 9 to 20 samples [26,27] or higher; 25 samples should be taken for each 4 Ha, following a W pattern, according to a recommendation cited in the literature [28]; according to another study, a range between 20-40 samples/Ha is sufficient to study heavy metals in soils [29]. Gomez distinguishes samples according to the surface to be sampled and the type of soil. Therefore, if the plot to be studied is smaller than 5 Ha, 12 individual samples are sufficient if the soil is homogeneous, but up to 20 samples are necessary if the soil is heterogeneous. For plots with areas between 5 and 20 Ha, it is recommended to carry out preliminary sampling and to try to subdivide the total area into plots smaller than 5 Ha. For plots larger than 20 Ha, a complete study should be carried out [30].

Some authors report other equations for non-Gaussian distributions [6]. For instance, this would be the expression for Poisson distributions, in which the variance is not considered:

$$n = \frac{t_\alpha^2}{R^2 \cdot x}$$

When the distribution has a variance larger than the mean value, a k factor is defined as the index of clumping, and a new expression is then used:

$$n = \frac{t_\alpha^2}{R^2} \cdot \left[\frac{1}{x} + \frac{1}{k}\right]$$

k and x are estimated for preliminary measurements on the system.

7.2.5 Composite samples

Samples can be analyzed individually or mixed. Use of composite samples enables the frequency of sampling to be increased, and the total number of samples analyzed to be reduced. The final result is a reduction in variability as a result of the heterogeneity of the sample population, but also in the information about the real variability in the plot, as peaks arising from individual samples disappear. High dilutions may also lead to false negatives. The different numbers of individual samples required to prepare a composite sample, deduced from previous experience, can be found in the literature; in a study by Fleming et al., composite samples were prepared from 20 individual cores [7]. Another paper describes a composite sample formed from 100 single samples, of 10 g each, giving a composite sample, to be analyzed in the laboratory, of about 1 kg [31]. However, some authors report an expression which can be used to calculate the number of individual samples that can be mixed to form a composite sample. This number cannot be greater than the ratio between a prefixed level and the detection limit of the analyte to be determined with a given technique [14].

7.2.6 Sample depth

The depth of a soil sample should be related to the objective of the work, taking into account the possibility of dilution effects in different soil layers which may lead to error in the final assessment. When studying the influence of sewage sludge on agricultural practices (e.g. amendment), with regard to trace element availability and downward migration, it has been recommended to sample down to 75 mm, for permanent grasslands, but down to 150-200 mm for arable lands [28]. In any case, sampling down to 20 cm, with hand augers (or drill-blows), is the most usual practice [29,31]. This depth is further justified because of the roots of the main crops. If trace metal transfer to fruit trees is being studied, a deeper layer (20-40 cm) should be sampled [30].

For soil survey and mapping, a soil profile down to 1 m should be considered. Furthermore, to assess the degree of penetration and trace element migration, the complete profile is better sampled in successive layers of about 10 cm [32]. The depth of sampling is most important for studying atmophile elements, such as Cd, Zn, and Pb. The mass transport of these elements is greater through the atmosphere than through streams. Thus, there may be accumulation in the uppermost layers of the soil underneath the canopy and surrounding the trunks of trees. Therefore, conventional soil sampling may fail to show such enrichment because of dilution effects after taking less contaminated soil from deeper horizons. This is shown in the literature in a sampling where thin layers (0-0.1 cm, 0.1-0.3 cm, 0.3-0.5 cm, 0.5-1.0 cm, 1.0-2.5 cm, 2.5-5.0 cm, 5.0-10.0 cm, and 10.0-20.0 cm) were taken beneath and outside the canopy, and results were compared with those from 20-cm layers [33]. It was shown that when conventional sampling was used, canopy influence over atmophile elements was masked by variability.

7.2.7 Materials and sampling techniques

Analysis of soil to study the distribution of contaminants requires a large number of samples to be taken, often over an extended area. Several mechanical devices have been developed to facilitate such sampling.

Care should be taken when choosing equipment, since the sample may be contaminated during extraction. The abrasive nature of soils may incorporate metals from equipment and/or containers, especially if the materials are stainless steel, aluminum, plated, galvanized, or carbon steel. Teflon plastics are suitable materials, but their structural strength is insufficient for some sampling applications [30,34].

Soil sampling techniques can be divided into the following categories [14,34]:
- manual surface grab sampling (only a few centimeters deep): spoons, hand augers, and shovels;

- manual shallow subsurface coring (down to 30-40 cm). Hand augers, metal push tubes, and split barrel samplers are used to carry out this sampling. As before, care should be taken to avoid materials which could contaminate soil samples as well as contamination coming from the overlying strata;
- deep subsurface coring with heavy equipment. The possibility of down-hole cross-contamination should be considered.

Sampling should progress from the least contaminated to the most contaminated areas, to reduce the potential for cross-contamination of samples. If this is not possible, care should be taken to ensure that the sampling equipment is cleaned between samples and that materials and samplers' hands (covered with gloves) and/or clothing do not contaminate the samples. Furthermore, plastic sheeting or aluminum foil should be used to keep the equipment off from the ground.

Careful cleaning of sampling equipment and sample containers is necessary to avoid cross-contamination. Several cleaning procedures, using detergents, steam cleaning, or high pressure washing, followed by use of nitric acid and distilled water, are described in the literature [34].

The most suitable material for inorganic species is high density polyethylene bags. The use of metal staples should be avoided [32]. It is usually recommended not to use glass or metallic containers, in order to avoid trace metal contamination. However, the plastic container should be chosen carefully since only some Teflon and linear polyethylene containers are free from trace metal contamination (especially Zn), leading to better results after a cleaning procedure involving HCl and HNO_3 [35]. Paper containers should be avoided, because moist soils may extract impurities from the paper, which may contain up to 0.2 % of Cr, Cu, Mn, Ni, Pb, and Zn. Metallic containers should also be avoided during transport, especially when Cu and Zn are to be studied [36].

Samples should be labeled. Gummed paper labels or tags are suitable. They should include, at least, sample number, name of collector, date and time of collection, and place of collection [3]. Labels should be affixed to sample containers prior to or at the time of sampling, and they should be filled in at the time of collection.

7.2.8 Sources of uncertainty

The error associated with sampling and subsequent steps in soil analysis has two components: random (which affects precision defined in terms of variability) and systematic (bias). Error and statistics theory in sampling is becoming more and more important and its complete description is beyond the scope of this chapter. Detailed studies can be found in the literature [37-39].

There are two main factors in the total variance of data due to random errors: those from the measurement and those from the sample [40]:

$$\sigma^2_{total} = \sigma^2_{measurement} + \sigma^2_{sample}$$

$$\sigma^2_{sample} = \sigma^2_{sampling} + \sigma^2_{population}$$

The variance arising from the measurement step can be monitored by a specific quality assurance program. Random errors originating from the sample come from the sampling process (in the study of soils, the error arising from this step is often greater than any other in the analytical process) and from the heterogeneity of the soil composition (population component in the variability [41]). According to a study carried out by James and Dow and quoted in the literature [42], soil variability can be divided into three categories: microvariation (variability between points in the soil separated by fractions of an inch), mesovariation (variability between points separated by up to a few feet) and macrovariation (scale larger than a few feet). To reduce this variability it is necessary to measure a large number of randomly selected samples.

Some studies have been focused on the spatial variability (in terms of relative standard deviation, RSD) of trace elements in soils. For instance, a range of 10-15% of RSD is found for Cu, Zn, Ni, and Pb total content in one-hectare scale, whereas the RSD goes up to 35% for Cr and Cd [29]. A range of 10-20% is found for the EDTA-extractable Cu in samples taken over three different years [32]. Another study has used data analysis by geostatistical techniques to study the spatial variability of several heavy metals (Ni, Cu, Zn, Cd, and Pb) at the one-hectare scale, in order to compare it with analytical (measurement) variability [43]. This study showed similar spatial and analytical variability for Zn and Pb, but clearly higher spatial variability for Cu, Ni, and Cd. Except for Cd, whose spatial variability was quite high, a limited number of samples was sufficient to assess a mean value for heavy metal content. Furthermore, after the application of nested sampling, with edge centered squares, it was clear that because of the heterogeneity at a microscopic scale, the variability at the center was not statistically different from the variability of samples collected throughout the entire plot.

Bias in the different steps (sample collection, handling, subsampling …) may be arise from contamination. To minimize the uncertainty resulting from this, all the processes involving changes in the samples in this step should be monitored; these include absorption or reaction of soil components and metals being studied with the container materials, or the use of sampling equipment that may not be suitable for the sampling plan.

7.2.9 Quality control samples

Blanks and control samples are recommended as the most effective tool for assessing and controlling sources of error and contamination of a soil sample, thus controlling bias in results and assessing precision. Contamination during collection arises mainly from the equipment and apparatus used, handling, preservation, ambient contamination, or sample containers [44]. Therefore, sampling that considers a suitable quality assurance program is needed to reduce systematic and random errors. The choice of the different quality control samples depends on the problem being assessed. It is evident that the need to reduce the incidence of errors in results will depend on the level of trace metal contamination in the soils studied (hot spots or, instead, if the levels are near to the detection limits of the analytical techniques used).

7.2 Soil sampling in trace element analysis

The following blanks and samples are those most used as field quality control samples [1,45,46]:

- field blank: a sample "free" from the analytes being studied. "Free" means here that samples are expected to have negligible or unmeasurable amounts of the analyte of interest. This blank is preserved and processed in the same manner as the associated soil samples. Distilled and deionized (DDI) water is often used as a field blank, although soils are more recommended since they have a similar analytical matrix. One field blank should be enough per sampling team per day and per item of collection equipment;
- equipment blank (field rinsate blanks): a sample of field blank that is used to rinse sampling equipment prior to sampling. It is useful to decontaminate the sampling equipment. Such a blank is recommended every 10 routine samples;
- trip (transport) blank: a sample of field blank taken from the laboratory to the sampling site and returned to the laboratory unopened. It is used to document contamination related to shipping, handling, or container. Although highly recommended for study of volatile organic compounds, it may be necessary to take it into account in assessment of trace metal pollution. One transport blank is necessary per day and per type of sample;
- field duplicates: independent soil samples collected as close as possible (in space and time) to the same point of sampling. They are considered to be identical and, as they go through the same analytical process, they may indicate the overall precision of sampling and subsequent analytical process;
- background sample (site blank): a soil sample taken from a location close to the site of interest, which is thought to be uncontaminated or unaffected by human activity and which can be useful for documentation of baseline or historical information.

An example of this is reported in literature [46] dealing with Pb determination in soil samples, in which DDI water used to rinse the soil collection equipment was used as a field blank. In this study, none of the 76 field blanks analyzed had a Pb concentration higher than 250 μg/l, which was the detection limit for the procedure which used atomic absorption spectroscopy as final measurement technique. This concentration limit suggested that equipment-cleaning procedures were adequate. However, other studies would indicate other recommended procedures for washing of equipment [44]. In that case, significant trace metal contamination (mainly of Mn, Cd, Cu, and Zn) was found using polyethylene collectors washed with deionized water. After acid-washing, the concentrations of these elements were 50-fold lower.

Besides the use of field quality control samples, internal quality control samples should be used, like blind blanks and spikes, especially when there are many samples to be analyzed, a new laboratory is going to be used, or when analytical problems are suspected [34].

Even though the ratio between quality control samples and routine samples should be at least 5%, it should in any case be defined according to the data quality required. Furthermore, if the data needed to calculate the minimum number of samples are not available at the time of sampling, empirical approaches may be followed and an equal number of field samples, field blanks, and spiked blanks is recommended [47].

7.2.10 Sampling report

A chain-of-custody record must be completed containing all the information related to sampling, and it should accompany every sample. This form should include sample number (sample identification), signature of the collector, date of sampling (year/month/day), site data (coordinates, utilization of the site), sampling techniques and description (depth sampled, boring and drilling tools), conditions of transport and prestorage (material of container, duration and temperature of transport), signature of persons involved in chain of possession, and inclusive dates of possession [3].

7.3 Soil pretreatment and storage conditions

Pretreatment and sample preservation techniques have to be considered as an important part of the overall analysis in trace element determination. In the determination of metals in soils these steps may condition the results obtained when applying extraction methods. Different metal species can be present in soils according to the mechanisms of soil interactions with metals and in most cases the conditions to preserve these forms are ignored or the risk associated with loss or contamination of the sample are not reported. The lack of methods assessing the integrity of the trace metal during pretreatment and storage can create serious risks of discrepancies since changes in humidity, temperature, oxidizing conditions, or biological activity can produce important changes in the original mechanisms of binding and consequently in their further release when applying extraction methods [48]. Although these aspects of sample handling are well known, in many studies of metal or nutrient determination in soils or in sediments the storage conditions are not reported and only few papers describe systematic studies on the effects of the main variables to be considered before extraction which can determine the final result. Moreover, studies on soil sample storage are mainly devoted to purposes other than trace metal determination, e.g. changes in nutrient elements, changes observed in the soil sample when a specific extractant reagent is applied, or changes in the solubilization of the iron and manganese content.

7.3.1 Drying processes

The results from the determination of elements in soils are usually reported on a dry mass basis. Moreover, dried samples are easier to handle and more appropriate for subsequent analytical steps, like mixing and sieving. However, drying methods can cause important changes in the sample since the equilibria between trace metals and the matrix is disrupted during the process and the experimental conditions must be strictly controlled to ascertain which will be adequate for the final information to be obtained.

Changes in soil composition during drying have been reported in the literature, especially the increase in surface acidity or the amount of organic matter, which becomes more soluble after drying [49], producing changes in the extraction of some metals, e.g. copper, directly related to their binding with organic matter [50]. Effects of drying on the further selective solubilization of trace metals from soils and sediments, beyond of the scope of this study, have recently been described [51]. During drying a decrease is also possible in the amount of iron extracted by the reagents used to solubilize amorphous forms of this element [50]. The most noticeable changes seem to occur in manganese; drying processes increase its solubility and studies of the dependence of these effects on the different drying methods used before metal extraction are reported in the literature [52,53]. There are studies on the variation of the properties of soils, e.g. redox behavior during drying. Thus, relationships have been found between the oxidizing properties of manganese-containing soils, compared with elements such as chromium, since no oxidized forms of this element are detected in dried soils. In such cases chromium can be considered as a test to demonstrate the oxidizing changes occurring in the soil during drying [49]. Drying temperature can also influence further solubilization of mineral phosphorus and nitrogen [54,55].

The changes in soils during drying also depend on time, temperature, and light, and these changes can also continue during storage [49]. As a consequence it is important to define and establish the conditions used for this treatment.

For analytical purposes, air-drying or oven-drying, using desiccators and vacuum or conventional ovens, are the methods most widely used [53,56]. No drying period is well established for soils. Berndt compared the results obtained when soil samples are air-dried at 28 °C or oven-dried at 60 °C, and pointed out that the two drying systems caused an increase in manganese solubilization and the increase was greater for the higher temperature [53]. These increases may change in relation to different soil composition, and no correlation can be established with parameters such as pH, organic carbon, or initial moisture.

In general, air-drying is preferred in studies related to extraction of heavy metals in soils [30,57-61]. Heat-drying is sometimes proposed before metal determination. Brümmer mentions air-drying for several days or drying at 60 °C for 24 hours as an alternative method [4]. However, drying at relatively high temperatures can cause losses of elements such as mercury or selenium.

In some studies on interlaboratory comparison, where the material to be analyzed by a relatively great number of participants has to be prepared under stringent conditions, the drying temperature is reported. Carlton-Smith et al. described the preparation of a sludge-treated soil and used a draught-oven not exceeding 30 °C for drying the sample [62]. Crosland et al. report an interlaboratory comparison for the determination of trace elements in two soils and use air drying before sieving [63]. Pickering mentions that air-drying or freezing a sample prior to storage can produce irreversible changes and affect the complexation state of trace metals [64].

Freeze-drying or infra-red heating are scarcelly reported in the literature for soil samples [31]. Bartlet and James report that freeze-drying does not offer advantages over air-drying since the changes during storage of the soil after drying are not avoided [49].

It is of interest to establish well-defined drying conditions in the analysis of inorganic components of soils, and to be able to compare results and to carry out this pretreatment properly and in reproducible form, because significant changes of soluble forms of metals during drying can occur, especially at high temperatures. The ISO guide on soil quality and pretreatment of samples for physico-chemical analysis recommends both air-drying and oven-drying [65]. For air-drying ISO recommends spreading the soil in a layer not thicker than 15 mm, on a tray which does not absorb any moisture and which does not cause contamination. Direct sunlight must be avoided. For oven-drying a thermostatically controlled oven, with forced ventilation capable of maintaining a temperature of 40 ± 2 °C, is recommended. A layer not thicker than 15 mm is also advised. The oven-drying time for sandy soils is normally not longer than 24 hours, for clay soils no longer than 48 hours and for soils with high organic matter content, a duration from 72 to 96 hours may be required. These norms also consider freeze-drying as an optional system. A temperature of 40 °C is reported by several authors for drying and it is recognized that for most parameters the changes upon drying at temperatures below 40 °C are minor [5].

Regarding the elements and organometallic compounds present in soils and their subsequent analysis, however, it should be borne in mind that for some volatile species, such as Hg, As, and Se compounds, freeze-drying of the sample is recommended [5].

7.3.2 Fractionation and subsampling

The metal contents of soils, as of sediments, varies with the particle size, and the highest contents are found in the smallest grain size particles [50,64,66-68].

In metal determination in soils it is assumed that the fraction to be selected from the whole sample is < 2 mm [4,50,54,64,66-68]. Some authors state that metal contents are lower if the coarse material is considered, as in field conditions [55].

Before grinding, soil aggregates must be broken up; this is a delicate operation which must be performed in a manner such that no soil particles are damaged. The ISO norms recommend removal of stones, fragments of glass, rubbish, etc, larger than 2 mm, by sieving or by hand picking, before crushing. To break up the aggregates the norms recommend crushers, mills, mortars and wooden or other soft-faced hammers.

This pretreatment has been already discussed [5]. It was recommended that the sample lumps should be gently rolled, and it was emphasized that pre-crushing procedures should be investigated and possibly standardized for agricultural soil analysis. It was asssumed that if pre-crushing is successful, further sieving at 2 mm is sufficient to homogenize the samples. Moreover the need to investigate suitable crushing procedures for calcareous soils was also stressed [5].

After pre-crushing it is recommended that the sample is sieved through 2 mm mesh. This operation is usually performed by means of nylon sieves, and it can be done by hand or by using a mechanical shaker. In all case, precautions must be taken to avoid the contamination of the sample by metal-containing materials.

Subsampling is necessary for analysis, and grinding, mixing, and homogenization are carried out to ensure the representativity of the laboratory sample. The size of the particle to be obtained will depend on the particular analytical purpose.

When small subsamples are required, grinding is recommended before homogenization. Different criteria are reported in the literature: 0.315 mm size for subsamples below 2 g [30], 40-mesh for subsamples less than 1 g [4], 250 μm for samples smaller than 2 g [55]. Depending on the nature of the soils other sizes are recommended. Grinding to obtain particles smaller than 0.250 mm, or grinding to 100 μm, is recommended when soils to be homogenized contain organic particles and sandy fractions [5]. For total metal content determination, a table adapted from Jackson [69], calculates the minimum and optimum sample masses for different mesh sizes. In the table a bulk soil density of 1.3 g/cm^2 is asssumed (this bulk density of soils can vary from 0.5 to 1.6 g/cm^2) [56].

Kramer *et al.* describe different grinding techniques for soils and sediments, and compare particle size distribution [70]. In this study jet milling is used for large scale (up to 200 kg) and includes a closed classifier to avoid any contamination of the workplace environment by the finest powder; planetary milling is used for medium scale (from 20 to 80 g), and mixer milling, using, e.g., the Retsch mixer is used for small scale (from 1 to 2 g).

From the comparison it is concluded that the three grinding techniques provide remarkable differences in distribution for the finest particles, and the highest percentage of these particles was obtained by using the jet milling process.

It should be noted that it is not always appropriate to obtain the finest particles during grinding of soil samples, e.g. for the preparation of reference materials. Indeed, grinding too finely may lead to unrepresentative samples since fine-grained soil particles will have larger specific areas and the extractability of trace elements will be higher than from coarser grain fractions.

Mixing in order to obtain homogeneous samples for analysis is the next step after sieving. Water can be added to the sieved sample to make a slurry, in order to obtain more homogeneous subsamples by volume measurement [49]. This method is also described as a simple, accurate way of preparing samples for analysis in the field, for the determination of soluble and exchangeable ions [71].

Representative subsampling of the sieved material can be achieved by different methods, but when the purpose of the subsequent analysis is the determination of trace metals some of the apparatus currently used for solid homogenization, e.g. mechanical riffles, may not be suitable because their materials can introduce contamination.

Subsampling can be performed by hand, by using a sample divider, or mechanically. Coning and quartering [65,55,56] is the most widely used procedure when subsampling by hand and requires simple apparatus. In this method the material is placed on a clean sheet, preferably of polythene to avoid contamination, formed into a cone which is divided into four quarters using a cross made of sheet aluminum, and then the opposite quarters are combined and mixed to furnish a subsample of approximately half the original weight. This subsample is again coned and quartered, and the process is repeated until the desired sample weight is obtained.

Sample dividers, usually riffle boxes or table samplers, are also used for subsampling, and different designs and dimensions are described according to the particular purposes, such as particle size and the number of portions to be obtained by splitting the bulk sample [65,56,72].

A wide range of equipment can be used to carry out mechanical subsampling. ISO describes an apparatus operating by rotation in which the bulk sample is introduced into a funnel and the different subsamples are collected in appropriate bottles [65]. The process can be repeated on the content of one or more subsamples until the desired amount of soil is obtained. An automatic device is described for supplying representative subsamples from large amounts of soils, sediments, and plant material for proficiency testing [73].

Homogeneity tests have to be applied after subsampling, by measuring properties selected according to the final purposes of the samples. An homogeneity test based on loss after ignition at 550 °C to determine organic matter is a good indicator of homogeneity [73]. The measurement of the main inorganic components and trace elements is carried out as an homogeneity test for soils and sediments. However, it should be pointed out that in many instances the results can be influenced by the particle size obtained by milling, thus if the total metal content is determined the result will be independent of particle size [54]. Homogeneity is achieved within bottle by analyzing ten subsamples taken from two bottles of the lot and between bottles by analyzing ten subsamples from different bottles randomly selected from the lot [61,74-76].

7.3.3 Storage conditions

Wet storage is not recommended because microbiological activity can cause significant changes in the properties of soils, especially in the redox character, and, consequently, these activities may influence the forms and extractability of trace metals.

Different systems are used to ensure the stability of soils during storage: low temperature storage (4 °C), steam sterilization, freezing, or gamma irradiation.

Low temperature short-term storage at 4 °C is the method most widely used to minimize microbial changes [31]. On the other hand, Bartlett and James state that the simplest method for short-term storage is to keep the sample close to field capacity moisture in polyethylene bags, permeable to CO_2 and O_2; for long-term storage the authors recommend low temperatures, above but close to freezing [49]. A wide study is reported on the influence of the storage of dried and field-moist soil samples at 4 °C and at 20 °C on the further extraction of manganese, concluding that under dry conditions the increase in soluble manganese is greater for samples stored at the highest temperature and for soils stored under field-moist conditions [53].

It should be remarked that when steam sterilization is carried out noticeable changes can occur in the sample because of the high temperatures used in the process. Singh and Pathak report the relationship between the soluble forms of manganese in different soils (acidic, neutral, or alkaline) and the temperature used when steaming; an increase in the water soluble and exchangeable forms of manganese is observed, particularly in acidic soils, whereas in neutral and alkaline soils an increase is observed in the amount of reducible forms of manganese [52].

Gamma irradiation is sometimes used for sterilization [50,77] because the method does not cause physical and chemical changes in the soil constituents. Nevo and Hagin reported a study on changes occurring in soil samples during storage and, by subsequent inoculation of nutrient broth, described the gamma irradiation conditions and time of irradiation (from 12 to 16 h) which ensure the total sterility of the sample [78]. These authors also report the use of an alternative method to sterilize the soil samples, addition of ethylene oxide gas to the containers after evacuating the air; the authors found, however, that ethylene oxide may react with water to some extent. In any case, whatever the sterilization method used, it has to avoid changes in the chemical forms of the elements which would produce significant changes in the subsequent extraction. It is for this reason that the use of chemicals should be avoided [31].

Partial losses of metals can occur during storage [56]. Metallic containers should be avoided but if they are used it has to be taken into account that aluminum or aluminum alloys can contain copper and magnesium, and that aluminum sheets can be coated with zinc or cadmium. Stainless steel should be avoided. When the choice is plastic it must also be considered that many plastics contain different metal compounds as stabilizers: polypropylene and silicone rubber are the preferred plastics. Losses are usually less likely than contamination, but some metallic compounds, such as organic forms of mercury or selenium, can be lost if the soil sample is not refrigerated; polyethylene materials, for example, are porous to mercury vapor [79].

Various storage periods (short, medium, or long-term) are used in laboratories, and it would be good practice to control the integrity of the sample before choosing the proper preservation conditions, because structural changes, losses, and contamination could occur, in some cases to a considerable extent, after previous treatments (drying, freezing, heating) and in most cases these changes continue during storage. This control may include periodic determinations of metal content or metal species as well as other properties, such as organic matter content or pH. Thus strict control of the metal content over time is recommended.

7.3.3.1 Storage of soil solution

The storage of wet soils at low temperature minimizes soil solution changes and can be a good method for a limited period of storage, although some changes in anion content (e.g. nitrate), pH, or Al can be observed in some cases, especially when solutions have been separated from the soils by centrifugation [80]. For these samples the precautions usually recommended for water samples should be considered to avoid metal contamination and losses. Metal contamination is attributed mainly to the type of material used for storage whereas losses of metals can originate from changes in properties such as pH, redox potential, organic matter, microbial activity, temperature, ionic exchange, or adsorption processes between the solution and the container walls. All these mechanisms can produce different kinds of loss, also depending on the metal, by volatilization, adsorption, and coprecipitation, including, in some cases, the transformation of ionic forms into higher molecular weight species. All these effects influence the final analysis of the metal species present in the original soil solution sample [50,81,82]. In general, low temperature storage, lowering of the pH to avoid hydroxide precipitation, and the filtration of the sample are the conditions most recommended for storage of solutions in which the metal species content will be determined.

7.3.4 Pretreatment procedures used in the preparation of reference materials (RMs) and interlaboratory trials

The use of reference materials (RMs), to assess the validity of the analytical methods used in laboratories is widely recommended [83], and the availability of soil RMs is of interest in a variety of fields of research such as agriculture, geochemistry, plant nutrition, or environmental contamination. Different soil reference materials have been prepared for metal determination and analyzed in interlaboratory trials and the description of the preparation, homogenization, components analyzed, and comments on the results are briefly mentioned. soil reference material SO-1 for trace analysis is described [74]; the material is dried at 105^0C, after air-drying, and sieved through a 0.15 mm nylon sieve, the residual fraction (0.15 to 1.0 mm) is ground until it passes through 0.15 mm mesh, and homogenization is performed by mixing in a rotating plastic drum for 72 h. Preliminary homogeneity tests between bottles and within bottles are performed by X-ray fluorescence determination of Fe, Ti, Sr, and Zr. No differences are observed from the results obtained after sterilization by irradiation when repeating the tests. The results obtained from the determination of 41 elements, carried out by 30 laboratories, were as follows: seven elements were classified as reference values, eleven with acceptable confidence, and sixteen were used for information.

The preparation of soil materials for proficiency testing is described within the International Soil-Analytical Exchange (ISE) in The Netherlands [73,84]. The soil samples are dried at 40 °C, milled, and sieved to 0.5 mm; after mixing and subsampling a homogenization test is performed by analyzing a test portion from each container. Thirty four elements are determined as total content, and the determination of more reduced groups of elements by using different extraction procedures is also reported.

An interlaboratory comparison of the total metal determination of six elements in a sludge-treated soil is described [62]. The soil is air dried in a draught-oven not exceeding 30 °C and ground to 0.5 mm. The homogeneity is checked by analyzing a sample from each of twenty locations and comparing the variance with 20 determinations from a single location, using X-Ray Fluorescence and Atomic Absorption Spectrometry (AAS) after acidic attack. The study includes a wide discussion on the results and analytical methods for each element.

As a part of the BCR Programme of the EC to harmonize methods for extractable trace metal content in soils and sediments, The Measurement and Testing Programme supported the ETMESS project which included the certification of EDTA and acetic acid-extractable metals in two soils (a terra rosa soil and a sewage amended soil), the certification of EDTA- and DTPA-extractable trace metals in a calcareous soil, and the certification of extractable trace metals in a lake sediment. The soil candidates, i.e. terra rosa, sewage-amended, and calcareous soils were collected in Catalonia (Spain), Northampton (England), and in the South of Spain, respectively. The bulk materials were air-dried and sieved at 2 mm size; homogenization, gamma irradiation (to avoid any microbial activity), and bottling were carried out by Muntau [85].

Finally, homogenization and stabilization (gamma irradiation) procedures have been described for the preparation of certified reference materials of soil and sludge samples (calcareous loam soil, domestic sewage sludge, and sewage sludge of industrial origin) within the Measurements and Testing Programme; these CRMs were certified for their total and aqua regia contents of trace metals [86].

7.4 Conclusions

Soils are heterogeneous materials in which the determination of trace constituents can be considered a difficult task, especially when the results obtained in different laboratories have to be compared, since sampling, pretreatment and storage influence the amount of metal ultimately extracted with a selected reagent.

Sampling has to be accurately programmed and the program should include the design and sampling strategy, quality control samples, number of samples and size, as detailed as possible.

Pretreatment and storage conditions and the binding of trace metals to the matrix constituents have to be carefully considered.

Special attention must be paid to avoiding contamination of the sample and loss of analyte during sampling, transport, sample pretreatment, and storage. The materials of the containers have to be properly selected.

Sample storage must be performed carefully and under the most stringently controlled conditions in such a way that the integrity of the sample is preserved.

Further recommendations and standardization methods should be available on the conditions of sampling, sample pretreatment and storage.

The availability of soil CRMs of different composition, reflecting the compositions prevaling in different countries, is of main interest for the validation of standardized methods on pretreatment and storage, since differences in matrix composition can require modification of the methods.

Interlaboratory trials play an important role in such studies.

7.5   References

[1]   A.A. Liabastre, K.A. Carlberg and M.S. Miller, in: *Methods of environmental data analysis*, C.N. Hewitt (Ed.), Elsevier Applied Science, London, pp 259-300 (1992)
[2]   R.J. Mesley, W.D. Pocklington and R.F. Walker, *Analyst*, **116**, 975 (1991)
[3]   U.S.EPA, *Test methods for evaluating solid waste*, Vol. II. Field manual physical/chemical methods. Chap. 9. Sampling plan. SW-846. 1986 (revised 1992)
[4]   G.W.Brümmer, in: *The importance of Chemical "Speciation" in Environmental Processes*, M. Bernhard, F.E. Brinckman, P.J. Sadler (Eds.), Springer, Berlin. (1986)
[5]   *Workshop on QA in Sampling and Sample Pretreatment in Soils and Sediments*, Organized by BCR, Commission of the European Communities, and Department of Analytical Chemistry of the University of Barcelona. Barcelona, 7-10 May 1994. Report based on Contract NR 5588/1/9/000/93/08-BCR-E(10)
[6]   B. Kratochvil and J.K. Taylor, *Anal. Chem.* **53**(8), 924a (1981)
[7]   G.A. Fleming, H. Tunney and E.G. O'Riordan, in: *Sampling problems for the chemical analysis of sludge, soils and plants*, A. Gomez, R. Leschber, P. L'Hermite (Eds.), Elsevier Applied Science Publishers, Brussels, pp 6-17 (1986)
[8]   E.K. Triegel, in: *Principles of environmental sampling*, L.H. Keith (Ed.), American Chemical Society, Washington, pp 385-394 (1988)
[9]   L.H. Christensen, G. Ciceri, S. Facchetti and M. Duane, *Quim. Anal.*, **13**, S43 (1994)
[10] ISO/CD 10381/1.2.1, *Soil quality. Sampling. Part 1. Guidance on the design of sampling programmes* (1991)
[11] M.J. Barcelona, in: *Principles of environmental sampling*, L.H. Keith(Ed.), American Chemical Society, Washington, pp 3-23 (1988)
[12] R.G. Petersen and L.D. Calvin, in: *Methods of soil analysis*, Part 1, A. Klute (Ed.), American Society of Agronomy Inc, Soil Science Society of America, Madison (1986)
[13] G.U. Fortunati, C. Banfi and M. Pasturenzi, *Fresenius' J. Anal. Chem.*, **348**, 86 (1994)
[14] G.U. Fortunati and M. Pasturenzi, *Quím. Anal.*, **13**, S5 (1994)
[15] K.W. Jackson, I.W. Eastwood and M.S. Wild, *Soil Sci.*, **143**(6), 436 (1987)
[16] J.P. Okx, G.B.M. Hewelink and A.W. Grinwis, in: *Contaminated Soil-'90*, F. Arendt, M. Hinsenveld, W.J. van den Brink (Eds.), Kluwer Academic Publishers, Dordrecht, pp 729-736 (1990)
[17] T. Brun and C. Lopez, in: *Sampling problems for the chemical analysis of sludge, soils and plants*, A. Gomez, R. Leschber, P. L'Hermite (Eds.), Elsevier Applied Science Publishers, Brussels, pp 55-65 (1986)
[18] J.P. Okx, H. Leenaers and R.M. Krzanowski, in: *Geostatistics Tróia '92*, A. Soares (Ed.), Vol. 2, Kewer Academic Publishers, Dordrecht, pp 673-683 (1990)
[19] F. Colin, in: *Sampling problems for the chemical analysis of sludge, soils and plants*, A. Gomez, R. Leschber, P. L'Hermite (Eds.), Elsevier Applied Science Publishers, Brussels, pp 66-78 (1986)
[20] F. Colin, M. Jauzein and G. Grapin, *Quím. Anal.*, **13**, S80 (1994)
[21] L.L. Eberhardt and J.M. Thomas, *Ecol. Monographs*, **61**(1), 53 (1991)
[22] D.R. Silkworth and D.F. Grigal, *Sci. Soc. Am. J.*, **45**, 440 (1981)
[23] P. del Castilho, *Quím. Anal.*, **13**, S21 (1994)

[24] M.I. Litaor, *Water Resour. Res.*, **24**, 727 (1988)
[25] M.G. Cline, *Soil Sci.*, **58**, 275 (1944)
[26] J. Fons, J. Romanyà, J. Cortina, T. Sauras and V.R. Vallejo, *Catena*, in press.
[27] R.E. Carter and L.E. Lowe, *Can. J. Forest Research*, **16**, 1128 (1988)
[28] N. Harkness, in: *Sampling problems for the chemical analysis of sludge, soils and plants*, A. Gomez, R. Leschber, P. L'Hermite (Eds.), Elsevier Applied Science Publishers, Brussels, pp 18-26 (1986)
[29] K. Aichberger, A. Eibelhuber and G. Hofer, in: *Sampling problems for the chemical analysis of sludge, soils and plants*, A. Gomez, R. Leschber, P. L'Hermite (Eds.), Applied Science Publishers, Brussels, pp 38-44 (1986)
[30] A. Gomez, in: *3rd Symposium International of Processing and Use of Sewage Sludge*, Brighton, 1983. D.Reidel Publ. Co. Dordrecht - Boston - Lancaster, EUR 9129, European Commission, pp 132-139 (1984)
[31] A. Gomez, R. Leschber and F. Colin, in: *Sampling problems for the chemical analysis of sludge, soils and plants*, A. Gomez, R. Leschber, P. L'Hermite (Eds.), Elsevier Applied Science Publishers, Brussels, pp 80-90 (1986)
[32] M.L. Berrow, *Anal. Proceed.*, **25**, 116 (1988)
[33] A. Banin, J. Navrot and A. Perls, *Sci. Total Environ.*, **61**, 145 (1987)
[34] R.J. Bruner III, in: *Quality Control in remedial site investigation: hazardous and industrial solid waste testing*, C.L. Perket (Ed.), 5th volume, American Society for Testing and Materials, Philadelphia, pp 35-42 (1986)
[35] J.R. Moody and R.M. Lindstrom, *Anal. Chem.*, **49**(14), 2264 (1977)
[36] R.L. Mitchell, *J. Sci. Food Agric.*, **11**, 553 (1960)
[37] P. Gy, *Analusis*, **11**(9), 413 (1983)
[38] F.P.J. Lamé and P.R. Defize, *Environ. Sci. Technol.*, **27**, 2035 (1993)
[39] M. Thompson and M. Maguire, *Analyst*, **118**, 1107 (1993)
[40] J.K. Taylor, in: *Principles of environmental sampling*, L.H. Keith (Ed.), American Chemical Society, Washington, pp 101-107 (1988)
[41] P.H.T. Beckett and R. Webster, *Soil and Fertilizers*, **34**(1), 1 (1971)
[42] B. Kratochvil, D. Wallace and J.K. Taylor, *Anal. Chem.*, **56**(5), 113r (1984)
[43] M.C. Wopereis, C. Gasmal-Odoux, G. Bourrie and G. Soignet, *Soil Sci.*, **146**(2), 113 (1988)
[44] D.L. Lewis, in: *Principles of environmental sampling*, L.H. Keith (Ed.), American Chemical Society, Washington, pp 119-144 (1988)
[45] J.J. Van Ee, L.J. Blume and T.H. Starks, EPA-600/4-90-013, US EPA, Environment Monitoring Systems Laboratory, Las Vegas (1990)
[46] S.C. Black, in: *Principles of environmental sampling*, L.H. Keith Ed., American Chemical Society, Washington, pp 109-117 (1988)
[47] ACS Committee on Environmental Improvement, *Anal. Chem.*, **52**, 2242 (1980)
[48] E.A. Thomson, S.N. Luoma, D.J. Cain and C. Johansson, *Water, Air and Soil Pollut.*, **14**, 215 (1980)
[49] R. Bartlett and B. James, *Soil Sci. Soc. Am. J.*, **44**, 721 (1980)
[50] G.E. Batley, *Trace Element Speciation: Analytical Methods and Problems*, CRC Press, Boca Raton, FL (1989)

[51] A.M. Ure, *Quím. Anal.*, **13**, S64 (1994)
[52] M. Singh and A.N. Pathak, *Plant and Soil*, **33**, 244 (1970)
[53] G.F. Berndt, *J. Sci. Food Agric.*, **45**, 119 (1988)
[54] W.O. Enwezor, *Plant and Soil*, **26**, 269 (1967)
[55] V.J.G. Houba, I. Novozamsky and J.J. Van der Lee, *Quím. Anal.*, **13**, S94 (1994)
[56] R. Rubio and A.M. Ure, *Int. J. Environ. Anal. Chem.*, **51**, 205 (1993)
[57] S.K. Gupta and C. Aten, *Int. J. Environ. Anal. Chem.*, **51**, 25 (1993)
[58] A. Payá-Pérez, J. Sala and F. Mousty, *Int. J. Environ. Anal. Chem.*, **51**, 223 (1993)
[59] L. Orsini and A. Bermond, *Int. J. Environ. Anal. Chem.*, **51**, 97 (1993)
[60] M. Vidal and G. Rauret, *Int. J. Environ. Anal. Chem.*, **51**, 85 (1993)
[61] A.M. Ure, Ph. Quevauviller, H. Muntau and B. Griepink, *Int. J. Environ. Anal. Chem.*, **51**, 135 (1993)
[62] C.H. Carlton-Smith and R.D. Davis, *Wat. Pollut. Control*, **51**, 544 (1983)
[63] A.R. Crosland, S.P. McGrath and P.W. Lane, *Intern. J. Environ. Anal. Chem.*, **51**, 153 (1993)
[64] W.F. Pickering, *Ore Geology Reviews*, **1**, 83 (1986)
[65] ISO/CD, *Pretreatment of samples for physico-chemical analyses*, Revised ISO/DIS 11464 (1992)
[66] U. Förstner and G.T. Wittmann, *Metal Pollution in the Aquatic Environment*, Springer-Verlag. Berlin (1981)
[67] W. Salomons and U. Förstner, *Metals in the Hydrocycle*, Springer-Verlag. Berlin, (1984)
[68] U. Förstner, W. Ahlf, W. Calmano and M. Kersten, in: *Metal Speciation in the Environment*, J.A.C. Broekaert, S. Gücer, F. Adams Eds., NATO ASI Series, Springer Verlag, Berlin (1991)
[69] M.L.Jackson, *Soil Chemical Analysis*, Prentice Hall, New Jersey. USA (1958)
[70] G.N. Kramer, A. Oostra and J. Pauwels, *A Comparison of Grinding Techniques Used for Large and Small Scale Preparation of Soil and Sediment Samples for Organic Analysis at IRMM*, Report GE/R/SP/05/94 of the European Commission, Joint Research Centre, Institute for Reference Materials and Measurements, B-2440 Geel, Belgium (1994)
[71] J.J.Oertli, *Commun. Soil Sci. Plant Anal.*, **21**(13-16), 1151 (1990)
[72] T. Allen and A.A. Khan, *Chemical Engineer*, 108 (1970)
[73] V.J.G. Houba, *Fresenius' J. Anal. Chem.*, **345**, 156 (1993)
[74] B. Holynska, J. Jasion, M. Lankosz, A. Markowicz and W. Baran, *Fresenius Z. Anal. Chem.*, **332**, 250 (1988)
[75] B. Griepink, *Fresenius' J. Anal. Chem.*, **337**, 812 (1990)
[76] H.D. Fiedler, J.F. López-Sánchez, R. Rubio, G. Rauret, Ph. Quevauviller, A.M. Ure and H. Muntau, *Analyst*, 1109 (1994)
[77] A.D. McLaren, L. Reshetko and W. Huber, *Soil Sci.*, **83**, 437 (1957)
[78] Z. Nevo and J. Hagin, *Soil Sci.*, **102**, 157 (1966)
[79] A. M. Ure, *Anal. Chim. Acta*, **76**, 1 (1975)
[80] D.S. Ross, J. Richmond and J. Bartlett, *J. Environ. Qual.*, **19**, 108 (1990)

[81] M. Betti and P. Papoff, *CRC Critical Reviews in Anal. Chem.*, **19**, 271 (1988)
[82] J. Buffle, *Complexation reactions in Aquatic Systems, an Analytical Approach*, Ellis Horwood Series in Anal. Chem., New York (1990)
[83] B.Griepink, *Int. J. Environ. Anal. Chem.*, **51**, 123 (1993)
[84] V.J.G. Houba, J.J. Van der Lee and I. Novozamsky, *Analusis*, **19**, 144 (1991)
[85] H. Muntau, personal communication, European Commission, Joint Research Centre. I-21020, Ispra (Varese), Italy
[86] G.U. Fortunati, *Meas. Testing Newslett.*, **2**(1), 14 (1994)

8.

Quality assurance of sampling and sample handling for trace metal analysis in aquatic biota

K.J.M. Kramer
MERMAYDE, P.O. Box 109, 1860 AC Bergen, The Netherlands

Monitoring of the environment, whether carried out on water, sediment, or biota is no better than the combined result of all the efforts that lead to the final outcome. The concept of analysis is the integrated result of many actions, such as:
- sampling;
- sample pre-treatment (e.g. preparation and preservation);
- transport and storage conditions;
- sample treatment;
- instrumental analysis;
- calculation, and
- evaluation of results.

It may seem surprising that most quality assurance (QA) programs that try to improve the quality of the 'analysis' focus on those aspects that are related to laboratory activities. The reasonable to good analytical results that are produced today in intercomparison exercises between different laboratories (comparability), are mainly the result of a decade of intercomparisons, the use of CRMs [1-3], and attention to good laboratory practice (GLP) in the analytical laboratory. Unfortunately the tested procedures rarely involve the sample handling before the samples enter the laboratory [4]. Moody [5] formulated the following statement: "For many elements the control of contamination or sample stabilization, or both, may become the limiting factors in the accuracy of an analysis". The establishment of guidelines [6] helps the understanding of prelaboratory problems, but traditionally most chemical analysts will be trained in the instrumental analytical procedures only. They are often neither experienced nor involved in field-related activities such as sample collection and pre-treatment.

In this chapter emphasis is placed on the extension of quality assurance to those activities of the "analysis" that take place before the actual laboratory procedures, notably to the sampling and the sample handling of biological material collected in the aquatic environment (including contamination control, use of materials, sample preparation, homogenization, transport, and storage conditions).

Since at present the majority of biological monitoring programs focus on the chemical detection of pollutants in tissues (so actually chemical monitoring of biota, and not biomonitoring) this chapter will deal mainly with the handling of biota for these chemically oriented objectives. Where appropriate, the handling of biota intended for biological effect monitoring will be discussed.

Elements traditionally analyzed in biomonitoring programmes include Hg, Cd, Zn, Pb, and Cu. Fewer data are available for other elements, including As and Se, Cr, Ni, Ag, and Sn. When looking at the periodic table of elements one may wonder whether yet other elements should not be analyzed. Multi-element analytical techniques, like ICP-AES and ICP-MS, offer now the possibility of doing so, and at a financially attractive rate. Even elements considered by most as "obscure", e.g. rare earth elements (REEs), including the lanthanides, are increasingly used in industrial processes. Several are relatively highly toxic and thus it is expected that they will become part of monitoring programs in the future.

A special group form the organoelement species. Because of serious histological effects observed in marine gastropods (especially in the dogwhelk, *Nucella lapillus*), organotin compounds (mainly mono-, di- and tributyltin) are also included in several routine monitoring programs. Other organoelement compounds (of As, Se, Hg, or Pb) have until now mostly been analyzed in the framework of surveys and special case studies. These organic compounds are usually much less stable and as such they merit special attention in sample handling if biased results are to be avoided.

Although this chapter will focus on the analysis of trace metals in three distinctly different groups of organisms (mussels, fish and macrophytes), it will be evident that a comparable discussion on quality assurance can be followed for other groups of organisms. The routines will differ only slightly. Even when the methods are applied to the study of organic pollutants in these organisms, many aspects can be considered valid provided, of course, that appropriate adaptations are made for example in the use of certain materials.

8.1 Biomonitoring approaches

In toxicological studies relatively few (but specific) species are used, for example because of their ease of handling, their representativeness, the possibility of collecting or culturing, their relatively insensitivity, etc. [7]. This has led to a wealth of data on the effects of toxicants upon these few organisms. It would be interesting to use the same organisms in field studies also, but since the objectives of toxicological testing do not agree in all aspects with field biomonitoring studies, different (groups of) organisms are often used in biomonitoring studies.

Considering the aquatic environment, many results on the concentrations of trace metals in organisms have been published, using, e.g., algae and macrophytes, zooplankton, worms, bivalves, prawns and shrimps, fish, and mammals [8,9]. It will be obvious that some organisms are more suitable for the detection of the chemical content of certain pollutants than others. In many cases practicality will result in the selection of the monitoring species, but the objective of the monitoring study can also be the steering factor in the selection of species (or organs). If the objective of the program is that organisms are to be used as indicators of (local) environmental conditions, in principle any organisms can be used (provided they obey certain basic rules, see below). However, when the objective of a monitoring program for, e.g., the Ministry of Agriculture and Fisheries, is to identify the potential risk for the (human) consumers, the choice will be limited to those species that are normally available for human consumption, and probably even to the edible parts (fish meat, prawn tails, mussel tissue) [10].

8.1.1 Selection of species

Apart from population characteristics, the following criteria for species selection can be summarized for environmental monitoring with biota (after: [11,12]):
- ease of identification;
- adequate body size (not too small, not too large);
- longevity;
- distribution, abundance and availability;
- mobility and action radius, preferably sedentary;
- diet and habitat;
- ability to determine age, class and/or sex;
- annual cycle characteristics, stages of life cycle;
- availability over a major period of the year;
- understanding of seasonal effects;
- hardy enough to be handled;
- tolerance to environmental stress (salinity, suspended particulate matter, pollutants);
- place of species in the food chain;
- physiology;
- accumulation of compounds of interest;
- no or minor transformation and/or regulation of ingested pollutants;
- sensitivity to biological effects;
- no tissue elements that interfere with the chemical analysis;
- handling without major risks of contamination;
- standardization of (sampling and analytical) methods feasible;
- cost of collection and maintenance;
- importance to man (as food source, economic reasons, ..).

This rather long list has led to the selection of a relatively short list of taxonomic groups commonly used in biomonitoring studies: macroalgae, (bivalve) molluscs, and fish. Since these taxonomic groups represent organisms with totally different functioning and behavior, sometimes two or more representatives are used in the same monitoring program.

8.1.1.1 Bivalves

Most common species included in biomonitoring studies involve mussels of the genera *Mytilus (M. edulis, M. galloprovincialis, M. californianus)*, and *Perna (P. perna, P. viridis)* for the marine environment, and for freshwater *Dreissena polymorpha, Anodonta* spp. and *Unio* spp. Oyster species are also commonly used, particularly of the genera *Ostrea (O. edulis, O. gigas)* and *Crassostrea (C. virginica, C. gigas, C. angulata, C. commercialis)*. Some tropical species were proposed by Sarkar et al. [13]. Many more species have been used, because of scientific interest or because of availability. Cantillo [14] documented 81 species. For an overview of this list see O'Connor et al. [15]. Certified reference materials (CRMs) for trace metals in bivalves (mussel and oyster tissue) are available from various sources [16].

8.1.1.2 Fish

Most studies reported on fish species used for human consumption. The choice of the species is often based on availability and wide distribution and will inevitably change from area to area. Worth mentioning are plaice (*Pleuronectes platessa*), flounder (*Platichthys flesus*), cod (*Gadus morhua*), tuna (*Thunnus* spp. and *Katsuwonus* spp.), mullet (*Mullus barbatus*), sea bass (*Leteolabrax* spp.), blue-fish (*Pomatomus saltatrix*), billfish (*Makaira* spp. and *Istiophorus* spp.), and the (teleost) dogfish (*Scyliorhinus* spp.) (see e.g. [17-19]). Certified reference materials have been prepared from certain species (cod, dogfish, tuna, sea bass) [2,16].

Interesting "sampling" is performed from natural history collections. This option may provide a long-time series [20,21]. Interferences from the methods of preservation (added chemicals) and from neglect of contamination control may be expected, however.

8.1.1.3 Macrophytes

Species of brown or green macroalgae have been employed most frequently, but red algae and other macrophytes such as seagrasses have also been used as bioindicators. Marine species include the genera *Fucus (F. vesiculosus, F. serratus)*, *Enteromorpha* spp., *Ascophylum* spp. and *Ulva* spp. [22]. In the freshwater environment vascular plants and aquatic mosses e.g. of the genera *Fontinalis* and *Rhynchostegium* have been used (see e.g. [23-26]). Also employed is the floating duckweed (*Lemna* spp.).

8.1.1.4 Other groups of organisms

The dogwhelk (*Nucella lapillus*) and the periwinkle (*Littorina littorea*) have been shown to be extremely good indicators of organotin pollution [27].

Worms (polychaetes and others) are used infrequently used in monitoring studies. Examples include the ragworm (*Hediste (Nereis) diversicolor*) and the lugworm (*Arenicola marina*) [28,29]. In freshwater, tubificid worms are used [30,31]. Worms have the disadvantage that they are true sediment dwellers and that most ingest sediment particles. Since sediment usually contains relatively high levels of pollutants, depuration is essential for these organisms [32,33]; this leads to the risk that compounds of interest may also be eliminated.

Crustacea, shrimps and prawns (especially the genera *Penaeus* and *Metapenaeus*), barnacles, or crabs (*Carcinus* sp.), lobsters, and crawfish belong to the benthos but are certainly not sedentary. Many migrate and exhibit seasonal distribution. Only their edible parts are usually tested (e.g. [34]). However, in an investigation concerning the distribution of Zn, Cu, Fe, Mn, and Cd in prawns Peerzade et al. [35] distinguished between muscle tissue and various other organs (hepathopancreas, ovaries, and gills) and subdivided the batch into size classes based on carapace length. Interestingly the muscle tissue contained the lowest concentrations (on a wet-weight basis). Sivadasan and Nambisan [36] investigated seasonal variations in the concentrations of trace elements (Hg, Cu, Zn) in *Metapenaeus* species. In an active biomonitoring approach (see Section 8.2.2). Palmer and Presley [37] compared different sites for the Hg content of shrimps. Everaarts et al. [38] compared Cd in the tissues of shrimps (*Crangon* sp.) over the area of the North Sea. Since the highest Cd concentrations were not found near the polluted coastal zone they concluded that the food source was of major importance. These findings seem to limit the use of this group of organisms in biomonitoring studies.

Barnacles are attached to a hard substrate and thus sedentary. Since they are crustacea, they are an interesting companion to the bivalves most often used in biomonitoring programs [39,40]. (Hermit) crabs were compared in a few studies of the geographic variation in trace metal content [41]. Several studies report on their uptake and/or elimination of trace elements, e.g. Zn and Cd [42], or As [43], both in the shore crab (*Carcinus maenas*).

Everaarts et al. [38] also investigated starfish (*Asteria* sp.), with similar findings as for their studies on shrimps, mentioned above. The use of hydroids for monitoring purposes was studied by Stebbing [44] in field experiments; sponges (*Halichondria panicea*) were tested more recently [45].

Marine and freshwater mammals, including seals, porpoises, and dolphins, are top predators and as such may be a valuable target organism in biomonitoring programs. Because of the public awareness, mammals are not often actively caught for this purpose. Most investigations depend on individuals that are washed ashore or found entangled in fishing nets [46]. Species with a coastal habitat include the common seal (*Phoca vitulina*), the common or harbour porpoise (*Phocoena phocoena*), and the bottlenose dolphin (*Tursiops truncatus*). Tissues of marine mammals have been collected for specimen bank purposes [47]. Organs that are analyzed include the liver and sometimes the brain. For a treatment of the methodologies one is referred to Law et al. [48].

8.1.2 Approaches in the chemical monitoring of biota

Methods for chemical monitoring of biota are based on the ability of organisms to accumulate (in)organic compounds to levels far above the concentrations found in the surrounding water. This bioaccumulation is the net result of both uptake and loss (or elimination) processes. An important requirement for application in environmental monitoring programs is the existence of a simple and consistent relationship between external and internal contaminant concentrations. It has been found in laboratory and field studies that the internal concentration (C_i) through uptake or loss of contaminants reaches equilibrium with the external pollutant concentration (C_e), thus C_i/C_e is approximately constant. This ratio, which has to be defined for each species and for each compound, is known as the bioconcentration factor (BCF) [49]. Given the species and compound, C_i represents the (bioavailable) contaminant concentration in the water column [50].

As an example of a laboratory experiment used to illustrate the accumulation of the element dysprosium with time in mussels (*Mytilus edulis*), as function of the concentration of this element in water (Figure 8-1), batches of 25 mussels were exposed in separate large tanks to different concentrations, each concentration being kept constant by use of a flow-through system. Over a period of 6 weeks sub-samples of 25 individuals were collected. In Figure 8-1 the tissue concentrations are plotted as function of the exposure time. They seem to reflect nicely the differences in the concentrations present in the water column.

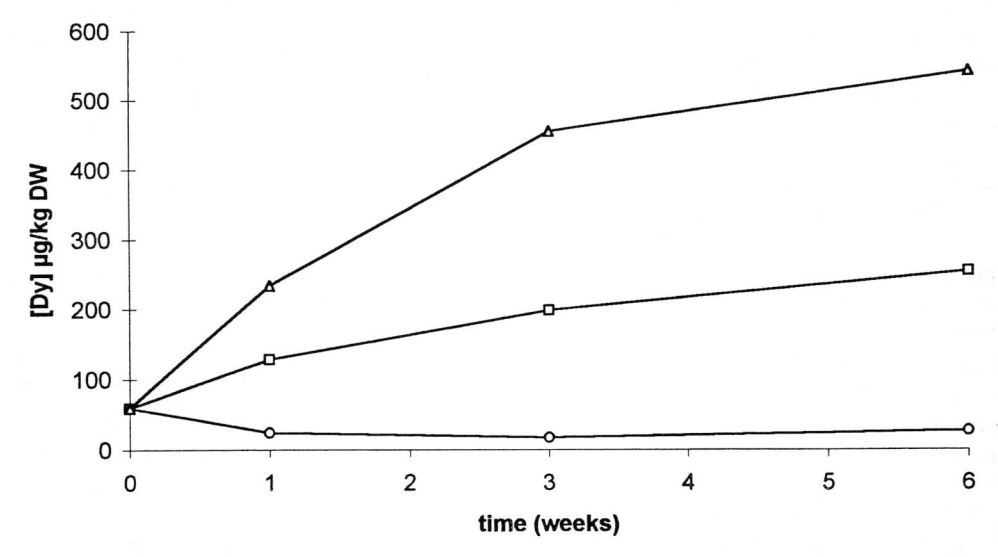

Figure 8-1: The uptake of ^{162}Dy by the mussel (*Mytilus edulis*) under controlled laboratory conditions, as function of time and the concentration of dysprosium in water (● = no addition; ■ = + 0.5 µg/l added; ▲ = + 1.0 µg/l added).

An important feature of this "sampling" approach is, that the tissue tends to be in equilibrium with the surrounding environment and represents not only the short period just before the moment of collection, but also includes an extended period before this moment. Thus this biomonitoring approach serves as an integrating sampler, representing a period rather than a moment, as in the spot sampling of water.

One may distinguish between two different approaches in the (chemical) monitoring of biota: the passive biomonitoring technique (PBM) and the active biomonitoring technique (ABM). Examples of both techniques are illustrated in a series of case studies using marine and freshwater bivalves [15,50,51].

8.1.2.1 Collection of wild populations

The technique of passive biomonitoring (PBM, in contrast to the active form ABM, see below) is characterized by the collection of organisms from their natural habitat at sites where a natural population exists. In this approach it is unavoidable that the samples originate from different (sub)populations. Not only will genetic differences exist, the natural history of the samples, in terms of, e.g., food availability and quality, temperature, salinity, suspended matter content or anthropogenic influences, will also be different from site to site. For example, the influence of these factors upon growth rate implies that similar shell lengths do not necessarily indicate the same age. As a result it has been shown that the between-site variability is significantly higher than the within-site variance [50].

PBM with mussels has become known as the "Mussel Watch" [12,14,15,52-54]. In a well designed monitoring program a protocol will be present that describes the practical aspects of sampling (site selection, sampling period, sampling method, tools, supporting field measurements) and sample handling (including packing, transport, contamination control) [15].

Because of natural variability it will not be sufficient to analyze one or a few individuals. Usually a batch consisting of 25-100 individuals is collected, depending on (species and site dependent) expected natural variability [55] and the size of the specimens. Some choose to analyze each individual separately, but to reduce analytical cost, most programs prefer to pool the individuals into a composite sample and to analyze a subsample after thorough homogenization. It will be obvious that information on the variation between individuals is lost by this procedure.

The PBM method is by far the approach applied most often, and has been used for the monitoring of almost all inorganic and organic pollutants of interest, both in fresh waters and in the marine environment (see, e.g., [15] for the US-NOAA list of compounds), including almost all the species mentioned above.

A major advantage of this approach is that the method is straightforward and relatively cheap. It seems especially suitable for situations where the intersite variability is not too critical, e.g. in the process of "hot-spot" detection.

8.1.2.2 Translocation of organisms (caging)

Many aspects of the PBM approach are applicable to the methodology of active biomonitoring (ABM). The major differences lie in the method of exposure.

For ABM studies one large batch of organisms (e.g. mussels) is collected from one preferably pristine site. This procedure ensures that all organisms are derived from the same (sub)population, with minimal genetic differences and a common pollution and environmental history. They are sampled and handled according to preset guidelines, e.g. describing their size range, water depth of collection, methods of transport and storage. The batch is then subdivided into subsamples consisting of 25-100 individuals of similar size. To keep the samples naturally distributed, it is important that a randomization procedure is used for this subsampling. Each subsample is placed in a cage or net, and suspended in the water column at the target positions defined by the monitoring program. Usually an exposure period of 6-8 weeks is applied, but a longer duration in general causes no problems.

At the start of the exposure period the organisms will be in equilibrium with their original (pristine) environment, but once they experience the new environmental conditions they will adapt to this new situation by uptake or elimination of the pollutants. After several weeks a new equilibrium will be reached. The time needed depends on the toxicant. The equilibrium will be a representation of the average C_e conditions the mussels have experienced over the exposure period in the water column. For detailed description, also of practical procedures, the reader is referred to De Kock and Kramer [50].

According to, for example, De Kock [56,57], Smith et al. [58] and Martincic et al. [59], major advantages of this ABM methodology are:
- the reduction of variability by employing similar groups of organisms leads to optimization of the resolution power for comparing chemical stress at different locations;
- monitoring stations can be selected independent of the (non)presence of naturally occurring populations;
- the exposure period is known;
- in addition to the detection of trace pollutants in the tissue, this method of exposure also offers unique possibilities for the investigation of biological effects (physiology, biochemistry).

Since one is not dependent on naturally occurring populations, selection of exposure sites can be target-oriented and the sites can be used for many subsequent years (trend monitoring). It is our experience that the logistically more complicated methodology is counterbalanced by the more useful results.

As for the PBM studies, this type of biomonitoring study has been applied to trace elements in marine and freshwater bivalves, e.g., [57,59-62], and organic pollutants such as PCBs and PAHs (e.g. [56,63-66]. Translocation of organisms for ABM studies is not limited to bivalves. Successful attempts have been described using fish, macrophytes, barnacles, and gastropods.

8.1.3 Approaches in biological effect monitoring

In contrast to the chemical detection of pollutants in biological tissue, biological effect monitoring aims to detect either behavioral or physiological (biochemical) changes. Many tests for biomarkers and other stress indicators have been developed over the recent decades at biological organization levels ranging from individual organism to subcellular [67]. Most have, however, so far only been tested under controlled laboratory conditions. Few have evolved to application in biomonitoring programs [68]. Notably the methods for detection of cytochrome p450 monooxygenase, as indicator for PCB/dioxin like compounds, and the metallothioneins as indicators of trace metal pollution seem well developed. Examples of field applications have been provided in a number of case studies [69,70]. The special case of the detrimental effects on tissues caused by tributyltin and related compounds has yielded a rather unique situation where a clear cause and effect relationship could be established both in the laboratory and in the field [27].

The characteristic of most biomarkers is that the detected compound is induced by one or more chemicals, and as such is an indicator of environmental effects and potential harm. These biomarkers are the result of physiological processes, and thus biochemical in nature. This implies the potentially unstable nature of the molecules, which gives reason for concern for the sampling and sample handling of such materials. In contrast to the chemical monitoring of biota, the risk of contamination is in these cases far less than the risk of degradation (or transformation) of the molecules. As a result of this, specialists in the field of biomonitoring have adapted their methods following strict procedures, which are probably better validated than for the analysis of common chemical pollutants in tissues. If we limit ourselves here to the (effects of) trace metals (analysis of metallothioneins), preferred methods include assays with freshly killed organisms. Since this is often not practical for field sampling, instead tissues (liver, kidney) are excised immediately upon collection. These tissues are wrapped in clean aluminum foil and immersed in liquid nitrogen. The low temperatures are essential to slow down deterioration processes. Subsequent storage at -80 °C is often applied. For detailed information of the analytical procedures one is referred to the literature cited in George and Olsson [70].

8.2 Contamination control

Most chemical analysts working in the field of trace element analysis are nowadays aware of the risk of contamination. Contamination will inevitably lead to increased concentrations of the trace element to be determined in the (biological) sample, leading to biased (too high) results with limited possibilities of proper interpretation. Contamination control is an essential part of every step in the analysis. Kramer [4] indicated that where specialist laboratory staff is usually well trained, this is often not the case with the sampling team. For curious management reasons, in many organizations and institutions the sampling team and the laboratory are strictly separated. It will be obvious that this is not an ideal means of combating contamination. The sampling team should be instructed by the laboratory staff where minimization of contamination is concerned. Instructions on what to do to minimize the risks of contamination should be provided. For example, handling of the samples with bare hands should be prohibited, subsampling or preparing organs should not be performed in the field but in a laboratory under controlled (clean room) conditions, no metal tools, etc.

8.2.1 Materials

For trace metal analysis only a limited number of materials should be used. Several plastics, without colorants or interfering stabilizers, with low metal content, can be sufficiently cleaned and are practical in use. Most used are polythene (the soft type, high density, conventional, or linear polythene, LPE), used for bottles, bags, tubing. Polypropylene (PP) is a good material also, but it has the tendency to become brittle after prolonged use. Teflon, preferably FEP quality, is very useful for bottles and tubing. It can be well cleaned, is temperature resistant over a large range, but suffers from high price. Polycarbonate also is used. Glass should not be used (many brands contain high Pb levels), but quartz glass is OK. Ampoules may be sealed and thus provide a really closed container. The disadvantage of quartz ampoules is their price and the risk of breakage [5,71].

8.2.2 Containers

Containers must be made of the materials mentioned above: polythene, Teflon or polycarbonate. For some special cases special materials must be used. Organotin compounds are preferred in glass or polycarbonate. Because of the volatility of Hg, samples for the analysis for this element must be stored in glass [72,73]. Wide-mouth types are most convenient for storage of tissue.

For storage of biological material over longer periods at very low temperatures (-196 or -80° C) Teflon seems to be the best choice. Other potentially suitable materials become brittle. Most containers are considered to be fully closed when the lid is screwed tightly. Some types appear to close not so well (even a "double seal cap" is no guarantee), thus testing is a useful precaution. A water loss of up to 3% has been observed with certain containers [71,73].

The size of the container should match the size of the sample, to optimize the volume to surface area ratio. The number of times a sample is transferred to other containers should be limited in order to minimize contamination.

Heavy duty plastic bags (two layers) are useful containers after sampling of larger batches. The bags may be sealed air-tight.

8.2.3 Tools

Tools for collection and preparation of biological samples should be made of materials not affecting the analysis. Plastics are usually preferred, but cutting and homogenization of other than soft tissues may require other materials.

To prepare specific organs or tissues, or to open the shells of bivalves, knives are essential. Of the nonmetallics (quartz or borosilicate) glass is sometimes used, but better are knives made of ceramic. Metal knives are possible when the element to be analyzed is not present in the knife. We use titanium, but titanium carbide, tungsten, or aluminum knives have been reported [74]. A "normal" knife, presumably of stainless steel, should be avoided [5,75].

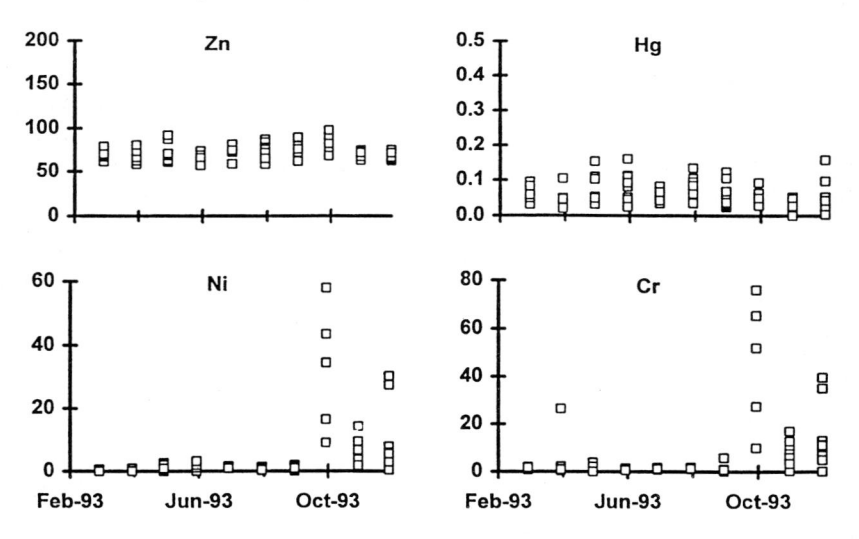

Figure 8-2: Monthly collected data for trace metals in prawns suddenly showed much higher concentrations and with a much larger variation. This was observed only for Ni and Cr, and not in the other elements analyzed. After a change in the laboratories involved in the preparation of the tissues contamination occurred during the homogenization step - the result of stainless steel knives.

In a homogenizer or blender the stainless steel blades and shaft should be replaced by noncontaminating material. An example of the effect of a change in methods is depicted in Figure 8-2. During the monitoring program involving the analysis of trace elements in prawn tissue, at a certain stage a different laboratory was involved. The increase in concentration (and in variation) of only Cr and Ni was a clear indication of the use of stainless steel in the process of homogenization.

Like all tools, sieves should preferably be made of plastic. Nylon sieves are available.

8.2.4 Clean room - clean bench

One of the important methods to combat contamination is the use of a clean room, or at least clean-bench techniques. This involves a working space where the air is passed over a so-called HEPA filter in order to achieve low particle content (Class 100). This reduction of particles, and the awareness that clean room techniques are essential, have reduced contamination considerably [76].

The use of a clean bench in the field is not always a realistic option, but we feel that the processing of the sample, cutting, dissecting, homogenization, etc., should be performed in a clean bench atmosphere. Portable clean benches are available. Otherwise, the preparation should wait until the sample reaches a (laboratory) place equipped with a clean bench.

Clean-room techniques require dust-free garments. This involves special coats (made of dust-free fabric) and talc-free plastic gloves (polythene). Ensure that the section of the lower arm between gloves and coat is well protected. A new set of disposable gloves is needed during sampling/preparation whenever the gloves have touched other than cleaned surfaces.

8.2.5 Cleaning of materials

All containers and tools should be properly cleaned. Removal of dust and plastic remnants is a first step. If necessary, a rinsing step to clean the surfaces from oil and grease may be added (think of collecting tools). The next process is soaking the materials in nitric or hydrochloric acid. Start with analytical grade, diluted by a factor in the range of 1:2 to 1:5, for a period of several days, followed by soaking the parts and filling the containers with high grade diluted acids (Supra Pur, Instra Analyzed, Ultrex, etc., ratio 1:10). The final cleaning step consists of washing with ultra clean distilled water [71,77].

After cleaning, proper packing until use is essential. Tools and bottles are packed in (cleaned) polythene bags until use.

8.3 Sampling

The objective of sampling is to collect a portion of the environmental matrix (organism, tissue) that is representative of that matrix in terms of location and time. The portion should be small enough that it can be treated, transported and analyzed conveniently. Representativeness is all that matters in sampling. Of course the action of sampling should not be in conflict with the other analytical requirement: that the concentration of the compound of interest should not change, neither be enriched (contamination) nor reduced (sorption to container walls, degradation, or transformation).

Sampling programs should be designed (long) before the actual sampling takes place [78]. Their lay-out should be discussed with all involved: policy makers, field staff, chemical laboratory, statisticians, and assessment staff.

Several authors have addressed the need for procedures and standard methods in sampling and sample handling methods [4,5,79-82]. In the following section the various aspects of quality assurance (QA) and good measuring practice (GMP) in these sampling and sample handling activities will be discussed in more detail for the specific needs of the most important groups of target organisms: bivalves, fish, and macrophytes.

8.3.1 Location, number of samples, frequency

The monitoring/sampling strategy should define the sampling locations, the number of samples, and the sampling frequency (including the number of independent replicates [51]). The analyst should provide information about the effectiveness (cost-benefit) of, e.g., increasing the number of replicates. In general the quality of the data in terms of precision will be better when more replicate analyses are performed. It has its price, however, and the accuracy will not be improved as this is dependent on bias or systematic error.

The sampling substrate should be limited to natural surfaces, unless special measures are taken to limit interferences with the objective of the monitoring program (e.g. exposure in plastic cages). Organisms should not be collected from surfaces that are covered with materials that interfere with the analysis. Metal or painted surfaces (buoys, piles) will contain metal oxides as colorant in the paint, preserved wooden structures could lead to unrealistically high concentrations of the preservatives in the tissue [83].

Samples should not be collected from public market places (unless the aim is to sample human food sources). No information is available on the area of origin and the living conditions of that area. Certainly not the least important, contamination control has not been applied.

8.3.1.1 Bivalves

General site selection methods for "Mussel Watch" studies are given by O'Connor et al. [15]. For ABM studies material from aquaculture farms can be used, provided they are from a relatively clean area and of desired size/age.

The dependence on water depth and/or submergence period has been discussed by several authors. Most prefer collection of samples which have been continuously submerged (subtidal). It may make a difference, however, whether samples are derived from subsurface or from near-bottom waters (see, e.g., [84,85]). Lobel et al. [86,87] used the bottom of a flat lagoon to enable collection of samples from the same depth (monolayer). They emphasized the use a relatively small surface area for collection (a few m^2), as differences may be found over larger areas differences. Our experience is that when samples come from subtidal areas of about similar depth, they will result in samples of sufficient homogeneity. Hand-picking, possibly using divers for deeper spots, is most useful. Dredging will result in an increased sediment content in the organisms (washed in during collection).

8.3.1.2 Fish

Fish are collected by a variety of methods: trawl surveys, seining, trapping, electro shocking. Chemicals should never be used to catch the fish [11]. Barghigiani and De Ranieri [17] used 30 hauls of 1 hour in trawl surveys according to a randomized stratified sampling design. Few papers seem to document the sampling strategy and give details on the method of collection. Seldom is the "sample" derived from more than one catch per location. Variability may be high, however [88]. Guidelines often follow traditional methods for fisheries investigations, although specialized instruction for chemical monitoring studies has been provided [89].

8.3.1.3 Macrophytes

Attention should be paid to the position of sampling on the shore (i.e. the site of the plants sampled with reference to changes in water level), as trace metal concentrations may vary with their position on the shoreline. Sampling should take place at similar shore levels with respect to tide marks at all monitoring sites in coastal waters and estuaries, or similar water depths in rivers and lakes [90]. For marine applications, the mid-shore level is generally considered the most appropriate [22,91].

8.3.2 Period of sampling, seasonal effects

Almost all organisms used in sampling programs are subject to seasonal variations. This may be a result of:
- migration, the organisms are only available during a limited period of the year;
- physiology and metabolism, as changes occur in feeding habits (availability of food, lipid content), processes linked to the reproductive cycle (spawning), etc.;
- growth regime; for example, in winter macrophytes have no or only minor parts exposed above the sediment.

This phenomenon of seasonality has been observed in the trace metal contents of tissues, and it is advised to sample at defined periods of the year. When several periods are to be considered, comparisons between different periods of the year must be evaluated with care. For most organisms literature data will reveal the best periods for sampling, i.e. periods when no major changes (spawning) take place. Do not copy months without understanding the biological reasoning behind the proposed selections. A series of monthly collected data will guide us to define the sampling period for the specific organisms in the given area(s).

8.3.2.1 Bivalves

Sampling for "Mussel Watch" programs fall in the winter period, thus in the northern hemisphere between mid-November and the end of March [15]. This is the period without reproduction and with reduced activity. Changes do occur within this period, however, and once the period has been established for a given location, the annual sampling should be within a period of three weeks. Some programs select two or more sampling periods. Practicality, such as ice coverage, will partly influence the definition of the period.

8.3.2.2 Fish

As (specific stages of) species will migrate, some fish can only be collected in a certain period. This period will depend strongly on the species concerned. As for the bivalves, spawning can seriously influence the concentrations of trace elements, and this period should be avoided. Since the start of the spawning is not linked to a given date, but to natural environmental conditions (temperature, light, food availability, ...) a substantial margin should be taken into account. If present at the monitoring locations, northern hemisphere species are collected in the winter period. Checking with the biological data on this aspect for the species and area concerned is most useful.

8.3.2.3 Macrophytes

The season of collection for aquatic plants is linked to the presence of the parts to be collected [91]. Since most programs advise sampling of younger tissues, the summer period is the best period of sampling. Care must be taken to collect tissues of similar age (exposure) for all samples.

8.3.3 Collection methodology

Samples are assumed to be collected randomly. This is often very difficult to achieve in a true sense. Sampling schemes may be directed by an approach based on random numbers generated by a calculator, e.g. in a grid approach, but this method is rather time-consuming for the collection of organisms as one can not be sure whether organisms will be available at these spots during sampling.

Unfortunately, most practical procedures for applied biomonitoring studies consist only of defining the location in terms of a dot on the map. No information is then provided on the distance from the shore line, whether the samples are collected at mid stream, at what water depth, etc. Certainly, samples collected just at the shoreline at 20 cm water depth or mid-stream at 5 m, will have a completely different history; this will probably be reflected in pollutant content. Only rarely does one observe detailed sampling instructions about how samples are to be collected and where. We believe that this neglect of quality assurance may be one of the reasons why relatively large variations are observed in pollutant tissue concentrations even when the sampling location was, so called, the same spot.

A strategy could be to collect along transects or grids, or using selected quadrants. For pelagic species, defined trawls of given time and speed or stretch can be applied. A series of sampling strategies and protocols has been summarized for estuarine work [6], but most techniques can readily be adapted to freshwater or coastal environments.

In many biological monitoring programs the outcome of the analysis depends strongly on the method of collection. This is true for the relative species abundance and biomass, but can also be true for trace pollutant content [6]. Hand-picked individuals, either on tidal flats or by scuba diver, may interfere with the principle of random collection (as one - subconsciously - tends to select on often irrelevant aspects); a dredge or nett will inevitably collect less selectively but may sample unnoticed in atypical areas. It is clear that no method is the best method for all organisms.

It is stressed that a consistent - and well thought out - sampling methodology is essential. The development of such a procedure should be the result of discussions between field staff, statisticians, and analysts. Whenever possible the methodology should be tested and validated for each monitoring program (species, area).

8.3.4 Selection of species, size and age, sex

Different organisms (species) may accumulate various elements at different rates (Figure 8-3). In the monitoring protocol it will be stated which organisms are to be used for the biomonitoring studies. This may be straightforward on paper, but less easy when sampling in the field. Depending on the group of organisms considered, it sometimes requires taxonomic specialists to identify the organism to the species level. A sampling team consisting of a set of nontrained technicians and the car driver may result in the inclusion of the wrong species. For example, it requires some skill to distinguish between *Mytilus edulis* and *Mytilus californianus* [58], and it can be quite a problem to identify the large variety of macrophytes. The identification of common brown seaweeds such as *Laminaria, Fucus*, or *Ascophylum* is relatively easy, however. For fish species the risk of mixing different species is also relatively small, because of their well defined (adult) characteristics. When there is a risk of doubt, written descriptions and a color photograph may assist the less experienced sampler. When in doubt about whether a given individual belongs to the correct species or not, discard it and collect the next one.

In the preparation of the monitoring outline the potential unfamiliarity with less well known species should be kept in mind: stick to easily recognizable species, even when specialists are to be used for the sampling. It could be necessary in the future to change (sampling) staff; this may influence the continuity of the sampling operation. It is strongly advised that old and new teams work together at least once to allow for an identical interpretation of instructions about selection of site, selection of species, sampling procedures, etc. This agreement on methods should also be confirmed when several sampling teams are used in the program.

The size of the individual is related to its age, especially in the juvenile and young adult life stages: the larger the older. The reverse is not necessary true: individuals of similar size, derived from different locations, are not necessarily of the same age. Different feeding conditions or low temperatures may have resulted in a retarded growth at one site. Different areas have different typical sizes. For example, Lobel et al. [87], at the east coast of Canada, used *Mytilus edulis* of 70-80 mm; we typically use 35-40 mm for the same species in North Sea waters. This implies that no general rules can be applied to whether a given size should be collected. The aspect of age is also important in that juvenile stages often differ in physiology from adult stages, including pollutant uptake and elimination rates. Whether young stages or adults need to be sampled depends on the species. If serious differences exist between juveniles and adults, it is best to refrain from selecting a size near the transition from juvenile to adult stage.

Practicality also plays a role in selection of the size. Usually a relatively large number of individuals needs to be pooled. When the individuals are quite large, the amount of tissue collected will prevent rapid freezing and will provide problems for transportation and storage. For the chemical analysis only a few grams will be required (for trace element analysis). On the other hand, many tiny organisms need to be sampled before sufficient (dry-weight) material is collected. The risk of contamination increases with the smaller sizes. When organs are to be excised, these observations are even more important.

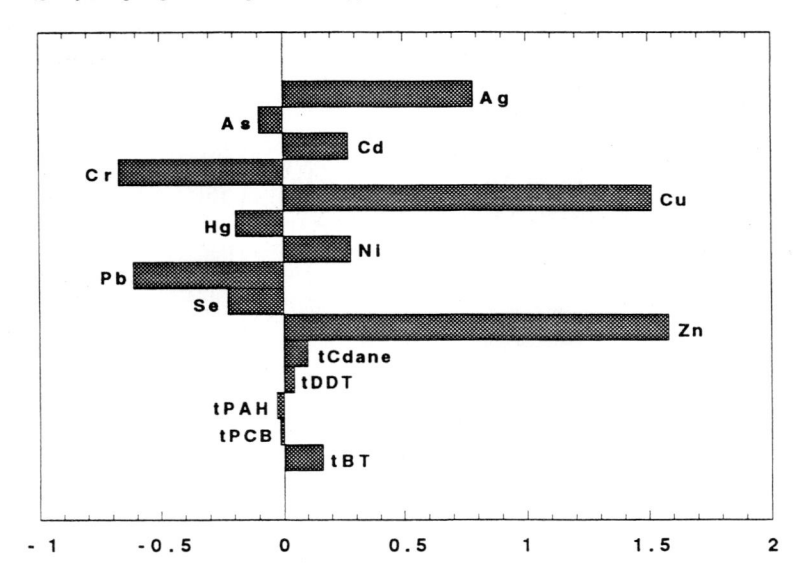

Figure 8-3: Logs of ratios of mean concentrations in oysters (*C. virginica*) to those in mussels (*M. edulis*). Data from the NOAA National Status and Trends Program [15]

Within a program one should stick to a relatively small size range, or series of ranges (possibly year classes). This range may change to some extent from year to year, as the growth rate may not be the same each year. Similar age prevails over the same size in this respect. Experience and field observations will teach what size relates to e.g. a two-year old specimen in the given area. With some organisms it is relatively easy to measure the age by using their growth marks (otoliths in fish, shells of molluscs), and this has been applied in some monitoring strategies [19]. Size alone is, however, usually sufficient for the purpose and length is the most used criterion.

For the given species and area the size range that can be pooled should be defined. In our monitoring work, we use 0.5 cm ranges for bivalve molluscs; 2.5-3, 3-3.5 or 3.5-4 cm (or several size classes). For fish a difference of 5 cm for upper and lower limit is often used. During sampling a measuring gauge, indicating the upper and lower limits, made out of an old ruler is most convenient. Since growth of an organism is not only reflected in the length, length/volume relationships may change with age. Instead of a linear scale, a set of log equidistant length classes has, therefore, occasionally been used [93].

Differences in pollutant uptake or elimination between the sexes have been found for a number of organisms [18,86], and sometimes this aspect is taken into account by selecting either males or females. In bivalve monitoring programs differences have usually been found to be small [94,95], and thus selection according to sex is not applied in most cases (also because of practical problems). In contrast, fish samples that are used for monitoring purposes are composed of either males or females. It can be discussed whether for trace metals the effects are substantial. In addition, for most organisms a trained biologist is required to detect the differences and usually each individual has to be dissected to reveal the sex organs. This implies an elevated risk of contamination by the additional handling.

8.3.5 Sample size and subsampling

In order to establish reliable mean levels of contaminant concentrations at a monitoring location, maximum effort is required to reduce the variance according to the differing individual contaminant concentrations, thus the natural variability. The sample should be of sufficient size for a reliable characterization of the pollutant level [96-98]. The larger the number of individuals, the better the coefficient of variation in sample series, the smaller the expected deviation from the population mean (Figure 8-4). This implies that one or a few big fish, which will each furnish sample material for analysis, may provide a false representation of the natural situation just because of the large variance in the individual samples. Pooling of these few individuals will not change this. A series of individuals can be analyzed separately until the variation becomes below a preset value (e.g. $\pm 10\%$ of the mean [80]); although this will result in a wealth of data, financial limitation will often prohibit such an approach.

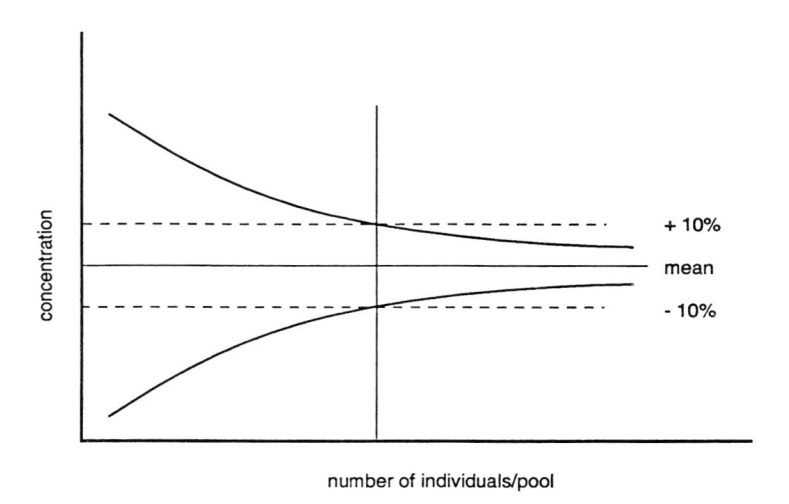

Figure 8-4: The higher the number of individuals per composite sample, the lower the variation in the analytical result.

Usually a sufficiently large number of individuals is pooled into one composite sample. For each species (not only for the group of species), for each pollutant (element) and possibly for each area, the number of individuals that is needed should be determined/validated. Numbers from the literature may not apply to other areas. However, when a sufficiently large number of individuals is taken, e.g. 50 or 100, one may assume that the natural variability is handled properly. As already discussed, owing to practical considerations, an optimum number of individuals per pooled sample will need to be determined, which will depend on the availability of the species and their size. It is our observation, though, that in most programs too few individuals are considered. The validation of this selection is almost never mentioned.

As was discussed above, the use of one population in the ABM exposure method has the inherent advantage that natural variability is reduced. This may never result in reduced sample sizes, however.

8.3.5.1 Bivalves

Especially in the case of smaller bivalves (e.g. *Dreissena polymorpha*) batches of 50-100 individuals are collected [50]. The US "Mussel Watch" program collects typical batches of 20 oysters and 30 mussels [15]. Replicates, e.g. three batches per site, will reduce the variability and enhance the interpretation, but increase the cost of the analysis considerably.

8.3.5.2 Fish

In many reports few (3-10) individuals were examined. Without validation that such a small amount of fish can indeed be used, we prefer larger numbers, e.g. 3 samples per site, consisting of ≥20 individuals [99,100].

8.3.5.3 Macrophytes

The number of seaweed individuals collected is rarely reported. One tends to collect until sufficient material is available. A batch should typically consist of 0.2-0.5 kg (wet weight). Some collect parts of ten individuals, perhaps at (a minimum of) five locations per sampling site.

Independent of the group of organisms, the total number individuals that need to be collected should be in excess of the final number pooled. When the pooling process starts, one often finds that some organisms belong to a different species, shells appear to be empty, or fish are of the wrong size class. It appears to be extremely inconvenient when insufficient individuals have been collected. A minimum of a 20-30% surplus is required (per sampling station). Only then can one be sure to be able to keep to the preset and validated methodologies.

To ensure statistical similarity between subsamples, our procedure for (sub)sampling requires that use is made of a randomization technique. From the batch of individuals that were selected on the basis of similar size, a series of single or multiple individuals is selected (portions of 1-5 organisms). A random table is used to combine certain batches into a composite sample. This procedure is of particular importance when the active biomonitoring technique (ABM) is used, as the intention of this procedure is to prepare subsamples for exposure that are as closely identical as possible. The procedure for preparation of the subsets and the use of a random table have been described [50].

8.3.6 Total organism or selected organs and/or tissue

In most monitoring studies, the chemical detection of contaminants is performed on whole soft tissue homogenates. A number of authors focused, however, on the analysis of specific organs, the main argument being that because some organs accumulate more pollutants than others, a better analytical performance could therby be accomplished (e.g. [101]). The distribution of trace elements in given parts may also show a lower variation. Unfortunately, not all elements (pollutants) are preferentially accumulated in the same organ or part. For lipophilic compounds, like organometallic compounds, tissues with a relatively high lipid content will contain larger amounts of these compounds than tissues with lower lipid content.

A second reason for selecting specific parts is the different exposure of these parts to the environmental conditions. Lastly, because of policy regulations, only the edible parts may need to be analyzed. This can be the whole (soft) tissue as in bivalve molluscs or a part of the fish fillet, thus muscle tissue only.

If higher concentrations in the tissue is the only argument for sampling selected organs, one should question the improved total analytical performance in routine monitoring programs. Not only is skilled staff required for such a procedure (one should be able to recognize the various parts) and the quantity of material for analysis is reduced, the preparation bears an elevated risk of contamination. Unless the parts can be sampled directly, as, e.g., in macrophytes, preparation of parts and/or organs in the field seems out of the question for this reason. But even under controlled conditions differences were found between various procedures. Clean-room techniques reduced the contamination considerably, as Schmitt and Finger [75] showed for eight elements in edible fish tissues.

8.3.6.1 Bivalves

A number of authors has addressed the question of subsampling specific organs. Organs of interest are: viscera, mantle, gonads, gills, retractor and adductor muscles, foot, labial palps and kidney. In general, highest concentrations are found in the kidney, viscera and gills [89,102,103]. Instead of soft tissue one may use byssus threads [104] or the shells [105,106]. Because of the yearly added "growth-rings" and the age of ten's of years that some bivalve species may reach, attempts have been made to use the shells as indicators of temporal changes in the concentrations of elements in the environment [107]. Such methodologies have not yet been often applied, however.

In the case of bivalves, unless applied for policy regulations it is doubtful whether the preparation of selected organs is necessary. Since most analytical techniques are sufficiently sensitive to measure most elements in total tissue, it seems no longer necessary to use the detection limits as an argument. It is our impression that in most cases the analytical risk in the sampling/preparation steps are not counterbalanced by the potentially better results.

8.3.6.2 Fish

Most fish studies are the result of edible tissue analysis. This results in the collection of (axial) muscle tissue (fillet). In flat fish one may select one side of the muscle tissue, usually the darker upper (dorsal) side, as this is in less contact with the sediment.

Organs of interest are the ovaries and liver, sometimes the kidney. For example, Hellou et al. [18] reported on 20 elements in the organs of Northwestern Atlantic cod.

8.3.6.3 Macrophytes

There are sound reasons for collecting distinct parts of seaweeds. The roots of macrophytes, such as seagrasses, are in close contact with the sediment whereas the leaves will exchange with the water column. In addition, older parts contain higher concentrations [22]. Bryan et al. [91] concluded that the younger tissue is to be preferred, as it best reflects recent conditions in the water column, whereas the older parts reflect conditions integrated over longer timescales. Distinct parts include the thallus, stipe, fronds, and tips (e.g. the last 10 cm). Although most monitoring programs use the parts above the substrate, roots and rhizomes also may be sampled. Stems, holdfast, and the older parts are usually discarded. Because of the variety in forms and sizes of the various species, no defined generally applicable procedures have been formulated.

Whenever it is decided, irrespective of the organisms investigated, to subsample for specific organs, the methods should be validated. This validation procedure should incorporate also the preparation methodologies and the number of organs to be pooled and analyzed. Sources of contamination should be investigated and, where possible, eliminated.

8.3.7 Cleaning of organisms

The objective of the sampling of biota is that, typically, whole organisms or selected tissues are to be collected. Particulate matter from adhered or ingested sediment and/or particulate matter may seriously bias the analytical result. This will be especially true for organisms (worms) living in the sediment and ingesting the particulates. Organisms living in the water column may also contain particulate material. Correlations found between biota and nearby sediments may be the result of the incorporation of sedimentary particulates in the analysis. Cleaning is therefore an important part of the sampling and sample handling process.

8.3.7.1 Bivalves

On sampling the adhering sand, silt, and other debris and the epibiota should be brushed off with nylon brushes. They should preferably be washed in the ambient water at the sampling site.

Most bivalves used for monitoring purposes are filter feeders. They filter particulates out of the water for food. This particulate matter may have relatively high contaminant loadings [28,32,33,108]. Depuration in water of similar characteristics (ambient water) for one to two days, is considered to be a solution to this problem [32,94,109]. These methods are, however, not easy to perform in a practical sense, and they are not easy to standardize. When organisms are exposed to ultra clean water for depuration, one should be aware of possible elimination of pollutants from the tissues and also the extra stress conditions. If depuration is applied, a period of 24 or 48 hours is considered sufficiently long.

Whether depuration is necessary will depend on the amount of particulates present or expected in/on the organism. When the amount is considered small, as is the case for mussels or oysters, this complicating methodology can be omitted [110]. For samples collected in the market place (human consumption) depuration cannot be applied as the original water conditions are unknown [111]. Normalization procedures based on elements, such as Al, Fe, or Mn, that are typical of particulate matter (clay), are not advised.

8.3.7.2 Fish

Fish need to be washed in the ambient water in order to remove adhering sand and dirt. For some species the scales are removed in the field, but unless this procedure is well controlled (e.g. the use of metal free tools) this is not advised. In cases where certain parts are to be sampled in the future, the skin is a reasonable barrier against contamination.

8.3.7.3 Macrophytes

The selection of an appropriate washing technique for collected macrophytes is most important, as surface contamination by a variety of organic and inorganic materials takes place. Suspended particulate matter deposited on the leaves offers a source of error. In addition, the leaf surfaces are covered with a mixture of epiphytic bacteria, diatoms, and protozoa, all capable of binding trace metals. The trace metal concentrations seem to vary with the washing technique applied [112,113]. Surprisingly little standardization has been achieved, however. Most studies speak of "removal" of adhering particles, without specifying the methods. Rinsing with filtered (ambient) water and/or with distilled water, scrubbing with a nylon brush, or application of ultrasonication, have all been reported [22,90]. Cleaning is essential, contamination control should be applied during this procedure, and the method used should be validated. Additional studies are certainly needed here.

8.3.8 Physical factors affecting sampling

Water levels may change over the year (seasonal effects) or even by the hour (e.g. tidal effects), and the simple instruction for sampling sedentary organisms "at one meter water depth" has no use if no reference level is stipulated. The definition of a defined shore level is preferred. In estuarine or coastal waters tidal effects can affect the degree of submergence of sedentary species. Differences in lighting regime, suspended particulate matter content, nutrient uptake and resulting differing growth rates are factors that may, as a result, influence the uptake/elimination processes of trace elements. Strong variations with water depth and/or shore level have been observed for different groups of organisms, such as bivalves and macrophytes [84,91,114]. It can be concluded that fixed depths are preferred, provided that they are chosen according to ecological principles.

Salinity, temperature, dissolved oxygen, etc. will affect the organisms to be tested. When atypical situations are present, e.g. close to an outlet or the merging point of rivers, these physical factors may bias the analytical results, as they may not be representative of the area considered.

Chemical speciation has been shown to influence the uptake (and possibly elimination) of trace elements [50,115,116]. Despite the evidence of its importance, speciation of trace elements is not taken into account in biomonitoring programs. Only defined organometallic species are identified separately.

8.4 Labeling, transport, and storage

8.4.1 Labeling

Although one would expect defined procedures to be applied to the labeling of (routine) monitoring operations, it is surprising that insufficient attention is often given to the labels and the sample code numbering system. The moment the sample is collected, it should be tagged with a unique identification. A simple "Number #1" will certainly not be unique and thus prove insufficient. It will depend on the monitoring organization whether numbering codes and labels are provided before the sampling trip starts, or the labels are to be composed at the sampling site. The latter procedure is not necessarily problematic (it offers more versatility), but it is subject to errors. Whatever method is used, it should be well documented and used throughout the organization.

It seems logical that the sample code comprises an indication of the type of sample, of analysis, etc. Minimum requirements to be included in the label information are:
- sample number, when necessary, including codes for replicates and/or subsamples;
- station number;
- date and year of collection;
- collector's name.

Additional information that may be included:
- ship and cruise number;
- codes for the matrix;
- information on the type and size of container to be used;
- codes for the type of analysis (trace metals, organics);
- (color) codes for the transport conditions (cool, deep-frozen, lighting conditions).

In addition to the labels on the containers, labels should be prepared for transport and storage.

Labeling can be performed with paper or plastic labels, which may contain preprinted information. With the introduction of bar-codes they may contain much more information. Be sure that the labels and marker pens used are of sufficient quality. The (self-adhesive) labels should stick to the container, even under damp conditions, and should stay on this container during the entire process of analysis, including cold storage! The ink should be water resistant, the labels should be able to withstand some wear during handling. For this reason one may put transparent tape over the label to protect it. Marker pens (black has the best light-resistance) should write on (damp) surfaces of bottles and packing material, be water-proof, and resist cold storage. All these aspects should be tested before actual samples are collected.

8.4.2 Forms

In addition to the sampling instructions, the sampling team should be provided with *"Sampling information forms"* and *"Field data inventory"* forms.

In case multiple samples, sometimes of a different nature, are collected during one sampling trip, the "sampling information form" will guide the sampler what actions are to be performed at a given sampling point. This form may contain (preprinted) information on location, (position, depth, shore level, etc), day of sampling, samples to be collected, storage containers to be used (material, size, and shape), number of individuals and replicates, specific instructions, etc. Each sample has a box on the form that may be ticked after completion of each task. This procedure helps the checking process when the analytical data are introduced in the data base. An example of such a form is given in Kramer et al. [6].

The "field data inventory" is used as a log-book, to note down additional information that may help later interpretation. This includes observations on the weather or water conditions, problems encountered during the sampling, reasons why certain samples have not been collected, or have been collected in duplicate, etc.

8.4.3   Transport conditions

As in all steps of the analytical process, transport conditions should in principle not interfere with the quality of the sample. This means dust-free packing and guarding against breakage. Once the samples are placed in their containers, they should be packed properly. First action is to protect the samples against dust. Plastic bags are most useful here. Seal or use tape to protect the samples during transport and storage. When samples are packed in plastic bags which are subsequently sealed, the receiver may check whether the samples have been opened by unauthorized personnel. This can be especially important when shipment of samples is required to other countries (customs).

Polystyrene "cool boxes" prevent breakage and are reasonably good insulators. There is no need to acid-clean these packing materials, but one should take care not to soil them.

Other aspects of the process of quality control include the temperature during transport and the time needed before the samples can be stored appropriately in the laboratory.

8.4.3.1   Temperature

Biological materials are subject to deterioration. The higher the temperature, the faster this process will take place or, conversely, cooling or freezing of the material will eliminate deterioration, minimize the interactions between container and the sample, and reduce bacterial activity. Storage conditions (see the next section) should, in principle, start directly after sampling. It seems rather useless to apply well controlled (cold) storage conditions when after sampling and during transport, the tissue has been subject to ambient or even elevated temperatures for days (exposed to solar radiation). Organometallic compounds are specially sensitive to elevated temperatures as they tend to be transformed easily. The faster the process of cooling/freezing takes place, and the lower the temperature, the better for maintaining sample integrity. Blast freezing is the technique applied for the preparation of foodstuffs, and this technique may be used in biomonitoring studies also [100]. Slow freezing will allow ice crystals to form in the tissue which will disrupt cells. As practical problems in the field may limit the possibilities, it is still essential to cool the samples as fast as possible to below 4 °C, after which deep freezing may take place after reception in the laboratory. An elegant method is to use a deep freezer of sufficient capacity (i.e. to handle the amount of tissue) to freeze to -20 °C within the hour, or to use dry ice (solid CO_2, -80 °C) or liquid nitrogen (-196 °C) to cool the material rapidly in a styrofoam insulating box. Put a surplus of dry-ice on the box, taking care not to seal the box completely. Several cool packs may help to keep the temperature low. For quality control reasons each shipment should contain a check on the temperature during transport, a simple min/max thermometer or, better, a digital recording device.

Bivalve and fish samples can both be deep-frozen and should be transported to the laboratory as soon as possible.

Macrophyte samples may be frozen, but because of the often fragile nature of many aquatic plants, the tissue will partly disintegrate as a result of the freezing process. Cleaning and washing thus becomes impossible after thawing. Therefore, in case the samples are to be washed later in the laboratory, cool (4 °C) transport, using cool packs, is preferred.

For ABM experiments it is essential to transport the organisms cool (not frozen) to the exposure sites in order to keep them alive. Cool packs in a cool box are very useful in this respect.

8.4.3.2 Transport time

The period between sampling and controlled storage should be as short as possible; in particular the reduced temperature conditions needed for fresh samples will not last for ever. Arrangements have therefore to be made if the sample has to be transported by post or public carrier (courier service). When properly packed, samples will stay cool for 24 hours. It is advisable to send advance warning to the laboratory that (and about when) the samples will arrive and to use an address label directing the samples to the laboratory (and not only to the institute, as they may be left at the front desk or the mail room for a considerable time). Stick a warning to the outside of the box that the samples should be stored cold after reception and that the laboratory should be informed immediately. The itinerary of the shipment should be such that the samples do not arrive at the laboratory in the evening, or worse, during weekends or public holidays. They may already have been spoiled by the time they finally reach the laboratory.

8.4.4 Storage

In principle there are two phases in the process of storage. The first is the storage of the samples as they are received from the sampling team. They have been subject to no or only minor sample treatment and need to be prepared before they may be stored for longer periods. The second phase involves the samples that are treated, preserved, freeze-dried, or treated in some other fashion (see next section). In both cases the storage conditions (temperature) should be controlled and be monitored. Power shortages should be backed up when they may occur for long periods. The rising temperature inside the freezer will define the time span that can be allowed without power.

In the first phase, directly after reception of the samples, they will in general be in a cooled or only slightly frozen condition. When the samples are to be further treated immediately (i.e. the same day) they are stored cool and allowed the thaw slowly. When the treatment will be performed at a later date, the samples are transferred to a freezer and stored at -20 °C or lower. This freezer should be different from the one used for the treated samples, as contamination by the potentially soiled samples should be prevented. Alternate freezing/thawing should be minimized as it may induce concentration gradients in the samples. Redistribution over the various organs has been observed during frozen storage [73,117].

For some compounds the light conditions also are important. Caricchia et al. [118] noted that, beside temperatures ≤ 20 °C, organotin compounds in mussel tissue were best preserved when light was excluded during storage, even for freeze dried material.

The phase of storage after sample treatment does not necessarily differ from the previous one. However, the state of the samples after treatment (method of preservation) and objectives of the storage influence this method of storage.

The strategy to be followed will also depend on the expected "shelf life". For use in the sequence of routine monitoring programs the analysis will be carried out in the weeks following sampling. Reference materials have a shelf life, or expected storage time, of several years, while "specimen banks" store fresh materials possibly over decades. Dried or sterilized material can be stored at ambient temperatures (e.g. CRMs), while fresh material for specimen banking will be best stored at deep frozen (-80 to -120 °C) under special conditions [11,119,120].

For routine monitoring programs most laboratories will store fresh material (thus deep frozen) until sample preparation starts; dissolution and instrumental analysis will follow the preparation step and no additional storage will be required. If subsamples are to be kept, e.g. for possible repetition of an independent analysis, this may be fresh homogenate stored at temperatures ≤ 20 °C or (freeze-) dried material at ambient temperatures.

8.5   Sample preparation, homogenization

This section discusses sample handling which results in a sample suitable either for long-term storage (e.g. in a specimen bank, as a reference material) or for handling over to the analyst who has to analyze the sample for trace metals. The items discussed include the important aspects of preservation and homogenization.

The analyses may be performed on wet, homogenized samples, and also on pretreated (e.g. dried) samples, where the dried material is first homogenized before analysis. There seems no reason to prefer one of these methods. Practicality will determine the procedures followed.

8.5.1   Tissue preparation

Once the samples have arrived at the laboratory for further treatment, either directly from the sampling operation or after some time of storage, they are allowed to thaw. This should be performed gently. The use of a microwave oven is advised against, as the tissue will not be treated homogeneously. Some parts may be cooked while others are still frozen. Leaving the samples overnight will in most cases be sufficient for thawing. The containers (plastic bags) must be kept closed to prevent evaporation of moisture.

After opening of the packing material/containers in a clean environment, the organisms are cleaned on the outside if this appears to be necessary to remove adhering sand or silt, epibiota, or other material. They are washed with high purity water and allowed to drip dry. Further treatment *must* be carried out in a clean bench atmosphere. The use of different vessels in the process of handling should be minimized.

8.5.1.1   Bivalves

Most bivalves will contain some water between the tissue (mantle) and the shell. This water is not really a part of the tissue and may be discarded. The posterior adductor muscle is severed by cutting with a (titanium) knife. The adhering water is allowed to drain for a standard period (20 min) on a slightly tilting glass plate. The soft tissue is removed from the shell and collected in an acid precleaned vessel. Selected organs may be collected in this stage. For pooled samples, all tissue from one batch should be collected for further treatment (homogenization, preservation). Total wet-weight is recorded.

8.5.1.2 Fish

Some homogenization techniques are based on the handling of frozen fish (see later). When selected parts or organs are to be collected, these parts are prepared using a noncontaminating knife. The tissue has to be collected from prescribed areas and be of defined size. Organs may be collected as a whole, but the fillet will result in a batch of tissue too large to handle effectively. Adhering water is allowed to drain, and the sub-samples are pooled according to their origin. The samples are ready for further treatment.

8.5.1.3 Macrophytes

After cleaning the parts of the aquatic plants are allowed to drain on tissue paper. It is essential that of all plants that were collected as one batch about the same amount is collected for pooling. As the plants may differ substantially in size, a sub sample will be needed. The sample-handling procedure should define, for the given species and batch-size, how much material has to be taken to form a sample. The parts are collected in an acid cleaned container for further treatment [121].

8.5.2 Preservation

The intention of preservation or stabilization is to convert the samples into such a form that they can be considered as relatively stable and thus able to be stored for a considerable time without disturbing changes in the trace metal content. The procedures followed should be carried out in such a way that the trace elements of interest are neither eliminated (loss by evaporation of volatile elements, transformation of organometallics) nor enriched (contamination). In addition to the method of deep freezing (-20 °C, -80 °C or lower), already discussed under the section on storage, a number of other options are available as preservation methods:
- addition of chemical preservatives;
- sterilization;
- oven drying;
- freeze drying (lyophilization);
- low temperature ashing;
- high temperature ashing.

These methods all enable the storage of samples after treatment without the need for a low-temperature regime. All these methods do change the matrix of the tissue, most by reducing the moisture content to <1 %. This may result in a matrix that is simpler to digest. For example, the carapace skin of crustacea is sometimes difficult to digest by standard methods. Treatment by oven-drying and homogenization in a mortar results in a matrix much easier to digest. However, only few methods appear to be suitable for volatile elements and/or organometallic compounds.

8.5.2.1 Addition of preservatives

Preservation by adding chemicals, e.g. formaldehyde (formalin) or glutaraldehyde, or fixation using "Helly's Fixative" or "Dietrich's Fixative" is intended to keep the tissue from deterioration [122]. It is a procedure useful for taxonomic or histological studies, but not for determination of trace elements in tissue. The chemical change the matrix and are of insufficient purity. No chemicals should be added to samples that are to be analyzed for trace metals (or organic micropollutants) [20,99]. If additional histologic or taxonomic studies have to be carried out, separate samples need to be collected and treated accordingly [6].

8.5.2.2 Irradiation sterilizing

Although the technique is hampered by the limited availability of irradiation equipment, sterilization by (γ)-radiation offers a useful technique when fresh samples are needed. This is especially the case when fresh samples are analyzed routinely. Comparison of the analytical performance with, e.g., that obtained with freeze dried material is not following the idea of analyzing similar matrices. A drawback of the method is that high level irradiation alters the structure of the tissue and may breakdown the more labile organic material, including organometallic compounds. For the latter group this method should not be used. Since the sterile phase will not be kept when the sample is further processed the preservation is only effective if applied when the homogenization has been performed and the samples are to be stored (or it should be performed before and after the homogenization step).

8.5.2.3 Oven drying

Although this can be considered the most common technique, also for the determination of the "dry-weight" (DW), there is surprisingly little agreement about the methodology. It will of course depend on the tissue to be dried, but even with the same type of organism or even species, differences are found in the temperatures used (e.g. 60 °C, 90 °C, 110 °C) or in the time allowed for the drying process (10 h, 24 h, 40 h, 48 h). The higher the temperature the greater the risk of losing volatile compounds. We would, therefore, not recommend this method for volatile elements (such as Hg) or organometallic compounds. Practicality often dictates the procedure, as the methods applied seem to aim at shortening the time span of the drying action. Differences do occur between the methods [123]. The rule is to dry until constant weight, but the procedure should be tested and validated for the given species (and amount of material).

8.5.2.4 High temperature ashing

This method is applied for the determination of the ash-free dry-weight (AFDW). The sample is placed in porcelain, fused silica or platinum crucibles. It is first dried following the DW procedure. Then the sample is incinerated at 500 °C in a muffle furnace for 4-8 h. Above a temperature of 650 °C carbonates are volatilized influencing the weight of the material [6]. This method is only advised for the determination of AFDW. For further analysis of trace elements in biological matrices this method is rather limited.

8.5.2.5 Freeze drying (lyophilization)

To safeguard against pollutant loss by volatilization at elevated temperatures it is possible to freeze-dry the material. The essence is to extract the water at reduced atmospheric pressure. Despite its elegance and proven applicability to many elements, (rapid) lyophilization may cause loss of the more volatile compounds (including mercury, some organometallic species, organics), thus careful testing of the settings of the instrument (vacuum, time) is advised [5,11,124].

8.5.2.6 Low temperature ashing (LTA)

LTA-treated samples are easy to composite and to homogenize. The method involves ashing procedures at reduced pressure, where (usually) oxygen is activated by a high-frequency electromagnetic field (27.12 MHz). The temperatures observed range between 100-200 °C. The ashing is rather slow but can be enhanced by using PTFE crucibles. For larger quantities the method is not very suitable. In an early study, over 30 trace elements were retained in the sample quantitatively. Exceptions were, e.g., Hg and Os [125].

8.5.3 Homogenization

8.5.3.1 Fresh tissue

8.5.3.1.1 Blender

In many laboratories the digestion of fresh tissue offers a simple technique which is also possible for the volatile compounds. The method can be used for most tissues that do not contain chitin like parts that hamper proper digestion.

Concentration on the basis of dry-weight is based on a separate DW determination and recalculation.

For fresh material, blenders such as Turrax, Polytron, Tissuemizer type are used. Since most rotating knives are made of stainless steel, they should be replaced by a high purity material that is not of analytical interest (Ti, W or other). The sample size and the blender should match. The time (and rotating speed) of the homogenization should be standardized, and be for at least 5 min to ensure homogeneity.

8.5.3.1.2 Cryogenic homogenization technique

An efficient and clean technique suitable in principle for the homogenization of practically all types of biological sample (including, e.g., whole fish and hard parts) was introduced by Iyengar and Kasperek [126]. In this brittle fracture technique (BFT) the sample is placed in a lidded Teflon container along with a Teflon-coated steel ball. The container plus sample are cooled with liquid nitrogen and subsequently vibrated at a defined high frequency and period of time. The method incorporates a number of desirable features: it protects heat- sensitive compounds, it yields a fine powdered material, the nature of the used materials is clean, and it enables a subsequent freeze drying step [127,128].

A variation is the use of a cryogenic Teflon ball mill, with similar advantages. The latter technique enables the use of larger quantities, such as for the preparation of reference materials (e.g. used for trace metals in fish muscle [2]). In order to have the samples in sizes that can be handled by the Teflon vibrator/mill, some use a cryogenic fracture technique for initial size reduction. The (large) sample is placed in a heavy duty polythene bag, immersed in liquid nitrogen and allowed to reach a low temperature. After retrieval, the bag is hammered with the result that the sample breaks up into useable blocks [129].

8.5.3.2 Dried material

Whether the material is from the process of oven/freeze-drying, or (low temperature) ashing, homogenization is easily performed by grinding using a Teflon or tungsten carbide pestle and mortar. The grinding should be carried out inside the clean bench. It will prove to be almost impossible to have a homogeneous grain size distribution at the end. Sieving of the ground material over a nylon sieve may remove the larger particles, but this may cause biased results as we can not be sure whether all particle sizes have the same trace metal content. The result of the grinding step, and whether additional measures are needed will depend on the original matrix. In case of doubt one can always analyze either the total sample or check on the differences in various subsamples.

8.5.3.3 Testing for homogeneity

The methods used for homogenization should be verified. Both the variation between and the variation within subsamples (bottles) should be tested. Ideally this procedure needs to be carried out for each (type of) organism. For the intra bottle homogeneity study 10 subsamples are collected from one bottle, for the inter homogeneity study 10 different bottles are analyzed. One way analysis of variance (ANOVA or F-test) should be used to evaluate the inter- and intra bottle homogeneity.

8.6 Additional measurements - Units of expression

From the preparation steps discussed above, the wet weight (WW), dry weight (DW), or ash-free dry weight (AFDW) can be used for the conversion of units either to grams of trace metal per gram tissue or to body burden (gram of trace element per individual). Differences in presentation can be interesting: a seeming decrease in concentration based on DW, may show a continuous uptake when plotted as body burden (Figure 8-5).

For the interpretation of the results additional information may be required, for example to relate (adjust) the trace metal findings to physiological effects. It should be realized that most additional variables need to be measured before tissue preparation starts. Since the organisms will be analyzed for trace elements later, contamination should be minimized and the measurements need to be carried out in the (clean) laboratory under controlled conditions.

8.6.1 Bivalves

Bivalve studies have shown a specially large variety of additional variables that are used for the interpretation of the monitoring results [86,102,130,131]. They include basic variables such as:
- shell length, used to separate into size classes;
- soft tissue weight;
- DW/WW ratio.

In addition the following measurements may be required:
- shell height and width;
- total weight (soft tissue and shell);
- flesh condition (mg soft tissue/g shell);
- condition index (DW/shell length);

8.6.2 Fish

For fish usually only the following variables are determined:
- length (and length classes);
- total weight or organ weight.

Additional are [132]:
- sex;
- DW/WW ratio;
- age (otoliths);
- lipid content (for organic micropollutant studies only).

The unit of expression for trace metals in fish tissue is per gram WW. The water content of fish tissue is much less affected by the condition of the fish than it is in mussel tissue. There seems no need to report on the basis of DW. Next to DW-based data, the WW-based results should be given (or the DW/WW ratios).

8.6.3 Macrophytes

Aquatic plants are sampled in total or only parts are taken. Typical analyses include:
- (blotted) wet weight;
- dry weight.

The water content of macrophytes may change considerably, therefore results are to be expressed as per unit DW.

8.7 Validation and protocols

It will be evident from the previous sections that many (combinations of) methods are applied in the sampling and analysis of biological material. Not only do techniques differ between various groups of organisms, large differences can even be found between methodologies used for the same species. It is most likely that this will lead to differences in the results. In view of the ongoing internationalization of monitoring programs, and consequently the need for comparison of results, this situation is rather unfortunate.

In a context of adaptation to proven methods, everyone will agree that one should use the same method for a given aspect of the analysis, provided only that the method selected is his own method. Although one may try to define internationally accepted "standard" procedures, the reality will be that universal acceptance is very hard to attain. The other, probably more realistic, option is to intercompare methods and techniques. When different techniques produce the same results there is no problem, and validation can be achieved.

Another question is the training of staff, in the laboratory but certainly also in the field. Similarly to the laboratories operating under GLP procedures, there should be a quality control system for the sampling and sample handling steps. Procedures and protocols should be documented, and the documentation should be available to the staff collecting and/or handling the samples. Today contamination (and general quality) control is probably more of a problem during the initial steps in the analysis than in the often well developed laboratory procedures. A general awareness of GMP - good monitoring practice - for the entire stretch of the sequential steps in the analytical process, will undoubtedly result in better data that can be better interpreted and thus fulfil better the goals and objectives of a monitoring program.

8.8 References

[1] E.I. Hamilton, *Mar. Pollut. Bull.*, **22**, 51 (1991)
[2] Ph. Quevauviller, G.N. Kramer and B. Griepink, *Mar. Pollut. Bull.*, **24**, 601 (1992)
[3] S.J. Reddy, M. Froning, C. Mohl and P. Ostsapczuk, *Sci. Total Environ.*, **140**, 437 (1993)
[4] K.J.M. Kramer, *Mar. Pollut. Bull.*, **29**, 222 (1994)
[5] J.R. Moody, *Phil. Trans. R. Soc. Lond.*, **A 305**, 669 (1982)
[6] K.J.M. Kramer, U.H. Brockmann and R.M. Warwick, *Tidal estuaries: Manual on sampling and analytical procedures*, Balkema Publ., Rotterdam, pp. 314 (1994)
[7] E.E Kenaga, *Environ. Sci. Technol.*, **12**, 1322 (1978)
[8] A.P. Vinogradov, *The elementary chemical composition of marine organisms*, New Haven, USA, pp. 647 (1953)
[9] R. Eisler, *Trace metal concentrations in marine organisms*, Pergamon Press, New York, pp. 687 (1981)
[10] P. Hagel, *Environ. Mon. Assess.*, **7**, 257 (1986)
[11] N.-P. Luepke (Ed.), *Monitoring environmental materials and specimen banking*, Proc. Int. Workshop Berlin, 23-28 Oct. 1978, Martinus Nijhoff, The Hague, pp. 591 (1979)
[12] D.J.H. Phillips, *Quantitative aquatic biological indicators - their use to monitor trace metal and organochlorine pollution*, Applied Science Publ., London, pp. 488 (1980)
[13] S.K. Sarkar, B. Bhattacharya and S. Debnath, *Chemosphere*, **29**, 759 (1994)
[14] A.Y. Cantillo, *Mussel Watch worldwide literature survey - 1991*, NOAA Technical memorandum NOS ORCA 63, Rockville, MD, pp. 142 (1991)
[15] T.P. O'Connor, A.Y. Cantillo and G.G. Lauenstein, *In: Biomonitoring of coastal waters and estuaries*, K.J.M. Kramer (Ed.), CRC Press, Boca Raton, pp. 29-50 (1994)
[16] UNESCO/IOC, *Standard and Reference materials for marine science*, Rev. Ed. IOC manuals and guides No. 25, UNESCO, Paris, pp. 577 (1993)
[17] C. Barghigiani and S. De Ranieri, *Mar. Pollut. Bull.*, **24**, 114 (1992)
[18] J. Hellou, W.G. Warren, J.F. Payne, S. Belkhode and P. Lobel, *Mar. Pollut. Bull.*, **24**, 452 (1992)
[19] R.T. Leah, S.E. Collins, M.S. Johnson and S.J. Evans, *Mar. Pollut. Bull.*, **26**, 436 (1993)
[20] R.H. Gibbs, E. Jarosewich and H.L. Windom, *Science*, **184**, 475 (1974)
[21] B.B. Colette, *Specimen banking marine organisms. Monitoring environmental materials and specimen banking*, N.-P. Luepke (Ed.), Martinus Nijhoff, The Hague, pp. 165 (1979)
[22] D.J.H. Phillips, *In: Biomonitoring of coastal waters and estuaries*, K.J.M. Kramer (Ed.). CRC Press, Boca Raton, pp. 85-106 (1994)
[23] P.J. Say, J.P.C. Harding and B.A. Whitton, *Environ. Pollut.*, **2**, 295 (1981)
[24] D.C. Mortimer, *Environ. Mon. Assess.*, **5**, 311 (1985)
[25] N. Oertel, *Verh. Internat. Verein. Limnol.*, **24**, 1961 (1991)
[26] J. Mersch, and J.C. Pihan, *Arch. Environ. Contam. Toxicol.*, **26**, 353 (1994)
[27] P.E. Gibbs, and G.W. Bryan, *In: Biomonitoring of coastal waters and estuaries*, K.J.M. Kramer (Ed.), CRC Press, Boca Raton, pp. 205-226 (1994)

[28] P.T.E. Ozoh, *Environ. Mon. Assess.*, **29**, 155 (1994)
[29] J.M. Everaarts, *Neth. J. Sea Res.*, **20**, 253 (1986)
[30] A.M. Gunn, D.T.E. Hunt and D.A. Winnard, *In: Environmental bioassay techniques and their application*, M. Munawar, G. Dixon, C.I. Mayfield, T. Reynoldsen and M.H. Sadar (Eds.), Kluwer Academic Publishers, Dordrecht pp 487-496 (1989)
[31] T.B. Reynoldson, S.P. Thompson and J.L. Bamsey, *Environ. Toxicol. Chem.*, **10**, 1061 (1991)
[32] A.R. Flegal, and J.H. Martin, *Mar. Pollut. Bull.*, **8**, 90 (1977)
[33] P.B. Lobel, S.P. Belkhode, S.E. Jackson and H.P. Longerich, *Mar. Environ. Res.*, **31**, 163 (1991)
[34] V.D. Sidwell, R.L. Lomis, K.J. Lomis, P.R. Foncannon and D.H. Buzzel, *Mar. fish. Rev.*, **40**, 1 (1978)
[35] N. Peerzada, M. Nojok and C. Lee, *Mar. Pollut. Bull.*, **24**, 416 (1992)
[36] C.R. Sivadasan and P.N.K. Nambisan, *Mar. Pollut. Bull.*, **19**, 579 (1988)
[37] S.J. Palmer and B.J. Presley, *Mar. Pollut. Bull.*, **26**, 564 (1993)
[38] J.M. Everaarts, E. Otter and C.V. Fischer, *Neth. J. Sea Res.*, **26**, 75 (1990)
[39] H.M. Chan, P.S. Rainbow and D.J.H. Phillips, *In: Proc. 2nd Int. Marine Biological Workshop: The marine Flora and fauna of Hong Kong and southern China*, B. Morton (Ed.), Hong Kong Univ. Press, Hong Kong, pp. 1240-1268 (1986)
[40] D.J.H. Phillips and P.S. Rainbow, *Mar. Ecol. Prog. Ser.*, **49**, 83 (1988)
[41] L. Karbe, M. Dembinski, J. Gonzalez-Valero, M. Mueller and R. Zeitner, *Arb. dt. Fischereiverb.*, **48**, 95 (1989)
[42] H.M. Chan, P. Bjerregaard, P.S. Rainbow and M.H. Depledge, *Mar. Ecol. Prog. Ser.*, **86**, 91 (1992)
[43] J.L. Andersen and M.H. Depledge, *Mar. Biol.*, **118**, 285 (1994)
[44] A.R.D. Stebbing, *ICES Report*, **179**, 310 (1980)
[45] T.M.E. Olesen and J.M. Weeks, *Bull. Environ. Contam. Toxicol.*, **52**, 722 (1994)
[46] R.J. Law, B.R. Jones, J.R. Baker, S. Kennedy, R. Milne and R.J. Moris, *Mar. Pollut. Bull.*, **24**, 296 (1992)
[47] T.I. Lillestolen, N. Foster and S.A. Wise, *Sci. Total Environ.*, **140**, 97 (1993)
[48] R.J. Law, C.F. Fileman, A.D. Hopkins, J.R. Baker, J. Harwood, D.B. Jackson, S. Kennedy, A.R. Martin and R.J. Morris, *Mar. Pollut. Bull.*, **22**, 183 (1991)
[49] R. Nagel and R. Loskill Eds., *Bioaccumulation in aquatic systems. Contributions to the assessment*, VCH Verlag, Weinheim, pp. 239 (1991)
[50] W.C. De Kock and K.J.M. Kramer, *In: Biomonitoring of coastal waters and estuaries*, K.J.M. Kramer (Ed.). CRC Press, Boca Raton, pp. 51-84 (1994)
[51] K.J.M. Kramer, *Int. J. Environ. Anal. Chem.*, **57**, 179 (1994)
[52] E.D. Goldberg, *Mar. Pollut. Bull.*, **6**, 111 (1975)
[53] E.D. Goldberg, V.T. Bowen, J.W. Farrington, G. Harvey, J.H. Martin, P.L. Parker, R.W. Risebrough, W. Robertson, E. Schnieder and E. Gamble, *Environ. Conserv.*, **5**, 101 (1978)
[54] D. Claisse, *Mar. Pollut. Bull.*, **20**, 523 (1989)
[55] S.F. Thrush, R.D. Pridmore and J.E. Hewitt, *Ecol. Applic.*, **4**, 31 (1994)
[56] W.C. De Kock, *Can. J. Fish. Aquat. Sci.*, **40**, 282 (1983)

[57] W.C. De Kock, *Environ. Mon. Assessm.*, **7**, 209 (1986)
[58] D.R. Smith, M.D. Stephenson and A.R. Flegal, *Environ. Toxicol. Chem.*, **5**, 129 (1986)
[59] D. Martincic, Z. Kwokal, Z. Peharec, D. Margus and M. Branica, *Sci. Total Environ.*, **119**, 211 (1992)
[60] L. Karbe and M. Hablizel, *Mitt. Nieders. Landesamt. f. Wasser und Abfall*, Heft 1 (1990)
[61] J. Mersch and L. Johansson, *Environ. Technol.*, **14**, 1027 (1993)
[62] M. Camusso, R. Balestrini, F. Muriano and M. Mariani, *Chemosphere*, **29**, 729 (1994)
[63] D.R. Green, J.K. Stull and T.C. Heesen, *Mar. Pollut. Bull.*, **17**, 324 (1986)
[64] M.M. Boom, *Intern. J. Environ. Anal. Chem.*, **13**, 251 (1987)
[65] S. Hervé, P. Heinonen, R. Paukku, M. Knuutila, J. Koistinen and J. Paasivirta, *Chemosphere*, **22**, 1945 (1988)
[66] H. Hummel, R.H. Bogaards, J. Nieuwenhuize, L. De Wolf and J.M. Van Liere, *Sci. Total Environm.*, **92**, 155 (1990)
[67] T. Aunaas and K.E. Zachariassen, in: *Biomonitoring of coastal waters and estuaries*, K.J.M. Kramer (Ed.), CRC Press, Boca Raton, pp. 107-133 (1994)
[68] K.J.M. Kramer (Ed.,) *Biomonitoring of coastal waters and estuaries*, CRC Press, Boca Raton, pp. 362 (1994)
[69] L. Förlin, A. Goksoyr and A.M. Husoy, in: *Biomonitoring of coastal waters and estuaries*, K.J.M. Kramer (Ed.), CRC Press, Boca Raton, pp. 135-150 (1994)
[70] S.G. George and P.E. Olssen, in: *Biomonitoring of coastal waters and estuaries*, K.J.M. Kramer (Ed.), CRC Press, Boca Raton, pp. 151-178 (1994)
[71] J.R. Moody and R.M. Lindstrom, *Anal. Chem.*, **49**, 2264 (1977)
[72] G.L. Fisher, L.G. Davies and L.S. Rosenblatt, in: *Accuracy in trace analysis: sampling, sample handling and analysis*, P.D. LaFleur (Ed.), NBS special publication No. 422. NBS, Gaithersburg Md, pp. 575-591 (1976)
[73] R.A. Durst, in: *Monitoring environmental materials and specimen banking*, N.-P. Luepke (Ed.), Martinus Nijhoff, The Hague, pp. 198-202 (1979)
[74] M. Heit and C.S. Klusek, *Sci. Total Environ.*, **24**, 129 (1982)
[75] C.J. Schmitt S.E. Finger, *Arch. Environ. Cont. Toxicol.*, **16**, 185 (1987)
[76] C.C. Patterson and D.M. Settle, in: *Accuracy in trace analysis: sampling, sample handling and analysis*, P.D. LaFleur (Ed.), NBS special publication No. 422. NBS, Gaithersburg Md, pp. 321-351 (1976)
[77] D.P.H. Laxen and R.M. Harrison, *Anal. Chem.*, **53**, 345 (1981)
[78] W.A. Maher, P.W. Cullen and R.H. Norris, *Environ. Mon. Assess.*, **30**, 139 (1994)
[79] A. Speecke, J. Hoste and J. Versieck, in: *Accuracy in trace analysis: sampling, sample handling and analysis*, P.D. LaFleur Ed., NBS special publication No. 422. NBS, Gaithersburg Md, pp. 299-310 (1976)
[80] M. Stoeppler, in: *Monitoring environmental materials and specimen banking*, N.-P. Luepke (Ed.), Martinus Nijhoff, The Hague, pp. 555-572 (1979)
[81] UNEP/FAO/IAEA/IOC, In: *UNEP/FAO/IAEA/IOC Reference methods for marine pollution studies*, No. 7, Rev. 2, UNEP, Nairobi, pp. 15 (1984)
[82] EPA, *Methods for the determination of metals in environmental samples*, C.K. Smoley (Ed.), Boca Raton Fl. pp. 339 (1992)

[83] F.G. Lowman, in: *Monitoring environmental materials and specimen banking*, N.-P. Luepke (Ed.), Martinus Nijhoff, The Hague, pp. 392-396 (1979)
[84] S.A. Nielsen, *N.Z. Jl. mar. Freshw. Res.*, **8**, 631 (1974)
[85] D.R. Young, T.C. Heesen and D.J. McDermott, *Mar. Pollut. Bull.*, **7**, 156 (1976)
[86] P.B. Lobel, S.P. Belkhode, S.E. Jackson and H.P. Longerich, *Mar. Biol.*, **102**, 513 (1989)
[87] P.B. Lobel, C.D. Bajdik, S.P. Belkhode, S.E. Jackson and H.P. Longerich, *Arch. Environ. Contam. Toxicol.*, **21**, 409 (1991)
[88] L.H. Atsatt and R.R. Seapy, *J. Exp. Mar. Biol. Ecol.*, **14**, 261 (1974)
[89] Anon, *Guidelines to be followed for sample collection, preparation and analysis of fish and shellfish in the conduct of of cooperative monitoring*, ICES JP/MM C3a/C8d/C8h, Charlottenlund (1982)
[90] S.W. Fowler, in: *Monitoring environmental materials and specimen banking*, N.-P. Luepke (Ed.), Martinus Nijhoff, The Hague, pp. 247-260 (1979)
[91] G.W. Bryan, W.J. Langston, L.G. Hummerstone and G.R. Burt, *A guide to the assessment of heavy metal contamination in estuaries using biological indicators*, Marine Biological Association of the United Kingdom, Special Publication No. 4 (1985)
[92] E.P.R. Goncalves, H.M.V.M. Soares, R.A.R. Boaventura, A.A.S.C. Machado and J.C.G.E. Da Silva, *Sci. Total Environ.*, **142**, 143 (1994)
[93] J. Stronkhorst, *Mar. Pollut. Bull.*, **24**, 250 (1992)
[94] Y.D. Latouche and M.C. Mix, *Mar. Pollut. Bull.*, **13**, 27 (1982)
[95] D.J.H. Phillips, *Mar. Ecol. Prog. Ser.*, **21**, 251 (1985)
[96] M. Gordon, G.A. Knauer and J.H. Martin, *Mar. Pollut. Bull.*, **11**, 195 (1980)
[97] C.R. Boyden and D.J.H. Phillips, *Mar. Ecol. Prog. Ser.*, **5**, 29 (1981)
[98] D. Wright and J. Mihursky, *Mar. Environ. Res.*, **16**, 181 (1985)
[99] J.L. Ludke, in: *Monitoring environmental materials and specimen banking*, N.-P. Luepke (Ed.), Martinus Nijhoff, The Hague, pp. 397-402 (1979)
[100] R.T. Leah, S.J. Evans and M.S. Johnson, *Mar. Pollut. Bull.*, **24**, 544 (1992)
[101] J.C. Amiard, C. Amiard-Triquet, B. Berthet and C. Métayer, *Mar. Biol.*, **90**, 425 (1986)
[102] D. Martincic, M. Stoeppler and M. Branica, *Sci. Total Environm.*, **60**, 143 (1987)
[103] P.T. Lakshmanan and P.N.K. Nambisan, *Bull. Environ. Cont. Toxicol.*, **43**, 131 (1989)
[104] T.L. Coombs and P.J. Keller, *Aquat. Toxicol.*, **1**, 291 (1981)
[105] M. Koide, D.S. Lee and E.D. Goldberg, *Est. Coast. Shelf Sci.*, **15**, 679 (1982)
[106] B. Carell, S. Forberg, E. Grundelius, L. Henrikson, A. Johnels, U. Lindh, H. Mutvei, M. Olsson, K. Svardstrom and T. Westermark, *Ambio*, **16**, 2 (1987)
[107] M.J. Imlay, *Malacol. Rev.*, **15**, 1 (1982)
[108] R.J. Clifton, H.R. Stevens and E.I. Hamilton, *Mar. Ecol. Progr. Ser.*, **11**, 245 (1983)
[109] A. Mitra and A. Choudhury, *Mar. Pollut. Bull.*, **26**, 521 (1993)
[110] NAS, *The international Mussel Watch*, National Academy of Sciences, Washington D.C., pp. 148 (1980)
[111] J. Coimbra, S. Carraca and A. Ferreira, *Mar. Pollut. Bull.*, **22**, 249 (1991)
[112] J.D. Wehr, A. Empain, C. Mouvet, P.J. Say and B.A. Whitton, *Wat. Res.*, **17**, 985 (1983)

[113] M.A. Holmes, M.T. Brown, M.W. Loutit and K. Ryan, *Mar. Environ. Res.*, **31**, 55 (1991)
[114] P.B. Lobel and D.A. Wright, *Mar. Biol.*, **66**, 231 (1982)
[115] D.W. Engel and W.G. Sunda, In: *Biological monitoring of marine pollutants*, F.J. Vernberg, A. Calabrese, F.P. Thurberg and W.B. Vernberg Eds., Acad. Press (1981)
[116] G.E. Batley, in: *Trace element speciation: analytical methods and problems*, G.E. Batley (Ed.), CRC Press, Boca Raton FL, pp. 1-24 (1989)
[117] K.A. Francesconi, E.J. Moore and L.M. Joll, *Aust. J. Mar. Freshw. Res.*, **44**, 787 (1993)
[118] A.M. Caricchia, S. Chiavarini, C. Cremisini, R. Morabito and R. Scerbo, *Anal. Chim. Acta*, **286**, 329 (1994)
[119] R.A. Lewis, N. Stein and C.W. Lewis Eds., *Environmental specimen banking and monitoring as related to banking*, Martinus Nijhoff Publ. The Hague (1984)
[120] S.A. Wise, B.J. Koster, J.K. Langland and R. Zeisler, *Sci. Total Environ.*, **140**, 1 (1993)
[121] M.C. Freitas, R. Cornelis, F. Decorte and L. Mees, *Sci. Total Environ.*, **130**, 109 (1993)
[122] P.P. Yevich and C.A. Yevich, in: *Biomonitoring of coastal waters and estuaries*, K.J.M. Kramer (Ed.), CRC Press, Boca Raton, pp. 179-204 (1994)
[123] C. Mo and B. Neilson, *Wat. Res.*, **28**, 243 (1994)
[124] P.D. LaFleur, *Anal. Chem.*, **45**, 1534 (1973)
[125] G.J. Lutz, J.S. Stemple and H.L. Rook, *J. Radioanal. Chem.*, **39**, 277 (1977)
[126] G.V. Iyengar and K. Kasparek, *J. Radioanal. Chem.*, **38**, 301 (1977)
[127] R. Ziesler, J.K. Langland and S.H. Harrison, *Anal. Chem.*, **55**, 2431 (1983)
[128] T.W. May and M.L. Kaiser, *J. Ass. Off. Anal. Chemists*, **67**, 589 (1984)
[129] J.A. Nichols and L.R. Hageman, *Anal. Chem.*, **51**, 1591 (1979)
[130] H. Fischer, *Mar. Ecol. Progr. Ser.*, **12**, 59 (1983)
[131] C. Nolan and H. Dahlgaard, *Mar. Ecol. Prog. Ser.*, **70**, 165 (1991)
[132] A. Jensen and Z. Cheng, *Mar. Pollut. Bull.*, **18**, 230 (1987)

9.

Quality assurance of plant sampling and storage

B. Markert
Internationales Hochschulinstitut Zittau, Lehrstuhl für Umweltverfahrenstechnik, Markt 23, D-02763 Zittau, Germany

Plants have always been considered a particularly interesting and important subject of scientific research. This is even apparent in the works of Aristotle and Theophrastos on Greek biology, which deal purely with plant systems. Later this aspect was found in conjunction with systematic/evolutionary studies in the work of Linne, in yield management and agricultural chemistry aimed at producing food for man on the basis of Justus von Liebig's "law of minima", or as a general subject of basic biological/chemical research at the level of molecular biology as in the case of Calvin and Krebs. In the last 30 years of plant research, the interest in new scientific findings has increased exponentially. There are three main reasons for this:

1. Plants account for over 99 percent of the total biomass of the earth. In other words, the greater part of all life on earth consists of plants and not of 5 to 6 billion people as one might at first mistakenly assume. The protection and preservation of individual species and the diversity of species, especially in tropical and subtropical regions, has become one of the most urgent political and ecological demands at both national and international levels.

2. Plants are responsible for the most important reaction on earth: photosynthesis, by which water and carbon dioxide are converted into sugar molecules in a complex reaction using sunlight as energy, oxygen being produced at the same time. Without the breakdown of carbohydrates and the oxygen required for synthesis in the respiratory chain, life on earth would be impossible. A system degenerating into inbalance, aggravated by steadily increasing CO_2 emissions and simultaneous ruthless destruction of the tropical rain forest, in particular, is the cause of the greenhouse effect - a major ecological and economic challenge to mankind worldwide.

3. In the context of food chains, plants are an important link between the atmosphere and the soil on the one hand and consumers of the first to the "n"-th order (animals and human beings) on the other, in respect of both material and energy flows. Pollutant substances often enter the food chain through the uptake channel "plant", and by accumulation and exclusion processes cause irreversible damage to individual organisms or whole communities.

Point 3, especially, makes great demands on analytical chemistry. The development of more and more sensitive analytical methods has made it possible to quantify practically all the elements of the periodic table in vegetable material [1-3]. Collaboration with associated scientific disciplines such as geology, physics, and medicine has led to a tremendous upswing in biological trace element research as a whole. Roughly speaking it can be assumed that about every two years the status of one of the chemical elements in the periodic tables will change. It may be found that a certain element which was once thought to be non-essential in character is in fact essential, i.e. the function it exercises is necessary to life; or an element that is used to be classified as nontoxic now proves to be toxic to organisms; or an element that used not to be accessible in plants by earlier analytical techniques is now measurable. A classic example of this is the "history" of selenium. In 1930 selenium and its compounds were generally thought to be highly toxic; from 1943 they were also considered carcinogenic. In 1957 selenium was realized to be essential to the life of some organisms, and in 1966 certain selenium compounds were successfully used in the treatment of cancer. In 1990 the study group of Professor Brätter in Berlin unequivocally characterized and isolated a second selenium protein and described its function.

Through modern trace element research we know today that virtually all elements occur in plant organisms. We also find that the concentrations of certain elements are usually lower in plants than in the earth's crust, which may be regarded as a natural reservoir for most of these elements (Fig.9-1). Exceptions here are a number of nonmetals such as C, H, N, P, S and B, as these are often structural elements building up the plant organism. If we plot the concentrations in the earth's crust against those in plants on a graph, as shown in Fig.9-2, the elements are seen to fall into three groups which have emerged during the evolution of plants and animals. The elements in Group 1, for example Ca, Mg, Mn, Ni and Cr, are clearly of an essential nature. The elements in Group 2 (As, Pb, Cd, etc.) are those whose actual function in the organism has hardly been described adequately so far (with the exception of Se, Mo, and Sn), but which may have a toxic effect at low concentrations and in the form of certain compounds. The elements in Group 3, which include the lanthanides, are thought not to have a biological function in the plant organism.

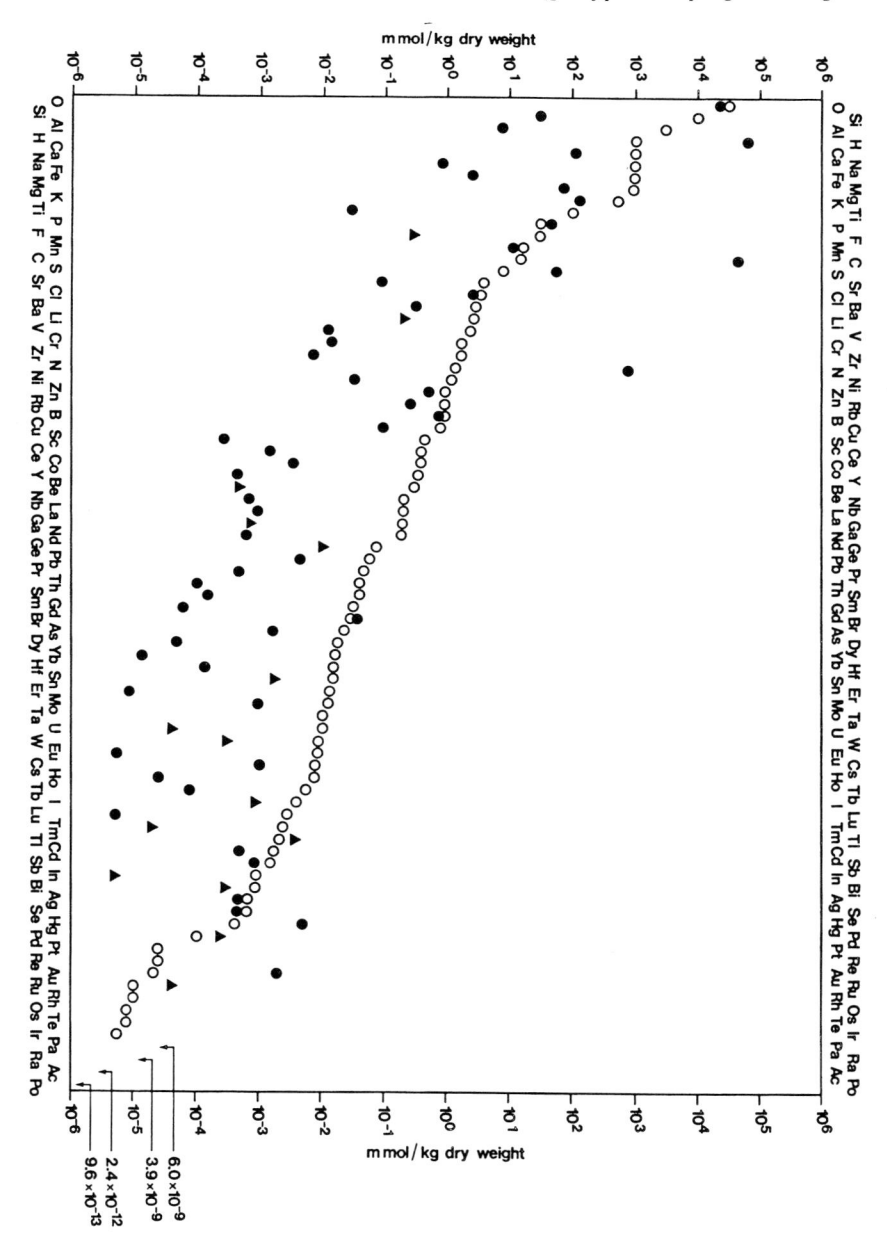

Figure 9-1: Concentration of 82 naturally occurring elements (excluding the inert gases) in plants and the earth's crust (from [65]). ○: concentration in the earth's crust (from [66]); ●: concentration in plants (after [67]); ♦: concentration in plant material (after [66]).

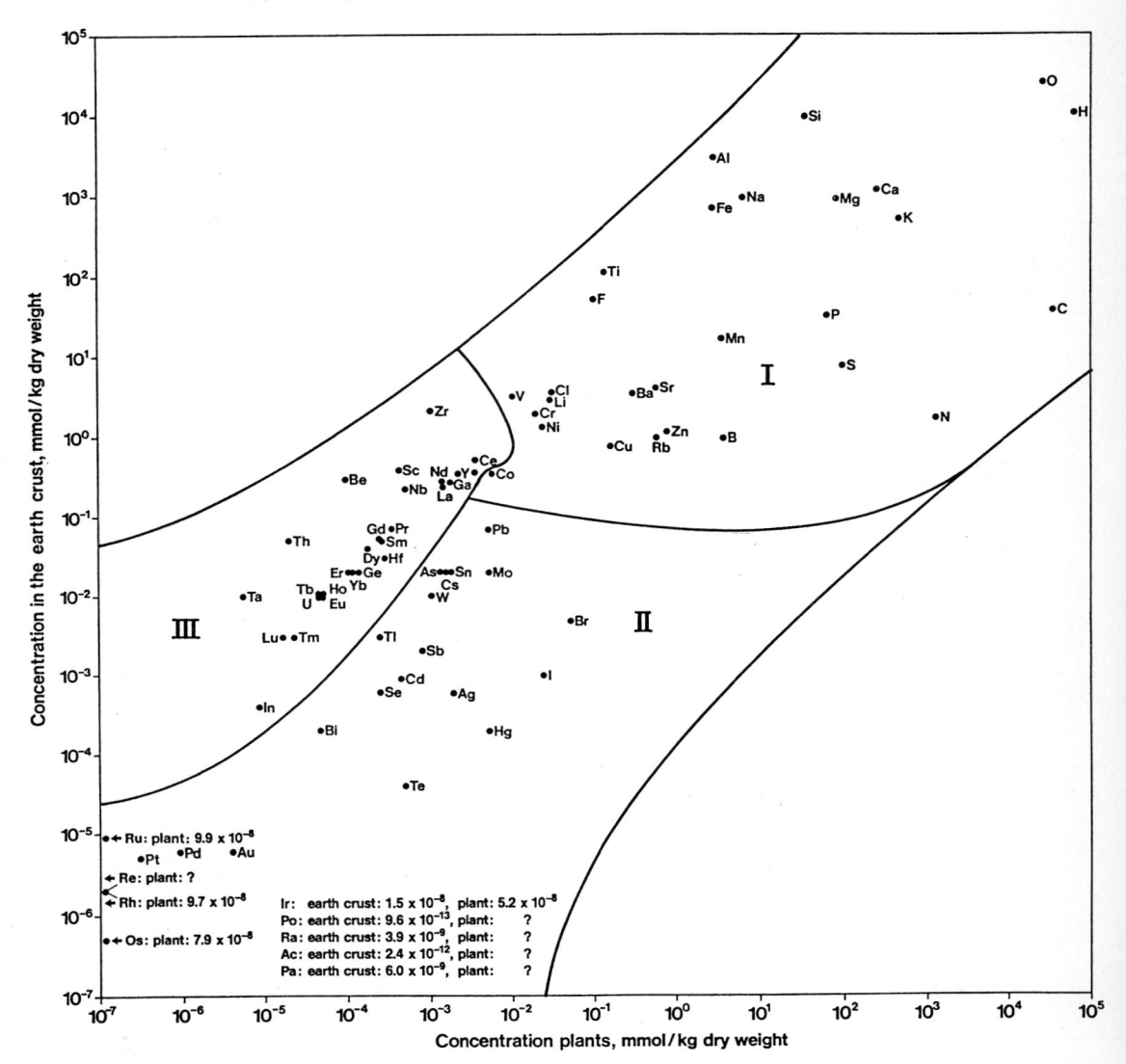

Figure 9-2: Natural concentrations of the chemical elements in the earth's crust plotted against the natural concentrations of the elements in plants. Group 1 consists of elements that the history of evolution plainly shows to be of an essential nature. The elements of Group 2 are those whose physiological function is, with the exception of Se, Sn, and Mo, still largely unknown to us but which may often be toxic at low concentrations (depending on their chemical forms). Group 3 consists of elements that have no known physiological function in evolution at the moment.

Highly sophisticated analytical methods have now placed us in the position of being able to quantify practically all the chemical elements in the table. We are now able to produce results that would have been unthinkable a few years ago. These include calculations of the total content of individual elements in the world plant biosphere (Table 9-1), a chemical fingerprinting technique comparable to the development of the genetic fingerprint (Fig. 9-3), and an initial attempt to establish a Biological System of Elements (BSE) drawing mainly on the latest findings from multielement analysis of plant specimens (Fig. 9-4). Whatever problem of plant analysis is to be tackled, the first practical step is always representative sampling geared to this problem. There is already a wealth of literature on the subject, some of which has been compiled by Hannappel [4]. Special attention is drawn to references [5-40].

Table 9-1: Estimation of the total element content in the world plant biosphere in tons [65].

Ac	?	Hf	$9{,}2 \times 10^4$	Rb	$9{,}2 \times 10^7$
Ag	$3{,}682 \times 10^5$	Hg	$1{,}841 \times 10^5$	Re	?
Al	$1{,}47 \times 10^8$	Ho	$1{,}472 \times 10^4$	Rh	$1{,}84 \times 10^1$
As	$1{,}841 \times 10^5$	I	$5{,}523 \times 10^6$	Ru	$1{,}84 \times 10^1$
Au	$1{,}841 \times 10^3$	In	$1{,}841 \times 10^3$	S	$5{,}523 \times 10^{10}$
B	$7{,}3640 \times 10^7$	Ir	$1{,}841 \times 10^2$	Sb	$1{,}841 \times 10^5$
Ba	$7{,}3640 \times 10^7$	K	$3{,}497 \times 10^{10}$	Sc	$3{,}682 \times 10^4$
Be	$1{,}841 \times 10^3$	La	$3{,}682 \times 10^5$	Se	$3{,}682 \times 10^4$
Bi	$1{,}841 \times 10^4$	Li	$3{,}682 \times 10^5$	Si	$1{,}841 \times 10^9$
Br	$7{,}364 \times 10^6$	Lu	$5{,}523 \times 10^3$	Sm	$7{,}364 \times 10^4$
C	$8{,}19 \times 10^{11}$	Mg	$3{,}682 \times 10^5$	Sn	$3{,}682 \times 10^5$
Ca	$1{,}841 \times 10^{10}$	Mn	$3{,}682 \times 10^8$	Sr	$9{,}2 \times 10^7$
Cd	$9{,}2 \times 10^4$	Mo	$9{,}2 \times 10^5$	Ta	$1{,}841 \times 10^3$
Ce	$9{,}2 \times 10^5$	N	$4{,}602 \times 10^{10}$	Tb	$1{,}472 \times 10^4$
Cl	$3{,}682 \times 10^9$	Na	$2{,}76 \times 10^8$	Te	$9{,}2 \times 10^4$
Co	$3{,}682 \times 10^5$	Nb	$9{,}2 \times 10^4$	Th	$9{,}2 \times 10^3$
Cr	$2{,}7615 \times 10^6$	Nd	$3{,}682 \times 10^5$	Tl	$9{,}2 \times 10^4$
Cs	$3{,}682 \times 10^5$	Ni	$2{,}76 \times 10^6$	Ti	$9{,}2 \times 10^6$
Cu	$1{,}841 \times 10^7$	O	$7{,}824 \times 10^{11}$	Tm	$7{,}364 \times 10^3$
Dy	$5{,}523 \times 10^4$	Os	$2{,}7615 \times 10^1$	U	$1{,}841 \times 10^4$
Er	$3{,}682 \times 10^4$	P	$3{,}682 \times 10^{10}$	V	$9{,}2 \times 10^3$
Eu	$1{,}472 \times 10^4$	Pa	?	W	$3{,}682 \times 10^5$
F	$3{,}682 \times 10^6$	Pb	$1{,}841 \times 10^6$	Y	$3{,}682 \times 10^5$
Fe	$2{,}76 \times 10^8$	Pd	$1{,}841 \times 10^2$	Yb	$3{,}682 \times 10^4$
Ga	$1{,}841 \times 10^5$	Po	?	Zn	$9{,}2 \times 10^7$
Gd	$7{,}364 \times 10^4$	Pr	$9{,}2 \times 10^4$	Zr	$1{,}841 \times 10^5$
Ge	$1{,}841 \times 10^4$	Pt	$9{,}2 \times 10^1$		
H	$1{,}196 \times 10^{11}$	Ra	?		

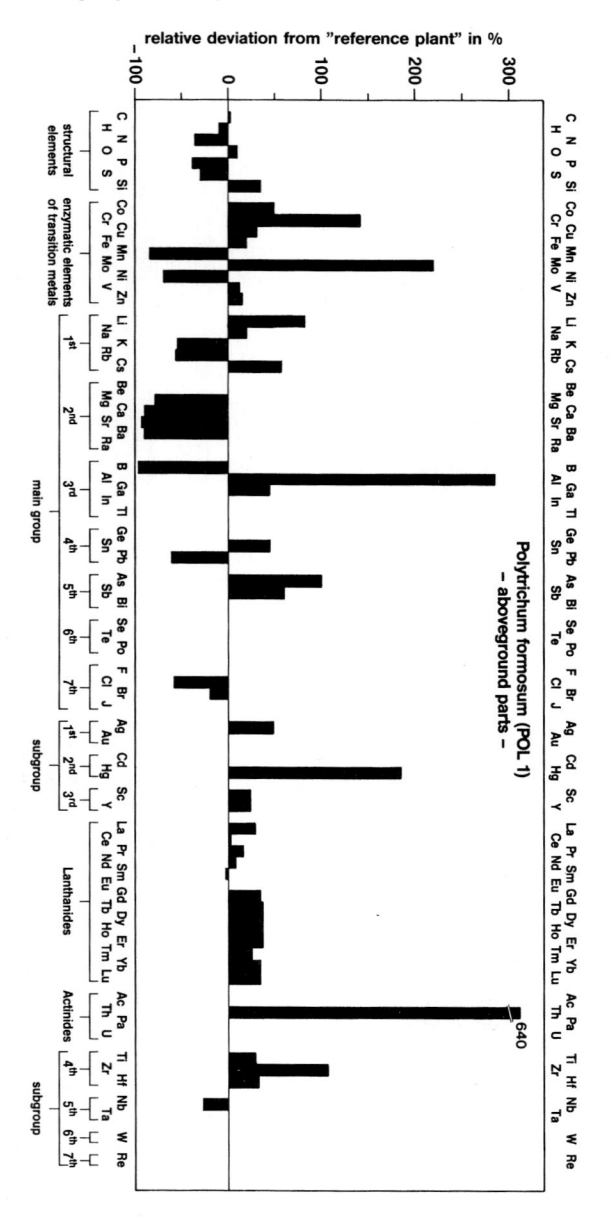

Figure 9-3: Chemical fingerprint of *Polytrichum formosum* (above ground parts) after normalization against "reference plant". The samples were collected at the "Grasmoor" near Osnabrück, North-West Germany (from [42]).

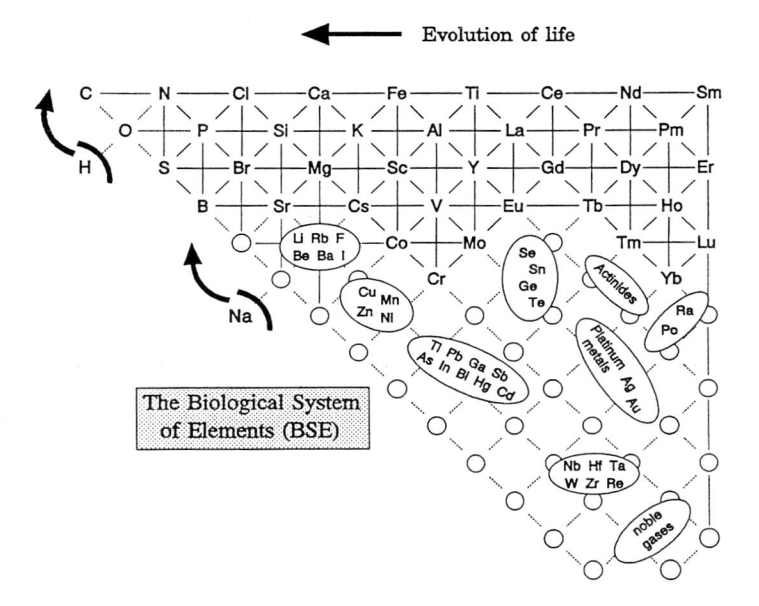

Figure 9-4: The Biological System of Elements (BSE) compiled from data on correlation analysis, physiological function of the individual elements in the living organism, evolutionary development out of the inorganic environment, and in respect of their form of uptake by the plant organism as a neutral molecule or charged ion (from [71,69]). The elements H and Na exercise various functions in the biological system so that they are not conclusively fixed. The ringed elements can at present only be summarized as groups of elements with a similar physiological function since there is a lack of correlation data or these data are too imprecise.

The problem areas dealt with in the following text refer largely to the discussion in Markert [42] of instrumental multielement analysis of plant specimens. As can be seen from Fig. 9-5, instrumental multielement analysis can be divided into the following analytical steps: analytical planning, representative sampling, sample preparation, instrumental measurement of the analytical signal, and data evaluation. In general, each analytical step involves errors. The degree of difficulty and also the susceptibility to error of each analytical investigation depends less on the physico-chemical properties of the substance or sample to be analyzed than on the concentration range in which the element to be determined occurs in the sample [42,43]. Even if analyses in the per cent and ppm range can frequently be carried out without difficulty as a matter of routine, analyses in the ppb and, in particular, the ppt range are frequently problematic and, indeed, in many cases impossible. The analytical error increases exponentially with the decreasing concentration of the element to be determined [41]. This is, on the one hand, because with a low element concentration the fundamental matrix may severely disturb the instrumental determination of an element, which is the reason why the accompanying matrix is usually separated before the actual instrumental measurement is made; and on the other hand the danger of contaminating the specimen by improper sample handling becomes greater. The concentration of mercury in laboratory air may reach such high levels that the mercury in body fluids can be determined contamination-free only under clean-room conditions. In extreme trace analysis the most inert materials possible should be used (e.g. quartz glass) along with highly purified reagents [43].

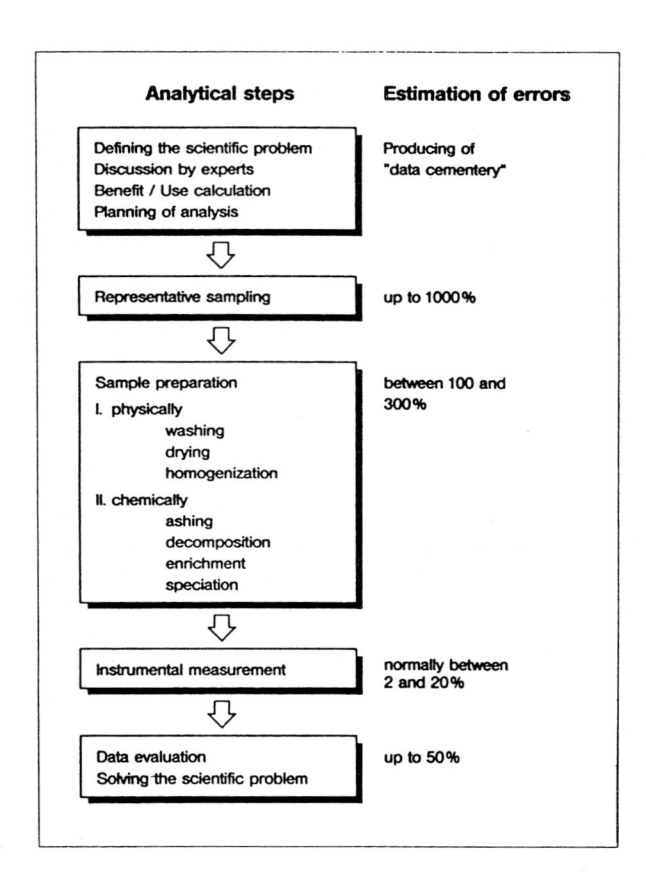

Analytical steps	Estimation of errors
Defining the scientific problem Discussion by experts Benefit / Use calculation Planning of analysis	Producing of "data cementery"
Representative sampling	up to 1000%
Sample preparation I. physically 　　washing 　　drying 　　homogenization II. chemically 　　ashing 　　decomposition 　　enrichment 　　speciation	between 100 and 300%
Instrumental measurement	normally between 2 and 20%
Data evaluation Solving the scientific problem	up to 50%

Figure 9-5: Simplified analytical flow chart for chemical multielement analysis. The full flow chart can be found elsewhere [42].

9.1 The individual analytical steps and their influence on the accuracy and precision of the analytical results

The analytical steps shown in Fig.9-5 are of varying significance in respect of their influence on the accuracy and precision of the analytical result. This may lead to misconceptions on the part of those not involved in the analysis [44]. The physical measurement of the analytical signal is then regarded as the most important, indeed often as the only, analytical signal and its relation to the actual concentrations no quantitative analytical result can be obtained. But with regard to possibilities of error, the physical measurement is one of the most precise and most easily reproducible analytical steps [44]. The analytical step most susceptible to interference is, however, sampling. If the first fundamental law of chemical analysis - namely that the analytical subsample measured in the measuring device should have exactly the same average composition as the total quantity of the analytical sample to be assessed - is infringed, analytical errors of much more than 1000% may occur [45]. Since the true elemental composition of an ecosystem or an ecosystem compartment is unknown and it is thus impossible to calibrate the sampling procedure, real quality control (how accurate or how representative is the sample?) is probably very difficult and in most cases impossible. Nevertheless, analytical and biological deviations from an average value can at least be directly compared by means of simple experimental approaches. Fig. 9-6 gives an example of the average manganese content of red whortleberry leaves. It is immediately apparent that the percentage biological deviation from the average is greater than the analytical scattering about the average. Furthermore, different average concentrations for the manganese levels in the leaves were determined after use of different sampling procedures.

In comparison, chemical sample preparation, the operational steps between sampling and instrumental measurement, such as washing, drying, homogenization, preparation of aliquots, ashing and decomposition, are less susceptible to interference although they represent the second greatest source of error. Ashing and decomposition may generate errors of between 100 and about 300% [46]. However, problems may also arise in data processing and data evaluation. It is generally assumed that data processing is error-free. But if, for example, the type of distribution of the analytical data obtained is not considered and a Gaussian distribution is assumed for the statistical data concentration when actually a logarithmic distribution is present, then with low trace element concentrations the averages of a parent population may contain errors of up to 50% [47].

224   9   QA of plant sampling and storage

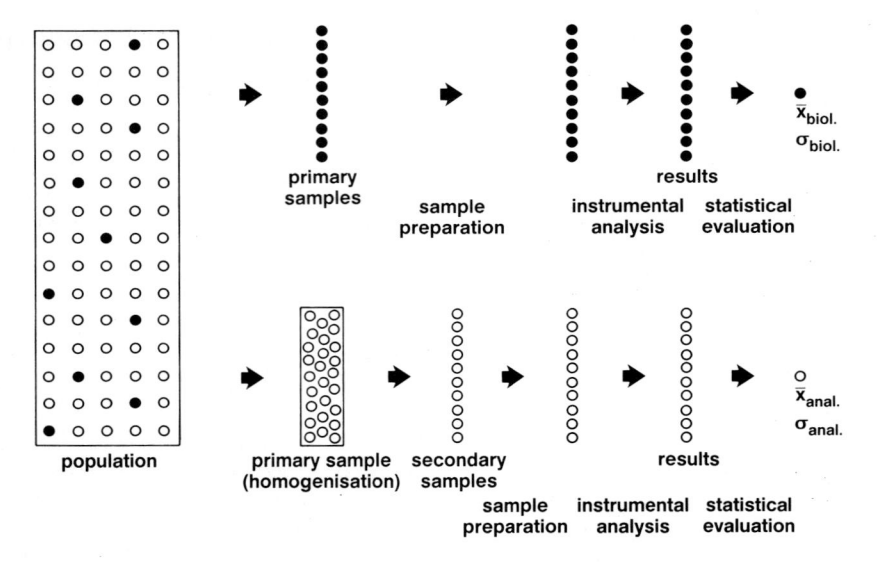

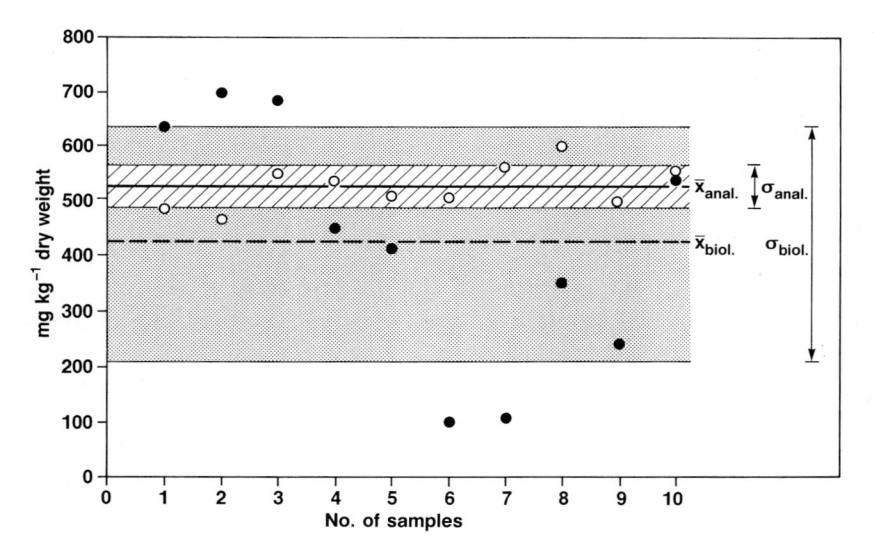

Figure 9-6: Top: removal of ten red whortleberry plants from a sampling area 3m x32m in size in the "Venner Moor". The leaves of each individual plant were analyzed separately to determine the biological scattering from plant to plant (●). A mixed sample of leaves from 100 individual plants was taken from the same sampling area in order to determine the analytical fluctuations (O). Ten subsamples were taken from the total sample and analyzed separately.

9.2 Identification of problems and analytical planning

Each environmental analysis presupposes the formulation of a specific problem; this frequently requires cooperation between several experts from different disciplines. The adaption of the analytical planning to the problem in hand is guided by practical requirements. In the following text a catalog of criteria will be given which may be applied in this or a similar form to the planning of new experiments. This catalog does not claim to be complete but merely represents an attempt to assist analytical planning.

9.2.1 Area of investigation

Which area is most suitable for answering the question in hand?

Will the area under investigation be available long enough for the planned project, or must it be expected that, as a consequence of public or private activities, the study area will be changed or destroyed (important for long-term studies)?

Is the area being studied in a protected area or is permission required from an authority or private person to undertake sampling?

Is access to the terrain possible by motor vehicle (this may be necessary to transport the samples) as far as the geographic conditions are concerned; and is it permitted?

Will the structure of the area under investigation be altered or its existence endangered by sampling or the accompanying studies?

9.2.2 Object of investigation

Which individual, which community of individuals, or which ecosystem compartment is most suitable for dealing with the problem in hand?

Is a sufficient quantity of study material available?

Must protected plants be included in the study?

9.2.3 Schedule and accompanying studies

How long will it take to investigate the problem?

How urgently required are the expected research results?

Are follow-up studies to be expected, or will the question be answered adequately for the time being?

Will further or accompanying studies be required in the area of investigation, apart from the actual sampling, in order to cover additional environmental analysis parameters?

Will further measuring instruments have to be set up and serviced for this purpose?

Have data relevant to the investigation already been acquired in other investigations (description of the soil structure, mapping of the flora and fauna, etc.)?

9.2.4 Chances of success and financial support

Is the sampling program designed in such a way that a solution to the problem can be expected with the aid of the data to be obtained?

Can the results of the study be compared with results from similar investigations by other working groups?

Does the expenditure justify a consideration of these scientific questions?

Are the expenses for instruments, personnel, and consumption of materials guaranteed for the entire analytical process from sampling to writing the final report?

Does financial support depend on attracting third-party funds and can the results to be expected possibly be used for problems arising in the current public debate?

This list of general decision-making criteria reveals that a specific catalog of questions must be drawn up for each investigation to be discussed with all those involved, namely scientists, technicians, and sponsors, and the list also demonstrates how closely intertwined the problem to be investigated and the analytical planning are. Planning of the investigation to be carried out thus takes up a considerable proportion of the analytical procedure, but careful implementation prevents the accumulation of inaccurate or false results.

9.3 Representative sampling of plant specimens

Two basic problems arise in taking plant samples from the field, and which counteract representative sampling adapted to the problem in hand: (1) the heterogeneous distribution of individual organisms in the overall system, and (2) the intrinsic dynamics of living organisms in respect of their chemical composition as a function of space and time.

The population to be sampled consists of more or less heterogeneously distributed individuals which have to be combined into an overall sample. The heterogeneity of the overall population arises, amongst other aspects, from the arbitrary distribution of the individuals in space, limited in the case of plant organisms in the spatial axis by the depth of the roots in a downward direction and by the growth height of the individuals in an upward direction. Laterally, the space is determined by the length of the side of the respective area of investigation or object of study. The heterogeneity of populations or individuals is furthermore promoted by time-induced changes in the metabolic-physiological processes within the organisms and communities of organisms. These changes are decisively influenced by seasonally alternating increases in biomass, differences in the balance of content substances as a result of age, and classic abiotic factors such as soil, light, precipitation, and wind conditions. Furthermore, the genetic predetermination of each individual may also be responsible for the element distribution pattern already discussed. The combination of individuals in a population to form an overall sample thus presupposes knowledge of a great number of individual variables, which only lead to a promising sampling strategy if combined exactly.

In general it is possible to differentiate between two procedures for representative sampling: (1) random sampling and (2) systematic sampling. Random sampling means the random collection of individual specimens from a total parent population (e.g. spruce needles from a spruce forest to determine average calcium content at time X). In the case of systematic sampling (e.g. the collection of birch leaves from various levels of a birch tree to determine the distribution function of calcium as a function of the tree height) a specific systematic study (the dependance of the calcium content of the leaves on the tree height) is coupled to a random sample (birch leaves). Systematic sampling mainly refers to problems of changeable values as a function of space and time, whereas random sampling is frequently aimed at determining an average X from a total parent population. The expression "random sampling" is misleading, since systematic sampling also works on the principle of taking random samples.

The following three "fundamental principles" are valid for both types of sampling.
- 1st principle: The sample taken from the system should have exactly the same chemical composition as the original sample;
- 2nd principle: The probability of being taken as a sample from a total population must be equally great for each individual; and
- 3rd principle: The greater the degree of dispersion of the individuals and the greater the number of individuals, the greater are the efforts required for sampling.

Practical difficulties during the actual sampling process make it virtually impossible to follow the first two principles. A sample taken in the field can never have exactly the same chemical composition as the original material; at best it will be very similar. One of the reasons is that often only a tiny fraction of the original material is actually analyzed (e.g. 100 mg of 100 kg of leaves from a forest ecosystem). The initial content of the original material can be altered by contamination or volatilization of individual constituents while the sample is being taken and during transportation. Moreover, it is scarcely possible to give each individual the same likelihood of being selected when such individuals are diffusely distributed over the ecosystem. The objective must rather be to come as close as possible to the first two principles by means of a carefully prepared sampling strategy. Some practical, basic rules may be helpful here:
- avoid contamination of the sample in any way with the equipment used, the containers, or by the person taking the sample;
- avoid any volatilization of elements as a result of microbial activity, absorption by the walls of the vessels in which the samples are kept or overheating of the samples during transportation and storage;
- take reasonably large samples, provided that there is enough material in the system and this is not subject to the nature conservation laws;
- take account of seasonal fluctuations in the composition of the original material and other parameters affecting its overall composition such as temperature, humidity, light etc.; and
- random division of the area to be investigated.

9.3.1 Classification and examples of element concentration differences and fluctuations in plants

Before discussing in detail the practical aspects of taking plant samples and the development of a sampling strategy, consideration must be given to the locations at which and times when special concentration differences and fluctuations in the plants' element balance must be expected. The analysis of a single plant specimen only resembles a snapshot illustrating the quantity of some element in some plant at some time.

The natural concentration differences and fluctuations in the element composition of plants may occur on five different levels (Table 9-2). The natural element concentration fluctuations in the plant organism must be taken into consideration when drawing up a representative collection program.

Table 9-2: Schematic classification of biological variations in the element content of plants. The element content of the plant body may be affected at all levels by anthropogenic influences [42].

BIOLOGICAL LEVEL	EXAMPLE	POSSIBLE REASONS FOR BIOLOGICAL VARIATIONS
1. Species	Plant species A ⇨ Whole plant kingdom ⇨ Plant species B	Genetic variabilities
2. Population	Population 1 (e.g. in Sweden) ⇨ Plant species A ⇨ Population 2 (e.g. in Germany)	Different climatic and edaphic conditions, genetic variabilities
3. Stand (within an ecosystem)	Stand 1 (e.g. on wet peat) ⇨ Population II (e.g. in Germany) ⇨ Stand 2 (e.g. on podsole)	Different edaphic and microclimatic conditions, genetic variability
4. Individual	plant 1 ⇨ Stand 1 (e.g. on podsole) ⇨ plant 2	Different microclimatic and microedaphic conditions stage of development (age of plant), seasonal changes, exposure, genetic variabilities
5. Plant compartments	Organ (leaf, shoot, root) Tissue Cell Organelle Plant 1	Transportation and delocalization of substances within the plant compartments

Within an ecosystem, edaphic and microclimatic factors mainly influence the observable fluctuations in the plants' element composition. Various detailed studies made in the area around Osnabrück provide information about the orders of magnitude of differences in element concentration to be expected within and between plant organisms. In the following, a few concentration distribution patterns will be presented for the "Grasmoor" near Osnabrück. The respective sampling parameters are summarized in Table 9-3.

Table 9-3: Origin of plant samples investigated. Sample amounts given in dry weight. The sampling of *Vaccinium myrtillus* was repeated after two months within populations 1 and 2 to investigate seasonal changes in the mineral content within leaves (from [42]).

		plant part	date of sampling	age of individuals (years)	weather conditions during sampling	height of individuals (m)	sample amount (dry weight)	soil type
TREES	*Betula alba*							
	Birch 1	leaves	15.05.87	30—40	rainy	19	total sample (300 g) + subsamples each meter	podsole
	Birch 2	leaves	28.05.87	30—40	rainy	19	total sample (300 g) + subsamples each meter	podsole
	Birch 3	leaves	28.05.87	10—15	rainy	7	total sample (300 g)	podsole
	Birch 4	leaves	03.06.87	25—30	rainy	15	total sample (300 g)	podsole
	Birch 5	leaves	17.06.87	10—15	rainy	7	total sample (250 g)	podsole
	Birch 6	leaves	27.06.87	15—25	rainy	10	total sample (250 g)	podsole
	Pinus sylvestris							
	Pine 1	needles	20.06.87	10—15	rainy	8	total sample (800 g)	podsole
	Pine 2	needles	09.07.87	15—20	rainy	10	total sample (600 g)	podsole
	Pine 3	needles	14.07.87	8—10	rainy	3	total sample (400 g)	peat
	Pine 4	needles	02.08.87	8—10	rainy	3	total sample (400 g)	peat
	Pine 5	needles	02.08.87	5	rainy	2	total sample (300 g)	peat
DWARF SHRUBS	*Vaccinium vitis-idaea*							
	Herd 1	leaves	05.08.87	diff. aged	rainy	diff. heights	total sample (200 g)	podsole
	Herd 2	leaves	07.08.87	diff. aged	rainy	diff. heights	total sample (200 g)	peat
	Herd 3	leaves	06.08.87	diff. aged	rainy	diff. heights	total sample (200 g)	gley
	Herd 4	leaves	08.08.87	diff. aged	rainy	diff. heights	total sample (150 g)	podsole
	Herd 5	leaves	08.08.87	diff. aged	rainy	diff. heights	total sample (150 g)	podsole
	Herd 6	leaves	04.07.87	diff. aged	rainy	diff. heights	total sample (250 g)	podsole
	Vaccinium myrtillus							
	Herd 1	leaves/fruits	03.06.87	diff. aged	rainy	diff. heights	total sample (200 g)/ total sample (5 g)	podsole
	Herd 2	leaves/fruits	17.06.87	diff. aged	rainy	diff. heights	total sample (200 g)/ total sample (5 g)	podsole
	Herd 3	leaves/fruits	20.06.87	diff. aged	rainy	diff. heights	total sample (250 g)/ total sample (5 g)	podsole
	Herd 4	leaves/fruits	27.06.87	diff. aged	rainy	diff. heights	total sample (250 g)/ total sample (5 g)	podsole
	Repetition							
	Herd 1	leaves	21.08.87	diff. aged	sunny	diff. heights	total sample (150 g)	podsole
	Herd 2	leaves	23.08.87	diff. aged	sunny	diff. heights	total sample (150 g)	podsole
HERBS	*Molinia caerulea*							
	Herd 1	leaves	02.08.87	grown this year	rainy	diff. heights	total sample (600 g)	gley
	Herd 2	leaves	02.08.87	grown this year	rainy	diff. heights	total sample (500 g)	gley
	Deschampsia flexuosa							
	Herd 1	leaves	09.07.87	grown this year	rainy	diff. heights	total sample (400 g)	podsole
MOSSES	*Polytrichum species*							
	Herd 1	above-ground parts	12.06.87	diff. aged	rainy	diff. heights	total sample (600 g)	peat
	Sphagnum species							
	Herd 2	above-ground parts	12.06.87	diff. aged	rainy	diff. heights	total sample (600 g)	peat

Site-specific differences in element composition can be demonstrated by various herds (site groups) of red whortleberries (Fig. 9-7). The range of fluctuation in the element composition of the leaves between the individual herds (interstand variations) is about 14 % for phosphorus and zinc and 31 % for aluminum. The differences in the element content can be attributed to different soil types (podsol, gley, and peat) and microclimatic differences within the overall ecosystem.

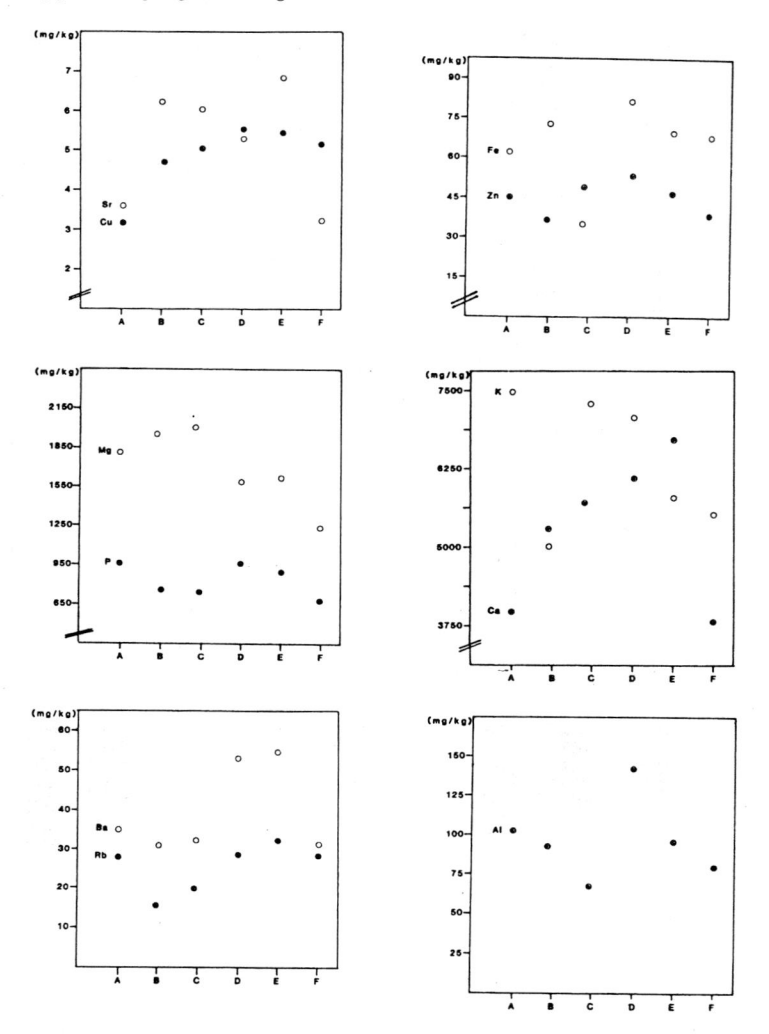

Figure 9-7: Different element contents of Cu, Sr, Zn, Fe, Mg, P, Ca, K, Ba, Rb, and Al in the leaves of *Vaccinium vitis-idaea* (red whortleberry) at six different stands. About 200 g of leaf dry substance was collected per stand (herd). Stands A, D, E, and F were on podzol soils, whereas stand C was on gley and stand B on peat (from [42]).

In Fig. 9-8, fluctuations in the element contents of leaves from four bilberry herds (interstand variations) are compared with the changes over time of the same stands within a period of two months (June - August). The range of variations in the element composition between the four individual stands of *Vaccinium myrtillus* are between 12.6% for zinc and 45 % for rubidium.

The concentration of potassium decreased slightly (- 1.69%) within the two months, and a drop in concentration of -14 to -32.6% was determined for the elements Rb, Fe, P, and Cu, whereas the element concentration for Mg, Ca, Sr, Ba, Al and Zn increased from 41 to 159 %. The temporary, seasonally conditioned differences in concentration thus exercise a greater influence on chemical composition for individual elements than is observed for the interstand fluctuations between individual populations.

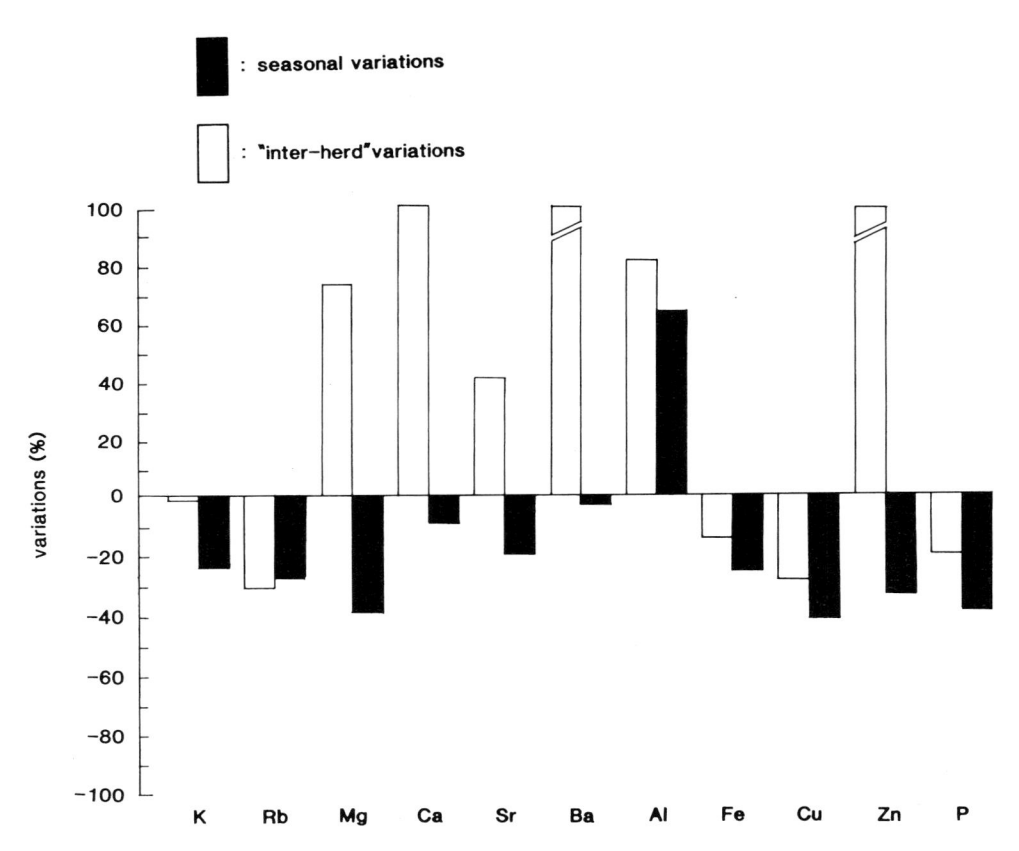

Figure 9-8: Percentage comparison of site-specific and seasonal fluctuations in the leaves of *Vaccinium myrtillus* (bilberry). All data in mg kg^{-1} dry weight (Ba: 159%, Zn: 105%; from [42]).

Four element-specific seasonally conditioned distribution patterns were determined for the moss *Polytrichum formosum* (Hedw.). Fig. 9-9 shows the trend for the element Cr. In 1985 to 1986 the elements Al, Cr, Fe, Mg, Pb, and Ti display element concentration fluctuations of about 80 % with maximum concentrations in the winter months and minimum concentrations in summer. The same distribution pattern, but not so pronounced, is shown by the elements Ba,

Ca, Cd, Cu, and Sr (about 50%), and Mg and Zn (about 30%). The decreasing concentration during the summer months is probably caused by a dilution effect as a result of increased biomass production by *Polytrichum formosum* in spring. It is interesting to note that in contrast to the other elements studied, the concentration maxima for K are found in the summer months and the lowest concentrations in the winter months. Engelke was unable to detect any, or only negligibly small, seasonal concentration changes for Cd and Pb in the moss *Dicranella heteromalla* (L. ap. Hedw.). This different behavior by the two moss species is possibly caused by their different water, and thus element, contents. *Polytrichum formosum* is an endohydrous moss in which the water uptake is supplemented by an internal transpiration flow and not, as in the case of the ectohydrous mosses (*Dicranella heteromalla*), exclusively *via* the moss surface. Differences in element composition may occur between individual plants of the same species even within a closely limited stand. These differences are caused by inhomogeneous soil structures (microedaphic differences within a very small area) and the different developmental stages of the individual plants (age). Fig. 9-10 shows the element concentration differences for the elements Cu, Fe, Mn, and Zn in ten individual red whortleberry plants growing close together in an area of 3 x 3 m². Compared with the analytical fluctuations, the biological fluctuations of the element concentrations determined for the elements Cu and Mn are significantly greater. Both the biological and analytical element concentration fluctuations are only slight for the elements Fe and Zn.

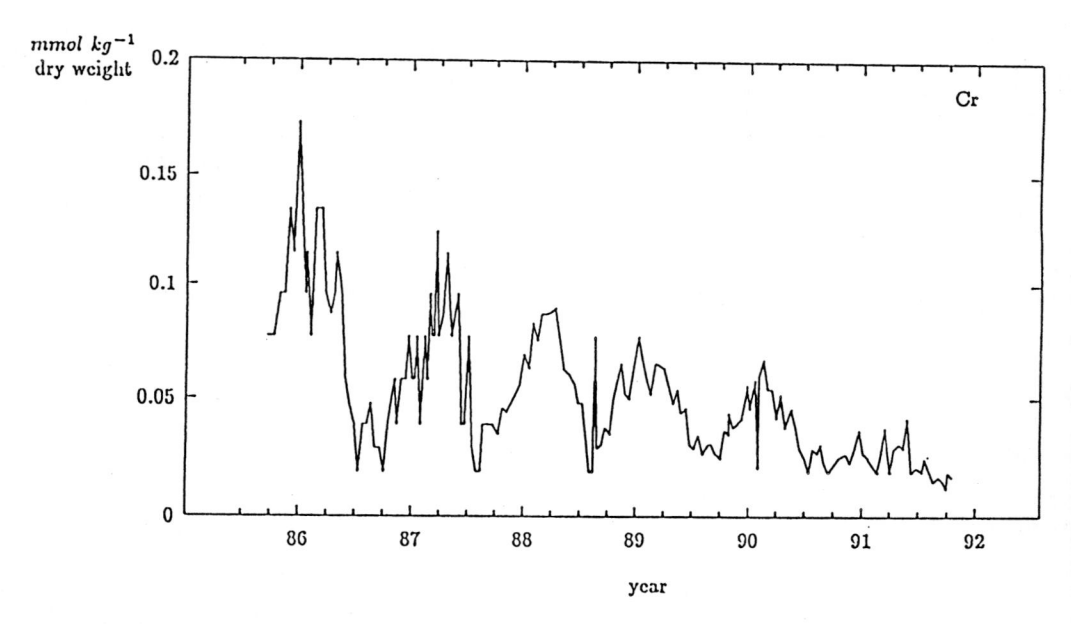

Figure 9-9: Concentration of Cr in *Polytrichium formosum* (above ground parts) collected from autumn 1985 to autumn 1991 (from [30]).

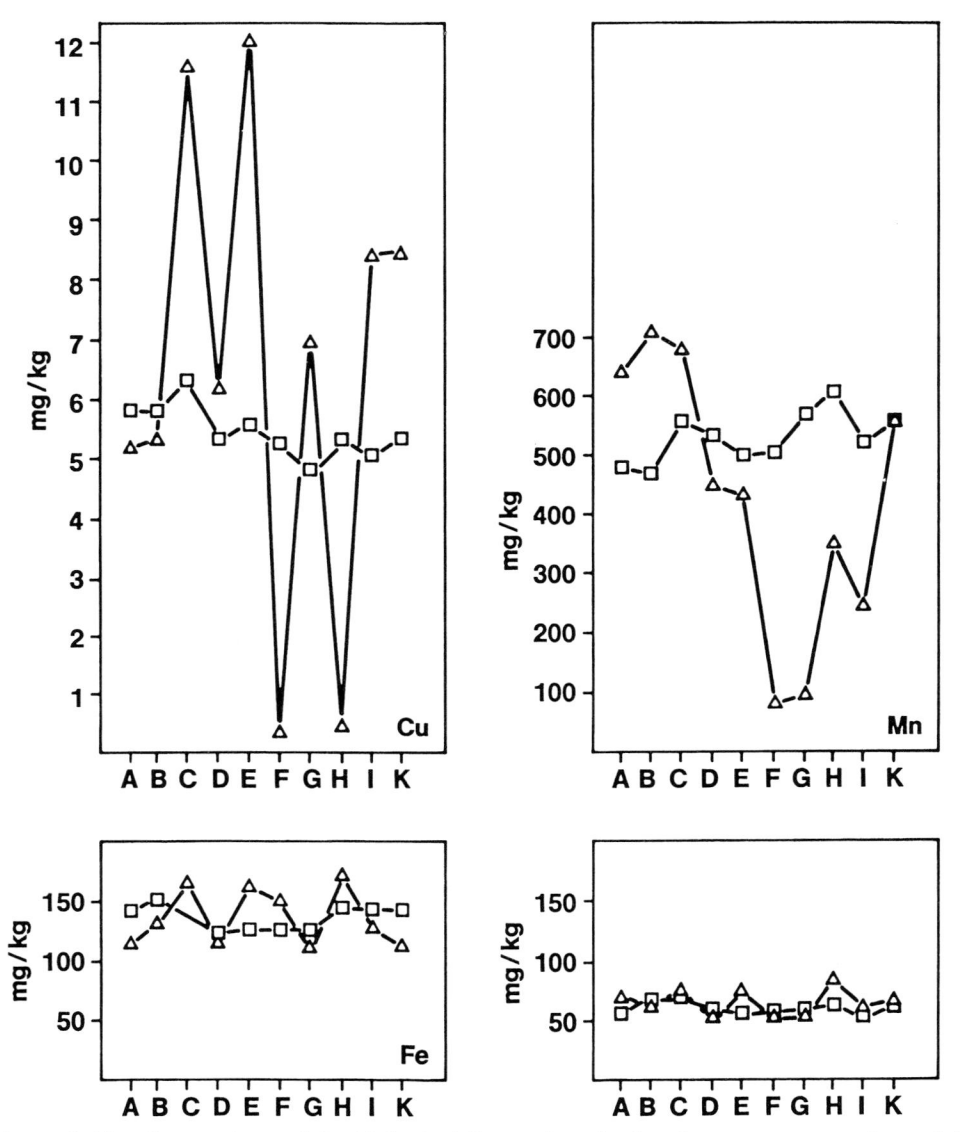

Figure 9-10: Comparison of the biological fluctuations in the element concentrations of 10 individual plants (leaves) of *Vaccinium vitis-idaea* with the fluctuations ranges of individual measurements of a total sample of red whortleberry leaves during instrumental analysis. A - K: 10 measurements from the analysis of 10 individual plants (Δ) and 10 measurements from a mixed sample of individual plants (□) (from [42]).

Heterogeneous element distributions within the same plant organ may arise in the investigation of an individual plant. Fig. 9-11 shows an example of this, the distribution of the element K in birch leaves taken from various heights of the tree [49].

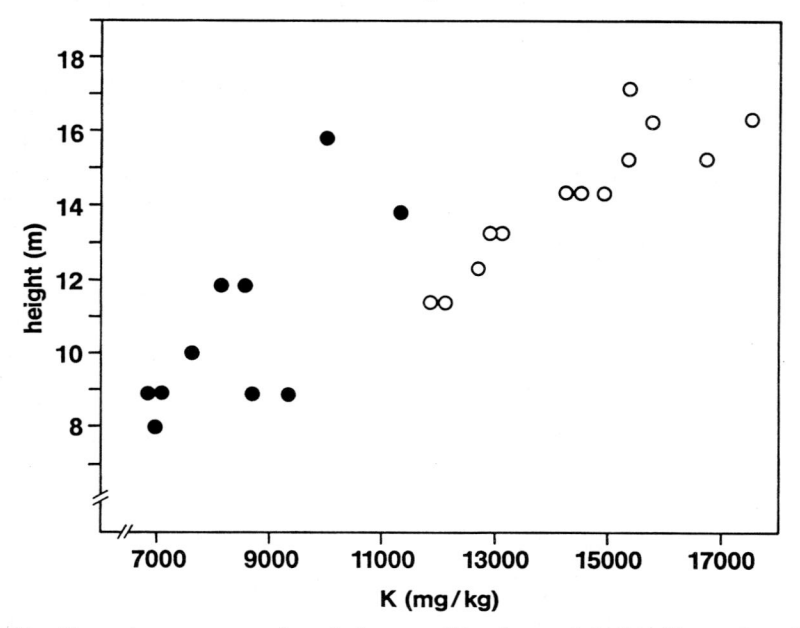

Figure 9-11: Potassium concentrations in leaves of *Betula pendula* (birch) as a function of tree height. Data in mg kg^{-1} dry weight. The leaves were taken from a horizontal position in the two trees (from [49]).

Inhomogeneous element distributions within individuals occur as a result of compartmentalization into root, shoot, and leaf. In the case of deciduous species and with the new formation of leaves (needles), shoots, and roots, translocation and transport processes, in particular, may result within a very short period; these may then alter the element concentrations in a plant organ in the short term. Such processes are described in detail in the literature.

In conclusion, it may be noted that the differences in element concentration in and between individual plant species are varied in nature. Differences occurring in element content are element- and plant-specific, that is to say, they can only be experimentally determined in a new investigation program and cannot be predicted. Nevertheless, in many cases the extent of element concentration differences probably decreases in the sequence: seasonal fluctuations > site-specific differences > analytical variances. In the following, an initial sampling strategy will be developed with the aid of the above insights.

9.3.2 Development of a general sampling strategy

As already emphasized on several occasions, the scientific problem and type of sampling must complement each other. For both random and systematic sampling it is often first necessary to divide the area of study into grid squares in order to give each individual the same chance of being studied.

The entire investigation area (Fig. 9-12) is first divided into four zones [50,51]. Zone 1 should be at least 40 x 40 m. No sampling takes place in this zone; it is intended for determining the usual environmental conditions such as light, temperature, precipitation, and wind. The measuring points must be set up at suitable locations in Zone 1. Zone 2 (buffer zone) serves as a protective belt separating Zone 3 from Zone 1. Mechanical alterations taking place in Zone 3 (for example the felling of trees) must not exert any influence on Zone 1 because this would bias the results obtained in Zone 1. The buffer zone should have a minimum width of 20 - 30 m. The samples required for subsequent chemical analysis are taken from Zone 3. Zone 4 may serve for the collection of zoological samples and may extend over several square kilometers. In order to obtain statistically random plant samples from Zone 3, this zone is divided into at least 100 individual squares and ten of these are selected by a random number generator. Sampling can now begin within these squares. If no plants relevant to the study grow within the selected square, then adjacent squares in a clockwise direction are examined sequentially. This concludes the division of the area to be studied.

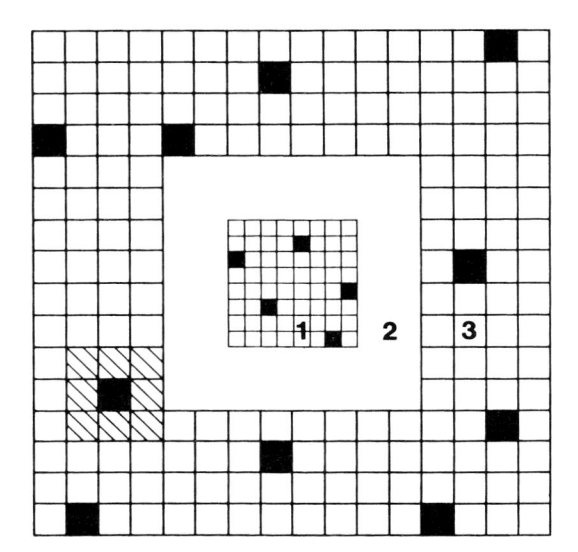

Figure 9-12: Systematic division of the area of investigation (modified after [50]). The inner zone serves exclusively for physico-chemical measurements (e.g. temperature and precipitation). It is surrounded by a buffer zone. Sampling takes place in Zone 3 after a random selection of sampling areas. If the randomly selected sampling site contains no plant material or insufficient for the study, recourse is taken to adjacent areas in a clockwise direction (from [69]).

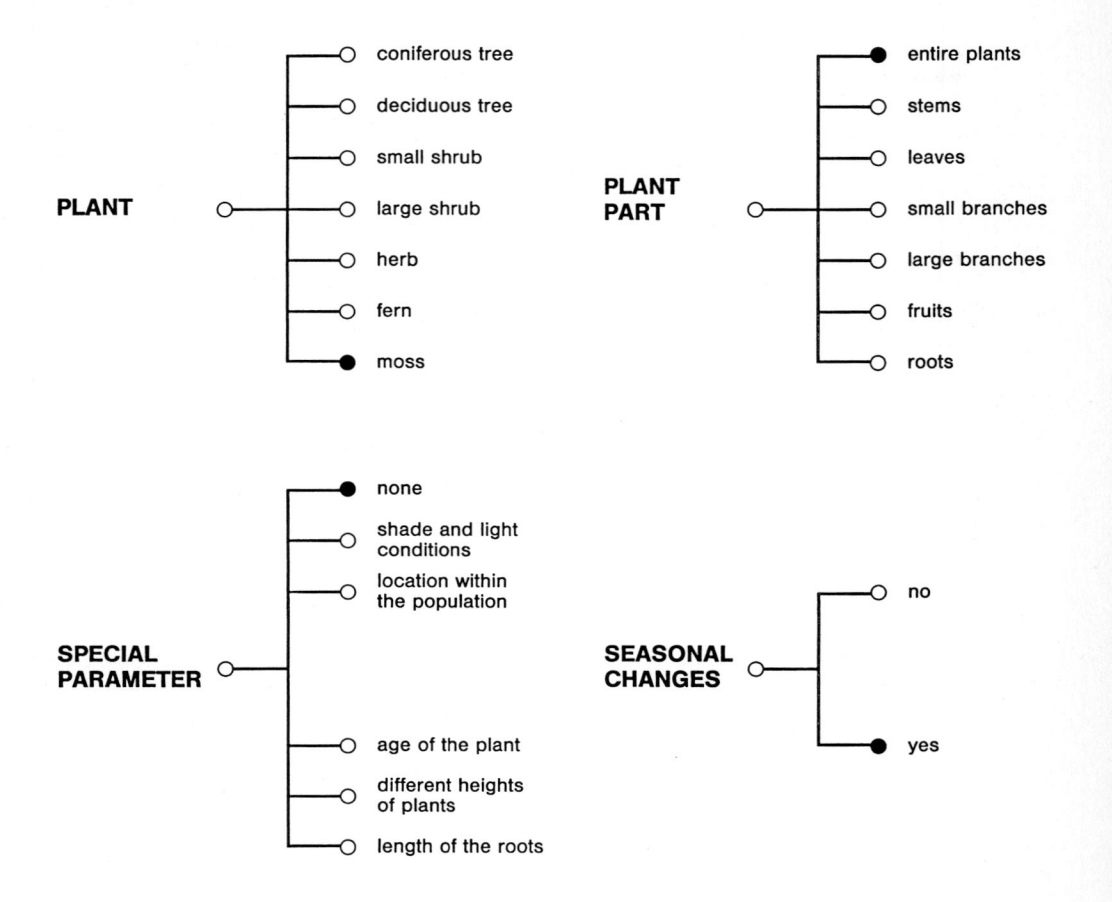

Selection of

```
                    ○ coniferous tree                          ● entire plants
                    ○ deciduous tree                           ○ stems
                    ○ small shrub                              ○ leaves
                                           PLANT
**PLANT**    ○──────○ large shrub          **PART**   ○────────○ small branches
                    ○ herb                                     ○ large branches
                    ○ fern                                     ○ fruits
                    ● moss                                     ○ roots


                    ● none
                    ○ shade and light
                      conditions
                    ○ location within
                      the population                           ○ no
**SPECIAL**                                **SEASONAL**
**PARAMETER** ○────                        **CHANGES**  ○─────
                    ○ age of the plant                         ● yes
                    ○ different heights
                      of plants
                    ○ length of the roots
```

Advice:
Choose a stand of moss after statistical division of your research area. Collect aboveground parts of single and different aged moss plants at weekly intervals (about 50 g fresh weight) over a period of 2 years at least. Transport the samples immediately to the lab to avoid changes in the mineral content produced by microbiological activities.

Figure 9-13: Combination of various individual parameters to establish a sampling strategy. The advice illustrated led to the results shown in Fig. 9-9 (from [28]).

As already described, the element concentration of a plant is essentially determined by the species, the compartment to be studied, seasonal concentration fluctuations, and the influence of various parameters such as soil type (site), light, plant age, height, etc. In many cases the factor constellations from the combinations of individual parameters provide a meaningful relationship between individual target parameters. Advice on sampling can be obtained for many parameter constellations from this experience, as shown in Fig. 9-13 using the example of taking samples of a moss in order to study seasonally conditioned element behavior. The practical implementation of this advice led to the results shown in Fig. 9-9. A method collection of over 300 pieces of advice for combining the most varied individual parameters is now available. The catalog of methods will not be discussed in detail here for reasons of space.

The development of this type of sampling strategy represents a first step towards developing a harmonized sampling system and may in future, if developed systematically, lead to comparable field analysis data from different working groups.

Ernst [52] has also recently published a general sampling strategy; this is shown in Table 9-4. In future it will be possible to apply this at various ecosystem levels.

Table 9-4: General sampling strategy for plant materials [52].

Sampling	Ecosystem budgets	Definition of the objectives: Community interaction (food web)	Population biology	Ecophysiology	Environmental monitoring and emission source identification
Sampling strategy	Random	Selected	Random/ selected	Random selected	1. Selected for emission source 2. Random for general changes
Sampled species	Dominant species in various ecosystem strata	Interacting autotrophic plants and heterotrophs	Species of interest; genotype selection	Species of interest; genotype selection	1. Same species along a transect 2. Several species common in the area
Sampled organ	All organs	At least affected organ	Selection due to objectives	Selection due to objectives	Mostly leaves
Sampling time	Once a year	During the interacting period	Throughout the year	Throughout the growing season	Mostly at the end of the growing season (maximum accumulation)
General attention:		Identification of individuals, development (age) stage			
Special attention	None	Impact on non-affected organ	Allocation changes during development	Allocation changes during development	Comparable age of the organ; comparable exposition; comparable environmental conditions except elements of the monitoring or identification purpose

9.3.3 Sampling for specific disciplines of plant analysis, e.g. biomonitoring trace metal status by means of plants

The use and analysis of living organisms to monitor the quality of our environment has greatly increased in recent years [42,53]. In this context a **bioindicator** is an organism (part of an organism or a community of organisms) that contains one or more pieces of information on the *quality* of the environment (or a part of this). A **biomonitor** is an organism (part of an organism or a community of organisms) that contains one or more pieces of information on the *quantitative aspects* of the quality of the environment (or a part of this) [42,53]. That is to say, in respect of both bioindication and biomonitoring an organism (part of an organism or a community of organisms) reacts to a variation in environmental conditions - for instance the effect of one or more noxious substances - with a change in its way of life in terms of morphology and/or the physiology of its metabolism. The nature of this change is visible or measurable.

The multitude of influences acting on a single organism in the course of its existence and the multitude of possible ways in which an organism can react to such influences show the difficulties involved in establishing and standardizing bioindicators. The accuracy and reproducibility of the results obtained depend to a decisive extent on the sampling procedure used for the organism analyzed. The following section describes two examples of sampling strategies for bioindicator organisms (poplar leaves and moss) which have already been used in some regional, national, and European measuring networks.

9.3.3.1 Use of poplar leaves (Populus nigra "italica") for biomonitoring of heavy metals

Wagner [54] has developed a method for extensive monitoring of environmental contamination with the metals lead, cadmium, and zinc using standardized leaf samples from Lombardy poplars (*Populus nigra* "italica"). Described below is the guideline for sampling summarized in Fig. 9-14. The record sheets prepared for this are included in Wagner's dissertation [54], which also contains initial bioindication results from the Saar region (see also [55]).

A. Choice of the sampling location

Look for at least 20 Lombardy poplars (*Populus nigra* "italica") in the proposed sampling area and select according to the following criteria:

1) Exposed position with no appreciable screening by buildings, other trees etc. Appreciable screening is assumed to exist if objects have a height of more than one-tenth of their horizontal distance from the tree above the level of 5 m at which the samples are taken, and obscure more than 6° of the horizon as seen from the sampling point.

2) The distance from public thoroughfares, railway lines, etc., should be at least 100 meters, and in the case of motorway flyovers or high-tension transmission lines at least 300 m or ten times the height of these objects above the ground at the sampling location.

3) The distance from water courses which may be polluted should be at least 25 m.

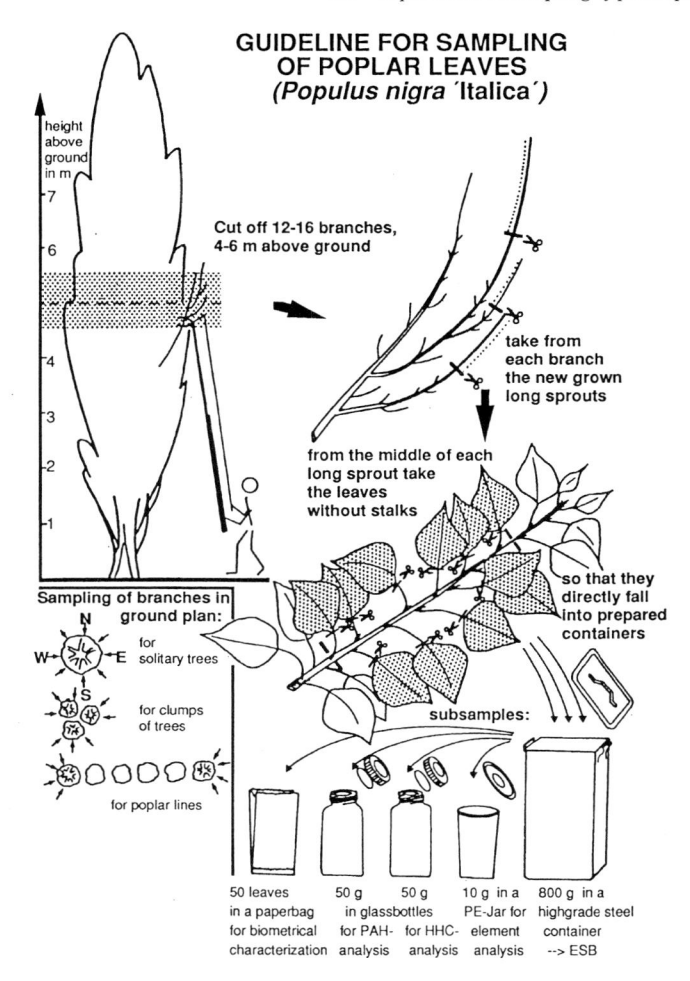

9.3 *Representative sampling of plant specimens* 239

Figure 9-14: Drawing to summarize sampling procedure for leaves of *Populus nigra* "italica" (from [54]). Thanks are due to Dr Gerhard Wagner, University of Saarbrücken, for preparing and submitting an English version of this drawing.

Enter in the record any potential emitters of pollutants and objects with a screening effect up to a distance of 1 km and major emitters (e.g. power stations, iron and steel works, chemical factories, airports, cities, etc.) up to a distance of 15 km, showing their direction and distance from the sampling site and their height above it.

To describe the sampling location, append to the record a section from a map or a drawing of the site on a scale of 1:500 or 1:1000 showing the permanently marked individual trees and their surroundings up to at least 50 m around the sampling points.

The population available must consist of at least twenty suitable Lombardy poplars and have long-term protection for sampling purposes and against disruptive influences. To ensure optimum comparability and reproducibility of the samples in the long term it is recommended that 20 to 25 Lombardy poplars be specially planted at a distance of at least 10 m from each other on a suitable site within the sampling area.

The trees will not be available for sampling until at least five years after planting. Only Lombardy poplars without visible signs of disease (e.g. dieback) and with average growth rates are suitable for sampling. All specimens with one or more of the following characteristics must be excluded from sampling:

1. Young poplars with a trunk diameter of less than 15 cm at chest height or with long sprouts averaging more than 35 cm at sampling height (4 to 6 m) and over-large juvenile leaves.

2. Aged or sickly specimens with very bent or stunted long sprouts averaging less than 15 cm.

3. Poplars with prematurely yellowed leaves, rust (more than 10 % of the leaves), severe infestation with aphids causing formation of "sooty mould" (crusts made up of aphid secretion, dust, and fungus on the leaves), or chlorosis, necrosis, or feeding damage affecting, on average, more than 5% of the leaf area.

B. Conduct of sampling

In the case of solitary trees, take the samples at a height of 4 to 6 m above the ground in all directions equally. With clumps of poplars, take the sample from the outsides of the trees all the way round the clump. If the poplars grow in rows, take the sample from the exposed side of the first and last tree (see Fig. 9-14).

For cutting off the twigs, use a pair of pruning shears with a telescopic handle than can be extended to 3 meters and a correspondingly long cord (gardening tool). At a height of 4 to 6 m in all directions equally around the edge of the crown, use the shears to cut off twigs about 50 cm in length with 3 to 5 long sprouts each and catch them before they reach the ground (see Fig. 9-14). As the leaves on the severed twigs quickly lose water it is advisable not to cut off more than four twigs at a time.

Until the leaves have been removed, the twigs should therefore be kept in the shade of the tree on untreated ground vegetation or on twigs that have already been cut off and used for sampling; on no account should they be placed on soil, on asphalt or concrete surfaces, car bonnets, or textiles (risk of contamination).

Take the leaves from the middle of the long sprouts only; throw away 5 leaves from the beginning and the end (the oldest and youngest leaves). Examine the beginning of the sprout, which is easily recognized by a ring of scars from the bud scales, for any leaves that have already dropped off (= semicircular scars); these must be included in the count. Do not use leaves from short sprouts. Use only average-sized, intact leaves without visible damage or impurities (e.g. fungus, attack by insects, aphid secretions, or bird droppings). Do not store the stems of the leaves.

About 2000 to 3000 leaves are required for a fresh weight of 1 kg.

To prevent contamination, remove the leaves without touching them with the hands, clothing, etc. Cut off the laminae at the leaf end of the stem with a clean pair of nail scissors in such a way that they fall straight into the prepared container without further handling. Make sure that the leaves are not creased or crushed.

Do not wash the samples; weigh them as soon as each container is full and then deep-freeze them in the vapor phase or liquid nitrogen. The temperature of the samples must not again exceed - 130° C before the analysis is carried out.

C. Equipment and facilities for sampling

To store the samples use stainless steel containers without a neck which have been washed according to instructions with dilute acids and extremely pure organic solvents and heated to 200° C. Secure the lids with an aluminum clip. For a total sample quantity of 6 kg at least 20 such sample containers (one for each 300 g per tree) and a suitable deep-freeze chamber (transportable LN_2-Dewars) will be required, as will a supply of liquid nitrogen sufficient to cover unexpected delays in transportation.

Furthermore:
Shears with a telescopic handle extendable to a length of 3 m
Clean nail scissors (stainless steel)
Disposable polyethylene gloves.

For characterizing the samples:
Paper bags (at least one for each tree)
Microscope or a good magnifying glass
Camera with a wide angle and macro lens and film for color transparencies
Scales (up to 3 kg ± 0.1 g)
Air thermometer and anemometer for measuring weather conditions.

D. Characterization of samples

Record the weather conditions and, if possible, the samples taken from each tree during the actual sampling. Fill in data not immediately available as soon as possible and pass them on to the data base.

To document conditions make color transparencies of the overall location, of each tree from which leaves were collected (appearance) and, where appropriate, of the branches cut (ramification, size, and color of leaves) and archive them along with a scale, to show the size, and a color chart.

To enable individual analytical characterization, determination of mean leaf weight, etc. take 50 randomly selected leaves separately from each tree, place in a previously weighed paper bag marked with the number of the tree, and weigh at once. After drying at 80° C they are used for determining dry weight and should be kept in a suitable manner for element analysis of individual trees.

When the first sample is taken from a newly selected or changed location, approximately 100 g more should be taken from each tree and stored. During homogenization, take subsamples for the proposed groups of substances from each tree for analytical characterization of the samples and send them to the analysts involved.

E. Sampling period

Sampling should be carried out in dry weather between 25 August and 7 September. Sampling should be interrupted if precipitation occurs. After dew formation at night or light rain, do not start or continue sampling until the leaves are completely dry. After lengthy wet periods of heavy rain, wait until at least three rain-free days have elapsed. Record weather conditions before and during sampling on a sample data sheet. Wait for at least two years before taking samples from the same trees again.

9.3.3.2 Use of mosses for large-area biomonitoring for heavy metals

In the late 1960s the Swedish scientists Åke Rühling and Germund Tyler discovered the usefulness of mosses for biomonitoring of heavy metals. They found that *Hylocomium splendens* and *Pleurozium schreberi* were especially excellent "catch organisms" for moist and dry deposition. Since then the use of mosses as biomonitors has been systematically extended. The study group Systems Research of the University of Osnabrück was requested by the Federal Office of the Environment in Berlin to set up a moss monitoring network covering the whole of Germany in the framework of the ECE. The objectives of this work are [70]:

- to determine the nature and quantity of regional atmospheric deposition of heavy metals with the aid of moss monitoring;
- to detect local sources of emissions;
- to determine the rate of atmospheric deposition on the basis of heavy metal analyses;
- to prepare easily readable maps showing heavy metal deposition in Germany and Europe; and
- to determine the degree of pollution in Germany in comparison with other European countries.

Very considerable help with the work of sampling and data evaluation was given by the study group Bioindication / Determination of Effects of the German Länder. Fig. 9-15 shows the record from sampling work, including explanatory notes, on which this was based. Fig. 9-16 shows the initial results for the distribution of lead [70].

9.3 *Representative sampling of plant specimens* 243

SAMPLING RECORD SHEET   Sample No. |__|__|__|
(Multiple responses possible)

Name of sampler:
Date of sampling:
Time of sampling:

Weather conditions:

- Sunny [ ]
- Cloudy [ ]
- Rainy [ ]
- Windy [ ]
- Windless [ ]
- Dry [ ]
- Other: _____

Species of moss:

- Hylocomium splendens [ ]
- Pleurozium schreberi [ ]
- Hypnum cupressiforme [ ]
- Polytrichium formosum [ ]
- _____ [ ]

Vegetation cover:

- Sparse [ ]
- Cushion-type [ ]
- Sward type [ ]

Sampling location

1. **Administrative unit**

   Federal state:
   County / district:
   Town / village:

2. **Geographic information on the sampling location**

   Hochwert: _____   Rechtswert: _____
   Longitude: _____   Latitude: _____   Height above MSL: _____ m

   Exposure: North [ ]   South [ ]   West [ ]   East [ ]
   Level ground [ ]

Figure 9-15: Sampling record sheet for large-scale collection of moss samples in the framework of the European ECE heavy metal biomonitoring system. With explanatory notes [70].

3. Ecosystem + use of land

Forest land ☐	Grassland ☐	Waste land ☐	Arable land ☐
Moor/bog ☐	Industrial site ☐		Trading estate ☐
Allotments ☐	Orchards and vineyards ☐		Parkland ☐
Traffic ☐	Leisure ☐	Residential ☐	Near-natural ☐
Not known ☐	Other ☐		

4. Sampling point

Types of plants - predominant species

| Tree layer | _____ | Shrub layer | _____ |
| Herbaceous layer | _____ | Moss layer | _____ |

Sample taken from   Ground ☐   Tree stump ☐

Type of soil (if known) _____

Other known soil parameters _____

Thickness of the humic layer _____

Distance to nearest tree	_____ m
Distance to nearest shrub	_____ m
Distance to nearest motorway	_____ m
Distance to nearest road	_____ m
Distance to nearest industrial site	_____ m
Distance to nearest residential area	_____ m

5. Other information

Nearest climatic measuring station _____ km

Figure 9-15 (ctd)

6. Other remarks

Explanatory Notes on the Sample Record Sheet

Sample No.: *This is the overall number for a whole sampling location. Repeat the number (with a waterproof marker) on the packed plastic bags and inside them (small slip of paper). Make out a separate record sheet for each overall sampling location.*

Date of sampling: *Enter day, month and year.*

Time of sampling: *Enter full hours*

Weather conditions: *To describe the **current weather** you may tick several boxes and enter observations of you own which are not listed.*

Species of moss: *Tick the box for the species of moss collected. If you do not find Hylocomium splendens (first choice) take the next moss, i.e. Pleurozium schreberi (second choice). If this is not available either, follow the sequence listed. If you do not find any of the four species, take the moss growing at the sampling point and identify it.*

Vegetation cover: *Rate the predominant density of the vegetation cover at the overall sampling location.*

Figure 9-15 (ctd)

1. **Administrative unit:** Enter the general location of the sampling point.

2. **Geographic information:** Enter coordinates with degrees, minutes and seconds. **Exposure** refers to a possible hillside aspect of a sampling area. If, for example, a slope or hillside faces south the exposure is said to be "south". if the slope faces south-west, tick the boxes "south" and "west". Level ground should be described as "level" on the record sheet. As far as possible the exposure should be the same for each location.

3. **Ecosystem + use of land:** Tick or name the ecosystem that specially characterizes the the sampling point. Also tick or describe the obvious or known use of the land at the sampling location. It is possible to tick several boxes.

4. **Sampling point:** Enter the obviously predominant species or, if known, the plant community. Examples: pine and birch for the tree layer and darnel for the herbaceous layer. Also tick the precise place from which the moss was taken (e.g. tree stump). If the moss was taken from both places, tick both boxes. Indicate this on the plastic bags. Do not take samples from living trees. For **type of soil** enter terms such as podzol, brown earth or gley. Also include other soil parameters such as the pH if known. State the thickness of the **humic layer** in centimetres. Under **distances**, exact information should be given on the distance of the sampling location from the individual objects listed.

5. **Other information:** Under this heading enter the climatic measuring station (name of the place) nearest to the sampling point. The same station may be entered for several sampling points.

6. **Other remarks:** Enter any other remarks that may be significant for later evaluation.

For details of procedure during sampling follow the instructions on the sheet **GUIDELINES FOR SAMPLING.**

Figure 9-15 (ctd)

Guidelines for Sampling

1. The samples should preferably be taken **in forest land**, from **the ground or from tree stumps**. Failing this, **open grassland, heath or bogs** may be sampled instead. The sampling points should be at least **300 metres** from the nearest **motorway / major highway**, the **nearest village** or the **nearest industrial site** and at least **100 metres** from **minor roads and houses**.

2. In **large conurbations, other densely built-up areas and city states** the mosses must be collected in **urban copses, parks and cemeteries** and not as described in Point 1. Here, especially, it is important to state **exact distances** from roads or industrial complexes (record sheet).

3. To avoid the direct influence of water dripping from the leaves and the filter effect of trees, the sampling points should be **in clearings**. Take the samples from **small open spaces at least 5 m from the nearest tree**. As mosses are often found associated with low bushes, care must be taken that these do not cover the mosses.

4. The overall sample from each sampling point should consist of **5 - 10 randomly selected subsamples** from an area of **50 x 50 metres**. If this is not possible, take the samples from **positions that seem similar** within a **radius of 500 metres**. Place the individual subsamples in separate plastic bags; these must contain **one species of moss only**. Then pack these **5 - 10 individual plastic bags** in one large plastic bag **(overall sample for the location)**. It is important that the bags should be **precisely marked**.

5. Take the sample of moss **in the form of a cushion**. The samples must not contain any impurities such as **litterfall, soil, humus or other material**. These must be carefully removed. The total sample taken from one location and consisting of 5 - 10 individually packed subsamples should have a **fresh volume of 1 - 2 litres** (5 - 10 g dry weight).

6. **Do not smoke** during sampling. Wear **disposable plastic gloves without talcum powder**.

7. Send the samples **in their fresh condition, protected against contamination and accompanied by the record sheets** to the University of Osnabrück, c/o Uwe Herpin **(by express delivery)**.

Figure 9-15 (ctd)

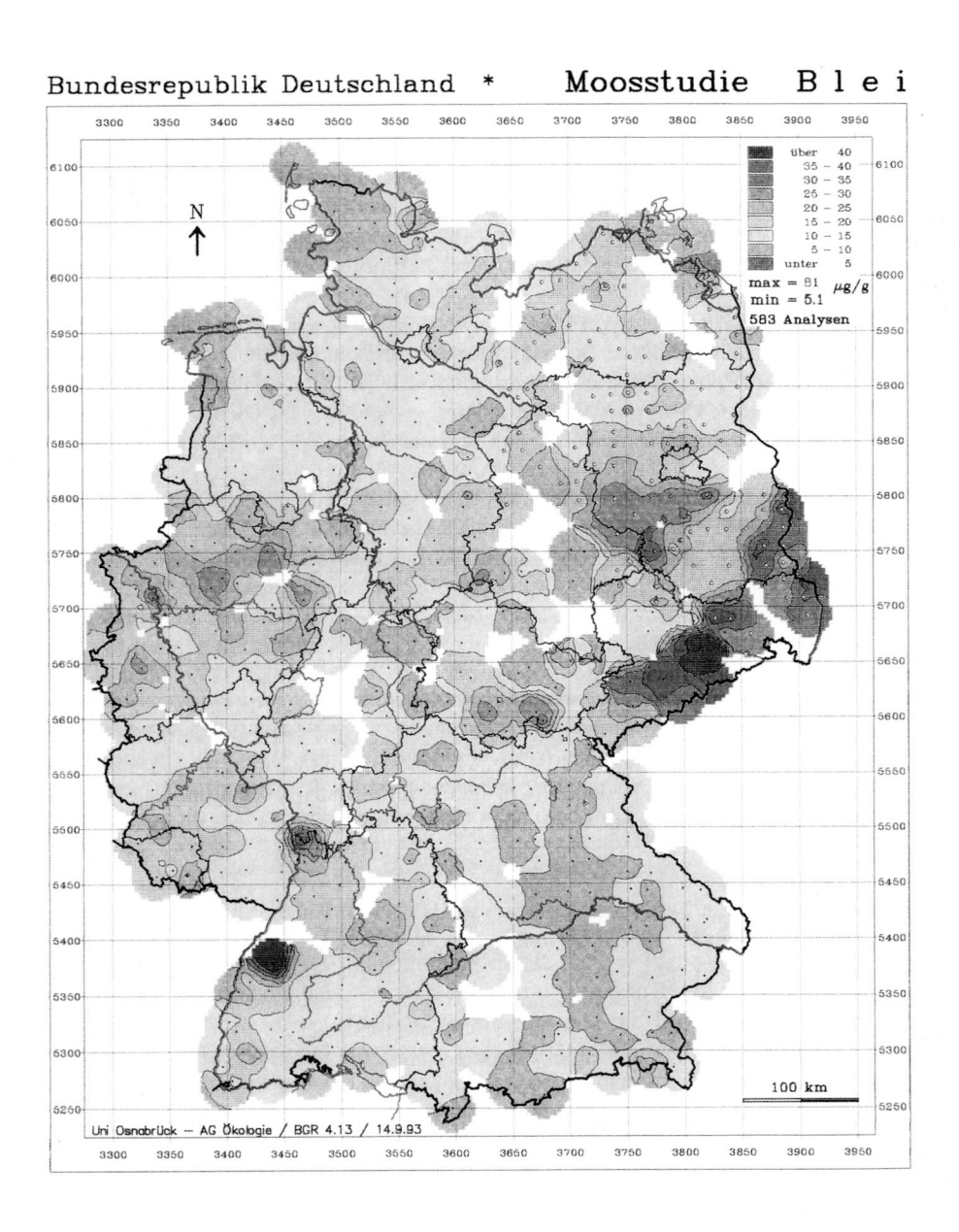

Figure 9-16: Map showing the distribution of the element lead over the Federal Republic of Germany as determined in moss samples [70].

9.4 Cleaning and drying the sample material

Cleaning of the sample material often represents the first of several physico-chemical steps through which a specimen will pass from its collection until instrumental measurement. In the case of plant samples, the question which must generally be asked is whether it is appropriate to clean the material before chemical analysis. In the case of studies concerning the food chain (for example in detecting animal uptake rates of individual elements *via* plant foodstuffs) in order to determine the total input into the animal, decontamination of the surfaces of plant samples is not required since animals do not clean their food before consuming it. In contrast, suitable cleaning procedures have to be carried out if the transfer rates of chemical elements from the soil into the plant are to be studied, for in this case the surface contamination of the plant would bias the uptake rates determined. In this connection, cleaning of the plant material is of significance if a high input of dust, flue gas, or similar is to be expected in the area studied.

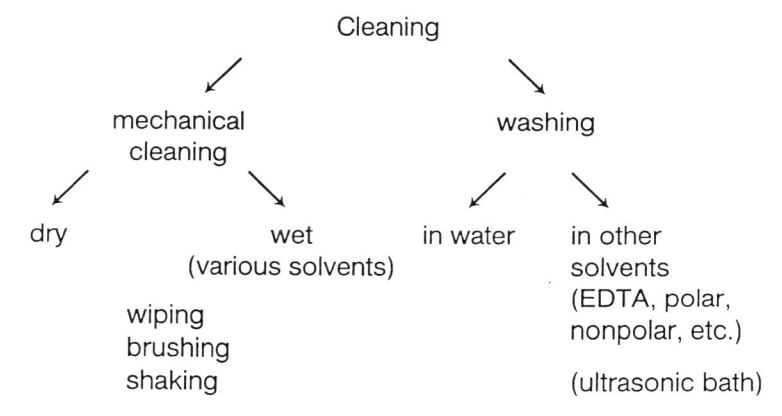

```
                           Cleaning
                         ↙        ↘
              mechanical            washing
               cleaning            ↙      ↘
            ↙         ↘      in water    in other
          dry          wet                solvents
                (various solvents)        (EDTA, polar,
                  wiping                  nonpolar, etc.)
                  brushing
                  shaking                 (ultrasonic bath)
```

Figure 9-17: Cleaning procedures for plant samples (from [42]).

As shown in Fig. 9-17, cleaning can include purely mechanical steps, for example the use of dry or moistened cloth, shaking, blowing out and brushing of the sample material, or the application of various washing procedures. In general, particularly if solvents are used, leaching of substances from the sample must be expected in addition to a removal of surface contamination. An unfavorable ratio of sample volume to solvent encourages the leaching of substances.

Various approaches are now available for developing suitable cleaning processes for plants and other biological materials. An intensive form of cleaning whereby conifer needles are completely cleansed of adhering surface contamination by removing the cuticle involves shaking the fresh needles in chloroform for one minute, then rinsing them twice with distilled water, and subsequently drying [56,57]. Although leaching of potassium was found to be negligible, removal of surface contamination by the elements Al, As, Fe, Sb, Sc, and V was considerable. Surface contamination accounted for about 80% of the total content of these elements in the needle. Porter [58] used various procedures for cleaning the leaves of *Ailanthus altissima*. As Fig. 9-18 shows, cleaning with 1% Alconox and subsequent treatment with 0.01 mol.l^{-1} EDTA solution achieves the most effective removal of surface contamination with Fe, Cu, Zn, and Mn while at the same time avoiding excessive leaching of Ca, Mg, K, Na, and Cl. Ernst [59] compared the lead content of washed and unwashed leaves with rough (birch) and smooth (Norway maple) surfaces in the course of a vegetation period. The washed leaves were characterized by lower lead concentrations, and this effect increased from the smooth to the rough leaf surfaces.

The quantitative separation of soil from root samples is also a problem. Even if only a small fraction of the rhizosphere soil is collected with the sample it may decisively bias the analytical data because the element concentration in the soil frequently exceeds that in the root by several orders of magnitude [60]. The danger of plant contamination from soil particles is therefore especially great. Since, in contaminated soil, the titanium content of a plant sample increases far more than that of many other trace elements, the titanium concentration may serve as an indicator of soil contamination.

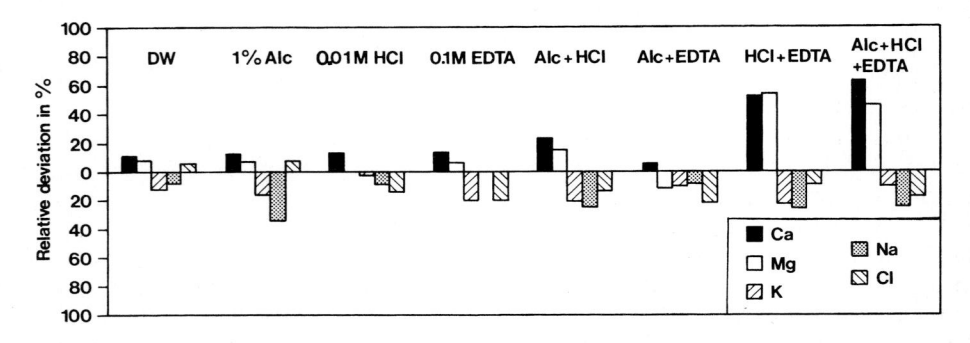

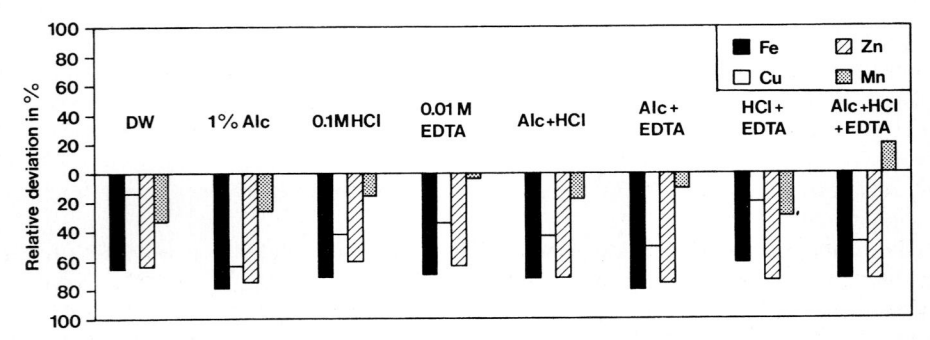

Figure 9-18: Various cleaning procedures for leaves from *Ailanthus altissima*. The most effective method is cleaning with 1% Alconox and subsequent EDTA treatment (highest leching rate for Fe, Cu, Zn, and Ti; lowest rate for K, Cl, Mg, Na, and Ca; altered after [58]). (from [69].

Precipitation during sampling is also significant with regard to the surface contamination of plant material. This precipitation washes off part of the dust adhering to the plant, which is not the case with plants gathered on days without precipitation. It is therefore necessary to keep an exact record conditions during sampling.

Drying of the sample material serves to protect the plant against microbial decomposition during subsequent storage and to ensure a constant reference value by determination of dry weight as opposed to fresh weight, which is difficult to quantify.

Careful drying in a vacuum (15 torr) at a temperature of 40 to 50 °C has proved to be a suitable method for the sample material. For the subsequent analytical determination of highly volatile elements or element compounds, drying without the use of reduced pressure should also be carried out at 40 to 50 °C. A further drying technique for trace analysis is freeze drying, in which moisture released from the sample after freezing and subsequent thawing is removed from the system by applying a vacuum. Samples which cannot be dried immediately can be stored for a few days in cold boxes or a refrigerator at 4 °C before being further processed.

9.5 Homogenization, aliquots and storage

The purpose of homogenizing the sample material is to achieve the most uniform possible distribution of all the elements in the sample. Grinding the specimen, however, may also lead to an additional risk of contamination due to abrasion from the grinding materials [61, 62]. In the case of materials particularly difficult to grind, previous embrittlement under liquid nitrogen has proved successful. Subsequent grinding in a vibration mill lined with Teflon and equipped with Teflon rods is the preferred procedure at the Environmental Specimen Bank [63]. The installation and operation of such mills involves considerable expenditure which few institutes are able to afford. As a rule, grinding equipment made of stainless steel, agate, zirconium oxide, tungsten carbide or titanium has to be used instead. If these materials are used, abrasion and thus a contamination of the sample must be expected as a result of the grinding procedure. With large sample quantities especially, it has proved advantageous to grind the plant material using agate equipment in disk vibration mills, although this may create problems with highly "resistant material" such as grasses and wood. More easily homogenized samples such as birch and red whortleberry leaves can be ground in planetary vibration mills with polyamide containers and Teflon balls [49]. Polyamide abrasion may be regarded as negligible in the subsequent instrumental measurement.

The homogeneity of the sample achieved by grinding can be verified without difficulty as a function of the weighed portion for analysis by applying the usual replicate measurements. After grinding, the sample can be stored for an indefinite period in clean plastic containers at room temperature and in the absence of light. Gamma irradiation, for example with the aid of a cobalt source, can be used to sterilize the samples. No change in inorganic chemical composition was observed at our institute in nonirradiated samples stored under normal conditions for eight years. In the case of finely homogenized red cabbage, several years of storage in glass jars had caused the material to become lumpy [64]; this may have been caused by too high a sugar content. To counteract possible grain size fractionation during storage, the sample container must be shaken several times before further subsamples are taken.

Acknowledgments

The author wishes to thank Mrs. Kathleen Kirschner for her valuable assistance in preparing the manuscript, Mrs. Marion Braase for the translation into English and my collaborators at the GKSS Institute for Inland Water Research for their untiring efforts.

9.6 References

[1] B. Markert, in: *D'ANS/LAX, Taschenbuch für Chemiker und Physiker*, Springer-Verlag, Berlin, Heidelberg, New York, 254-272 (1992)
[2] B. Markert, in: B*iogeochemistry of Trace Metals*, Adriano D.C. (Ed.), Lewis Publishers, Boca Raton, 401-428 (1992)
[3] B. Markert and W. Wtorowa, *Vegetatio*, **98**, 43 (1992)
[4] S. Hannappel, in: *Sampling of environmental materials for trace analysis*, B. Markert (Ed.), VCH-Publisher, Weinheim, New York, Tokyo, 493-509 (1994)
[5] S.E. Allen (Ed.), *Chemical Analysis of Ecological Materials*, Blackwell Scientific Publications, Oxford, 557 (1974)
[6] J. Benton Jones, R.L. Large, D.B. Pflederer and H.S. Klosky, *Crops Soils*, **23**, 15 (1971)
[7] D.H. Brown and R.M. Brown, in: *Element Concentration Cadasters in Ecosystems*, H. Lieth and B. Markert Eds., VCH-Publisher, Weinheim, New York, Tokyo, 55-62 (1990)
[8] G.W. Bryden and L.R. Smith, *Amer. Lab.*, **21**(7), 30 (1989)
[9] G.W. Bryden and L.R. Smith, *Amer. Lab.*, **21**(9), 19 (1989)
[10] U.M. Cowgill, in: *Sampling of environmental materials for trace analysis*, B. Markert (Ed.), VCH-Publisher, Weinheim, New York, Tokyo, 187-202 (1994)
[11] U.M. Cowgill, in: *Sampling of environmental materials for trace analysis*, B. Markert (Ed.), VCH-Publisher, Weinheim, New York, Tokyo, 347-364 (1994)
[12] Zueng-Sang Chen, in: *Sampling of environmental materials for trace analysis*, B. Markert (Ed.), VCH-Publisher, Weinheim, New York, Tokyo, 365-378 (1994)
[13] E.M. Dick, in: Sampling of environmental materials for trace analysis, B. Markert (Ed.), VCH-Publisher, Weinheim, New York, Tokyo, 255-278 (1994)
[14] R. Djingova and I. Kuleff, in: Sampling of environmental materials for trace analysis, B. Markert (Ed.), VCH-Publisher, Weinheim, New York, Tokyo, 395-414 (1994)
[15] W.H.O. Ernst, in: *Element Concentration Cadasters in Ecosystems*, H. Lieth and B. Markert (Eds.), VCH, Weinheim, 17-40 (1990)
[16] O. Fränzle, in: *Sampling of environmental materials for trace analysis*, B. Markert (Ed.), VCH-Publisher, Weinheim, New York, Tokyo, 305-320 (1994)
[17] A. Gomez, R. Leschber and P. L'Hermite (Eds.), *Sampling Problems for the Chemical Analysis of Sludge, Soils and Plants*, Elsevier Applied Science Publishers, London and New York, 94 (1986)
[18] H. Guhr and E. Weber, in: *Sampling of environmental materials for trace analysis*, B. Markert (Ed.), VCH-Publisher, Weinheim, New York, Tokyo, 223-248 (1994)
[19] G.V. Iyengar, *Anal. Chem.*, **54**, 554A (1982)

9.6 References

[20] G.V. Iyengar, *Elemental Analysis of Biological Systems, Biological, Medical, Environmental, Compositional and Methodological Aspects*, CRC Press, Boca Raton, 430 (1989)

[21] R. Jayasekera, in: *Sampling of environmental materials for trace analysis*, B. Markert (Ed.), VCH-Publisher, Weinheim, New York, Tokyo, 443-447 (1994)

[22] L.H. Keith (Ed.), *Principles of Environmental Sampling*, ECS Professional Reference Book, American Chemical Society, Washington, 458 (1988)

[23] H. Klapper, in: *Sampling of environmental materials for trace analysis*, B. Markert (Ed.), VCH-Publisher, Weinheim, New York, Tokyo, 203-221 (1994)

[24] M. Kovacs, in: *Sampling of environmental materials for trace analysis*, B. Markert (Ed.), VCH-Publisher, Weinheim, New York, Tokyo, 435-442 (1994)

[25] G. Kraft, in: *Sampling of environmental materials for trace analysis*, B. Markert (Ed.), VCH-Publisher, Weinheim, New York, Tokyo, 3-7 (1994)

[26] V. Maavara, A. Martin, A. Oja and P. Nuorteva, in: *Sampling of environmental materials for trace analysis*, B. Markert (Ed.), VCH-Publisher, Weinheim, New York, Tokyo, 465-489 (1994)

[27] B. Markert, *Verhandlungen der Gesellschaft für Ökologie*, Osnabrück, Vol. XIX, III, 345-373 (1991)

[28] B. Markert and N. Klausmeyer, *Toxicol. Environm. Chem.*, **25**, 200 (1990)

[29] B. Markert and V. Weckert, *Water, Air and Soil Pollut.*, **43**, 177 (1989)

[30] B. Markert and V. Weckert, *Toxicol. Environm. Chem.*, **40**, 43 (1993)

[31] A. Paetz and G. Crößmann, in: *Sampling of environmental materials for trace analysis*, B. Markert Ed., VCH-Publisher, Weinheim, New York, Tokyo, 321-334 (1994)

[32] R. Puls, in: *Sampling of environmental materials for trace analysis*, B. Markert (Ed.), VCH-Publisher, Weinheim, New York, Tokyo, 287-302 (1994)

[33] M.H. Ramsey, in: *Sampling of environmental materials for trace analysis*, B. Markert (Ed.), VCH-Publisher, Weinheim, New York, Tokyo, 93-108 (1994)

[34] R.W. Scholz, N. Nothbaum and Th. May, in: *Sampling of environmental materials for trace analysis*, B. Markert (Ed.), VCH-Publisher, Weinheim, New York, Tokyo, 335-345 (1994)

[35] R. Smith and G.V. James, *Anal. Sci. Monographs*, Vol. 8, Dorset Press, Dorchester, 191 (1981)

[36] Z. Tuba, in: *Sampling of environmental materials for trace analysis*, B. Markert (Ed.), VCH-Publisher, Weinheim, New York, Tokyo, 415-434 (1994)

[37] G. Wagner, *Element Concentration Cadasters in Ecosystems*, H. Lieth and B. Markert (Eds.), VCH-publishers, Weinheim, 41-54 (1990)

[38] J.G. Watson and J.C. Chow, in: *Sampling of environmental materials for trace analysis*, B. Markert (Ed.), VCH-Publisher, Weinheim, New York, Tokyo, 125-161 (1994)

[39] B. Zielinska and E. Fujita, in: *Sampling of environmental materials for trace analysis*, B. Markert (Ed.), VCH-Publisher, Weinheim, New York, Tokyo, 163-184 (1994)

[40] R.D. Zimmermann and W.E. Plankenhorn, *Allgemeine Forst Zeitschrift*, **41**, 33 (1986)

[41] W. Horwitz, L.R. Kamps and K.W. Boyer, *J. Assoc. Off. Anal. Chem.*, **63**, 1344 (1980)

[42] B. Markert, *Instrumentelle Multielementanalyse von Pflanzenproben*, VCH-Publisher, Weinheim, 266 (1993)

[43] G. Tölg, *Naturwissenschaften*, **63**, 99 (1976)
[44] B. Sansoni Ed., *Instrumentelle Multielementanalyse*, VCH-Publisher, Weinheim, New York, Tokyo, 782 (1985)
[45] O.U. Anders and J.I. Kim, *J. Radioanal. Chem.*, **39**, 435 (1977)
[46] B. Sansoni, *Fresenius' Z. Anal. Chem.*, **323**, 573 (1986)
[47] B. Sansoni, R.K. Iyer and R. Kurth, *Fresenius Z. Anal. Chem.*, **306**, 212 (1981)
[48] R. Engelke, *Dissertation*, Universität Hamburg, 101 (1984)
[49] B. Markert and R. Steinbeck, *Fresenius Z. Anal. Chem.*, **331**, 616 (1988)
[50] H. Lieth and B. Markert, *Aufstellung und Auswertung ökosystemarer Element-Konzentrations-Kadaster*, Springer-Verlag, Berlin, Heidelberg, New York, 193 (1988)
[51] H. Lieth and B. Markert (Eds.), *Element Concentration Cadasters in Ecosystems*, Methods of Assessment and Evaluation, VCH, Weinheim, 448 (1990)
[52] W.H.O. Ernst, in: *Sampling of environmental materials for trace analysis*, B. Markert (Ed.), VCH-Publisher, Weinheim, New York, Tokyo, 381-394 (1994)
[53] B. Markert (Ed.), *Plants as biomonitors - Indicators for heavy metals in the terrestrial environment*, VCH-Publisher, Weinheim, New York, Tokyo, 644 (1993)
[54] G. Wagner, *Beiträge zur Umweltprobenbank*, M. Stoeppler and H.W. Dürbeck Eds., 412, 225 (1987)
[55] P. Müller and G. Wagner, *BMFT-Forschungsbericht T 86-040*, 115 (1986)
[56] V. Krivan and G. Schaldach, *Fresenius Z. Anal. Chem.*, **324**, 158 (1986)
[57] A. Wyttenbach, S. Bajo, L. Tobler and T. Keller, *Plant and Soil*, **85**, 313 (1985)
[58] J.R. Porter, *Environ. Pollut.*, **12**, 195 (1986)
[59] W.H.O. Ernst, *Chemische Rundschau*, **31**, 32 (1978)
[60] H.J. Fiedler und H.J. Rösler (Eds.), *Spurenelemente in der Umwelt*, Ferdinand Enke Verl., Stuttgart, 278 (1988)
[61] J.D. Schladot, F. Backhaus and U. Reuter, *Spezieller Bericht der Kernforschungs-anlage Jülich*, **330**, 25 (1985)
[62] B. Markert, in: *Encyclopedia of Analytical Chemistry*, Elsevier, 1995 (in press)
[63] M. Rossbach, J.D. Schladot and P. Ostapzcuk (Eds.), *Specimen banking, environmental monitoring and modern analytical approaches*, Springer-Verlag, Berlin, Heidelberg, New York, 242 (1992)
[64] H. Muntau, personal communication
[65] B. Markert, *Vegetatio*, **103**, 1 (1992)
[66] H.J.M. Bowen, *Environmental Chemistry of the Elements*, Academic Press, London, 333 pp. (1979)
[67] B. Markert, *Jül. Spez.*, M. Stoeppler and H.W. Dürbeck (Eds.), **360**, 166 pp. (1986)
[68] B. Markert (Ed.), *Sampling of Environmental Materials for Trace Analysis*, VCH-Publisher, Weinheim, New York, Tokyo, 524 (1994)
[69] B. Markert, *Fresenius' J. Anal. Chem.*, **345**, 318 (1993)
[70] U. Herpin, B. Markert, U. Siewers and H. Lieth, *Monitoring der Schwermetallbelastung in der Bundesrepublik Deutschland mit hilfe von Moosanalysen*, Forschungsbericht 108 02 087 des Umweltbundesamtes Berlin, 161 (1994)
[71] B. Markert, *The Sci. Total Environ.*, **155**, 211 (1994)

10.

An holistic structure for quality management: a model for marine environmental monitoring

D.E. Wells[1] and W.P. Cofino[2]

[1] Scottish Office Agriculture and Fisheries Department, Marine Laboratory, POB 101 Victoria Road, Aberdeen AB9 8DB, United Kingdom
[2] Institute for Environmental Studies, Free University, De Boelelaan 1115, 1081 HV Amsterdam, The Netherlands

10.1 Introduction

The quality management and laboratory accreditation systems [1-5] (ISO9000, EN45001, BS5750) available at present do not comprehensively cover all aspects of the quality of information required for international marine monitoring programmes (MMPs). The current standards focus on specific quality needs and the overlap of systems is limited since they have been developed to support quite distinct needs. The three main areas covered at present are: (1) the quality management covered by the International Standards Organization, (ISO9000), the European Standard (EN29000), and the UK National Standard (BS5750); (2) laboratory accreditation covered by national organizations, such as NAMAS (UK), STERLAB (NL), SWEDAC (S), SAS (CH), and in Europe under EN45001; and (3) the Good Laboratory Practice (GLP) operated by the Organization for Economic Cooperation and Development (OECD). More specifically in the UK, GLP is operated by the Department of Health [6] for laboratories performing nonclinical laboratory trials in relation to the safety of regulatory compounds and their effect on the environment and man.

Quality management standards such as ISO9000 are generally regarded as horizontal systems which can be applied over a number of widely differing activities and functions, not confined to scientific or administrative applications. The whole series of these standards is *management* orientated (Fig. 10-1). Laboratory accreditation, on the other hand, is primarily *techniques* orientated and focuses specifically on the area of measurement and testing in the laboratory. The *management* aspects are limited to laboratory organization, audits, and staff audit and staff training to undertake the specific tests or measurements. It is by nature rather specific and precise in its application and execution. Laboratory accreditation focuses primarily on the traceability of measurement and testing to specified standards (Fig. 10-1).

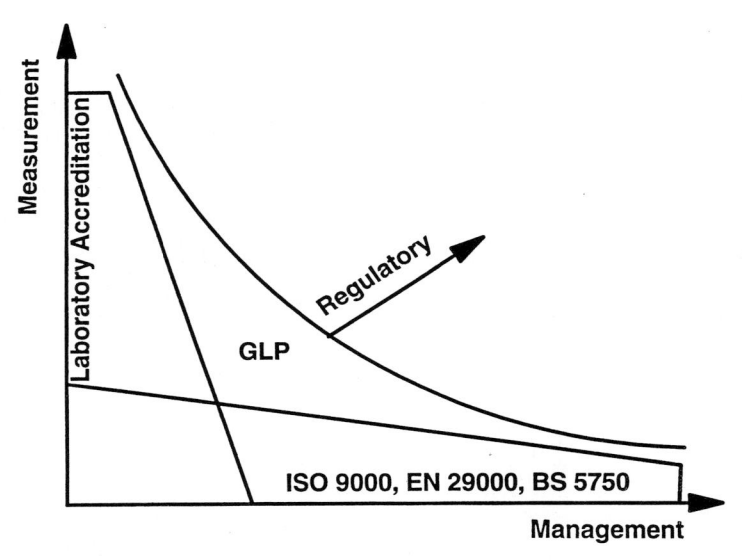

Figure 10-1: Relationship of management and measurement quality systems. GLP: Good Laboratory Practice. Ref: Scottish Office Agriculture and Fisheries Department (SOAFD) course on GLP, A. Waddell Inveresk Research, Musselburgh, Scotland.

GLP, however, includes both measurement and management in covering specific studies and filling some gaps left in the other two systems. The formal UK GLP is accepted worldwide, but is limited to determining the effect of regulatory compounds. The Environmental Protection Agency (EPA) and the Food and Drug Administration (FDA) in the US have their own internal GLP systems. Apart from the OECD program there is currently no European equivalent. More informal use is made by individual laboratories implementing their own quality system based on similar GLP procedures. However, these have no formal recognition and, therefore, different standards can exist between different laboratories. The standards, protocols, and operation of an informal quality system can suffer from a lack of conformity and are not always fully compatible between laboratories or countries.

The advantages of the ISO9000, EN29000, BS5750 standards, GLP, and laboratory accreditation (NAMAS, STERLAB, EN4501) is that they are well developed, internationally accepted norms; this creates a common platform in specific areas. The main disadvantage of such systems is that they do not cover all aspects of quality. Laboratory accreditation requires documentation of data of *known* quality, not necessarily of the highest quality. It is the responsibility of each laboratory to ensure that the quality is suitable for the purpose of the measurement. The standard of performance for each accredited measurement or test can, for example, be set by the institute at a relatively less stringent level of precision and bias than is actually possible, even on a routine basis. The laboratory goal can be simply to be accredited and conform to a standard of *known* quality, but have no incentive within the system to *improve* the performance of the measurement. Laboratory accreditation is clearly becoming a necessary part of quality assurance for those institutes undertaking chemical measurements in the marine environment, but is not on its own sufficient to ensure that data obtained from all laboratories are of a specific quality.

What is missing from these quality systems is an holistic framework (Fig. 10-2) which can include continuous assessment, stepwise improvement, and training programs in an international setting. In addition to the laboratory work, the quality control of measurements must be extended to field work, sampling and storage. Finally the quality objectives of the international programs should be included in this framework. Each of these additional elements of the holistic framework is now considered.

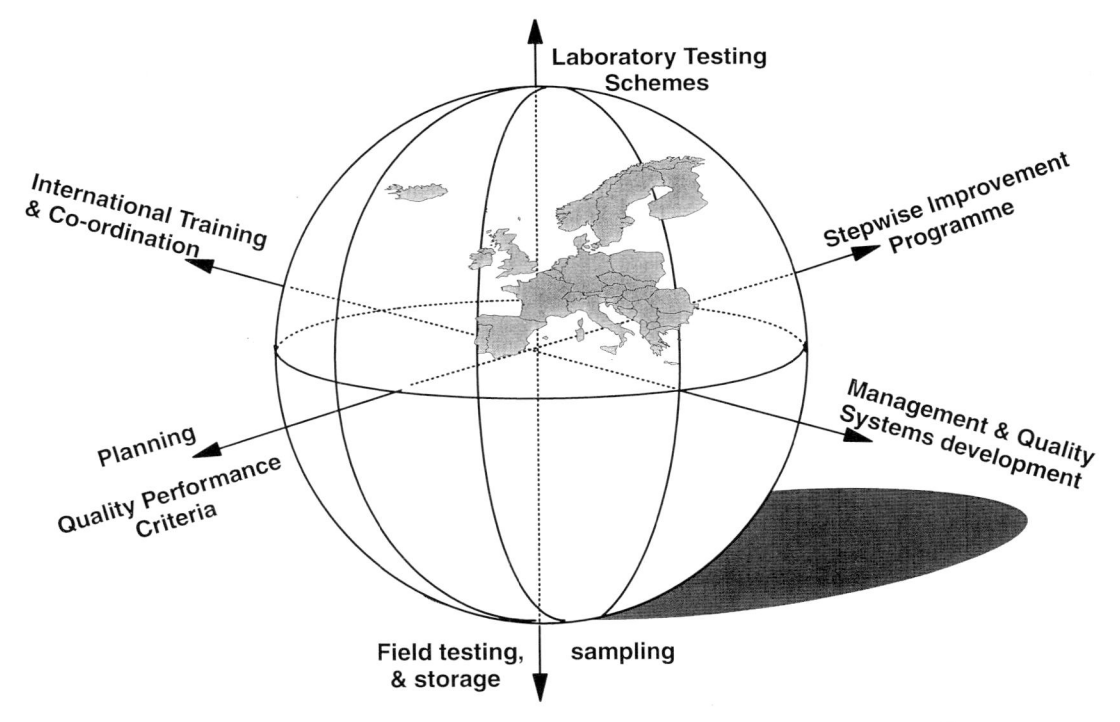

Figure 10-2: The extension of the elements of management and measurement quality systems to include: (a) continual external assessment of quality through laboratory testing schemes; (b) a stepwise improvement and learning program to reduce sources of error; (c) extension to national and international coordination; (d) development of QA in field measurements, storage and sampling; and (e) quality measurement as a function of the monitoring objectives.

10.1.1 Continuous assessment

Current standards do not automatically include a facility for assessing the quality of the information. Do these data achieve the objectives for which the information was gathered? Laboratory accreditation successfully documents the accuracy of the data and requires these data to be fully traceable. A laboratory can also include specific quality performance criteria in its own accreditation documentation, but this practice is not essential for accreditation. Normally, it is assumed that the institute will judge whether the information obtained will meet its own project objectives. However, the quality required for the chemical measurements to meet the MMP strategies is still a developing theme [7-9]. At the very least it is necessary to establish targets for the analytical bias and precision of the chemical measurements.

10.1.2 International comparative assessment

Quality management and laboratory accreditation systems relate to quality activities within a single laboratory and not between institutes, and there is no provision for a *comparative* assessment of QA data in cooperative programs. Laboratory accreditation is recognized between institutes in one country, for example through the National Accreditation of Certified Bodies (NACCB) in the UK, and between countries in Europe through the European Accreditation Multilateral Agreement and European Accreditation Certification. The documentation, accountability, and traceability operate to an agreed standard, but the actual level of performance of the measurement between laboratories can be different. So the precision and bias, for example, can be set at different target values for the same measurement in different laboratories, but *both* be accredited under international agreement.

Some intercomparison of analytical quality control (AQC) data is available through national and international schemes, such as the Food Analysis Performance Assessment Scheme (FAPAS) [10,11] operated by the Ministry of Agriculture Fisheries and Food (MAFF), UK, the National Marine Analytical Quality Control (NMAQC) scheme operated by the UK Department of the Environment (DoE), and the US National Oceanographic and Atmospheric Administration (NOAA) Status and Trends QA Program. Each of these programs focuses on the AQC which goes part of the way to achieving a more comprehensive coverage of the quality requirements.

10.1.3 Stepwise improvement

The third element which is absent in existing systems is a structured framework for **improvement**. Norms and standards are essential to describe, control, and document the quality of information, but these systems usually require a certain standard of attainment *prior to* participation and acceptance. This standard may be requested by an external customer to judge the quality of the data before using the information or placing a contract. However, these established quality systems do not include an improvement program for the new and developing laboratory, or for the good laboratory which seeks a higher standard of performance. In some instances the criticism could well be made that accredited institutes which operate under ISO9000 (or equivalent) may view their standard of performance as adequate and have no perceived need to improve.

10.1.4 Field studies and sampling

Factors other than AQC contribute to the quality of information and must be included in a comprehensive, *holistic*, scheme [12]. Sampling, sample handling, and storage are notorious for introducing artifacts and interferences which ultimately affect the final determination. Inadequate methods of data transcription, assessment, and storage increase the errors in the information base, along with poor data for cofactors, such as lipid measurement in biota or total organic carbon (TOC) in sediments.

Accreditation currently focuses on the *laboratory* and, with few exceptions, does not cover the broader aspect of QA relating to *field measurements*. Many problems are related to the sampling strategies, the techniques used, and the equipment and the manner in which it is deployed. Correct sample handling and storage methods prior to the material arriving in the laboratory are of paramount importance to the preservation and integrity of the material. In many cases e.g. nutrients in sea water, the whole cycle of sampling and analysis is conducted at sea in conditions quite different from those which exist in the shore laboratory. The analyst is in somewhat of a dilemma since attempts to preserve the sample and return it to the laboratory are fraught with as many, but different, problems as are encountered when making the determination on board ship often in less than ideal conditions. General guidelines [13,14] do exist to cover these aspects, but there are no controlling or unifying protocols beyond the laboratories own (good?) laboratory practice.

The criteria for data assessment and particularly the quality of the project objectives must be scrutinized to obtain a complete picture of the quality requirements for the information **as a whole.** Topping [15] outlined the role and application of QA in marine environment protection in which he lists 13 elements of QA practice for the individual laboratory. These principles encompass the requirements of the laboratory, but go far beyond just AQC to wider aspects which cover the quality of the whole process of obtaining the chemical information.

10.1.5 The holistic project "QUASIMEME"

The elements of QA required to monitor and improve the reliability of information have been developed in an holistic framework [12] by the EU Standards, Measurement and Testing Project **'QUASIMEME'** (Quality Assurance of Information in Marine Environmental Monitoring in Europe) (Fig. 10-2). The QUASIMEME initiative has focused on a stepwise improvement of the quality of management, operation and communication for the whole information cycle, from planning through to data assessment (Fig. 10-2) in support of international marine monitoring programmes (MMPs) in Europe.

These monitoring programs form part of the responsibilities of the new Assessment and Monitoring Group (ASMO) under the auspices of the Oslo and Paris Commissions (OSPARCOM), the Baltic Monitoring Programme (BMP) of the Helsinki Commission (HELCOM), and the Barcelona Convention in the Mediterranean. The ASMO group has subsumed the work of the former North Sea Task Force [16,17] (NSTF) and the North Sea Quality Status Reports. European marine monitoring programs are a multinational undertaking and those countries which contribute to these programs formulate a joint policy and agree on those measures necessary for the protection of the marine environment. Data on trace metals and trace organic contaminants in biota and sediment and nutrients in sea water form the core of the chemical information for these programs. These chemical data along with their cofactors and associated information e.g. sample location and conditions of sampling, must have a high and well defined level of reliability if the interpretation and subsequent marine protection policies are to have a sound basis.

Hence the quality of these data has a direct influence on the formulation of environmental policy. It also has a significant economic impact. Any chemical analysis undertaken to provide specific environmental information has a finite cost e.g. detection of a trend, say, 10% decline over 10 years may not be possible in economic terms unless the quality of data is very high. The *cost* of this *information* is directly related to the accuracy of the measurement. An imprecise or biased measurement will require a significantly greater number and frequency of samples to detect the required change with a given confidence. For many measurements the savings can be very substantial. Data which do not meet the standard required will ultimately mean that all of the resources invested are wasted.

There is also a diplomatic benefit. Improved quality will result in a greater trust in the data generated in different countries. Replication of the same analysis in each country can be minimized, and corporate, complementary studies can be developed in a harmonized, international setting.

10.2 The holistic philosophy

The underlying principles and practices of the holistic approach for an international program are now discussed using the QUASIMEME project to draw upon illustrative examples.

The term *quality of information* is used because the project covers much more than just the measurement or the analytical procedure. The quality of the data, e.g. the concentration of copper in sediment, will, for its environmental interpretation, be dependent upon a number of cofactors such as the aluminum and or total organic carbon content, the grain size and the location within the sediment layers, and the geographic position of the sample. Information on the manner of sampling, e.g. grab or core sample, and the subsequent sample handling techniques are all part of the integrity of the data on the sediment. All these data collectively produce the information necessary to make a true environmental assessment. The quality of each part of these data make up the whole, corporate picture.

10.2.1 Cooperative interaction

The holistic approach requires a clear framework and a focused set of objectives. It requires a cooperative, confidential atmosphere where the participants regard themselves as stakeholders in the venture. They own the collective information and share the success of the project. One way to create this cooperative atmosphere is to allow time for people meet informally. The *workshop* style of forum may appear to be an overplayed option, where most people are willing to travel at the expense of the project. However, the combination of active involvement in the discussion of a common problem in a relatively isolated, but pleasant venue will stimulate good, focused interaction. People meet with a unifying background, and a commitment is made, not to a system, but to a fellow worker in another aspect of the international project. This may seem idealistic, but it works! These practical and essential steps help participants to become stakeholders in the development of the program. The success of this type of project is entirely dependent upon the commitment of individuals in each of the contributing laboratories.

10.2.2 Interlinked resources

An holistic program should not repeat those aspects of quality management which are already in place, but should subsume them in the overall structure. Duplication in any area should be avoided since quality management and measurement can take as much as 15-20% of the staff resources in individual laboratories as well as the coordination and assessment teams for the whole program. Many countries, within the European Union for example, have their own accreditation bodies. Each participating institute is encouraged and supported to develop its **own** quality system within its **own** laboratory to the standards and specifications agreed with its accreditation body. The program should never be considered as a surrogate for accreditation or equivalent, nor as a replacement for the laboratory's own QA system.

National (e.g. UK NAMQC) and international initiatives have been developed to establish laboratory testing schemes and to improve the quality of measurement [18]. Each of these schemes is time-consuming for the participant, so it is essential that an holistic project such as QUASIMEME does not compete, but complements such studies. Where possible, closely related studies should either be developed in harmony or combined to form a larger program. Participants dislike replicate QA studies which provide effectively the same information to more than one organization. A single laboratory testing scheme for a specific area, e.g. QUASIMEME, can provide the necessary information to both the national and the international agencies responsible for the marine environment within European waters and the North Atlantic.

10.2.3 Advice and guidelines

Clear advice must be given and guidelines produced on laboratory protocols and practices which are specific for the area of interest so that all the institutes in each country are following the same common approach. The holistic approach should take into account **all factors** which impinge on the final outcome of the information in these guidelines [19].

For example, extractable lipid measurement has been used as a cofactor for the determination of persistent organochlorine compounds in biological tissues. Until recently laboratories have been using a wide range of methods from extraction with chloroform methanol water mixtures [20] to a simple extraction with a nonpolar solvent such as hexane. For some fatty tissues the difference between the data obtained by the different methods has been relatively small, but for low-fat fish muscle tissue the range of values in lipid measurement can increase the variability of organic contaminant measurements by 100-200%, thus completely negating the use of a cofactor as a means of reducing the variability of the data! Cooperative studies between participants have resulted in agreements on a methodology and recommendations on a total extraction method. Studies with the whole group of participants have:

- undertaken a stepwise improvement and intercomparison programme for lipid measurement;
- identified poor methodologies and recommended their cessation of use; and
- agreed on methods which give acceptable results and confirm their application in practice.

10.2.4 Flexible framework

The holistic program should not attempt to be a rigid structure which fixes the specific way a measurement is made or information gathered. The quality of the information required should be stated and the laboratory should provide evidence of the validated methods used [3].

Consensus agreement to use reference methods, identical method protocols, and standard operating procedures (SOPs) in all laboratories is difficult to implement particularly in an international setting. Apart from the practical problems of conformity, the fundamental weakness in using a reference method is the likelihood of producing biased data. Any chemical measurement is only the best estimate of the true value. If such measurements are produced using a single method it is difficult to demonstrate that these values are free from bias. However, this does not imply that any method can be used. Different, validated methods should provide the best means of producing data with a minimum acceptable bias which can be recommended in the MMP QA guidelines [19]. Above all these recommendations should exclude unsound practices, give good advice, but not necessarily prescribe a specific method/technique for use in detail.

A recurrent analytical problem associated with the digestion of trace metals from sediment shows the value of using different methods for the same determination. The *total* methods involve the use of hydrofluoric acid (HF) or nondestructive methods like proton induced X-ray emission (PIXE) and X-ray fluorescence (XRF), and the *partial* digestion methods use strong acids e.g. concentrated nitric acid. In some instances, especially when the overall concentration of trace metals is low, the values obtained by the two techniques are method- dependent, with the *total* method generally giving the higher values [21].

A laboratory may, where they are not fully equipped to use HF, select the partial method on the grounds of health and safety. In addition each MMP can recommend *different* methods. At present the Baltic Monitoring Programme of the Helsinki Commission recommends the use of partial digest methods and the ASMO group of OSPARCOM recommend the use of total digest methods. A QA program would need to support *both* monitoring programs, without necessarily making a value judgement on the current program in place. While the laboratories are required to submit data to both programs it is essential that these data are of known quality. Therefore a QA assessment of both partial and total extraction methods should be made independently. Where the data from the two methods are statistically different then these points should highlighted in the reports [21] .

10.2.5   Collaborative improvement

It is reasonable to assume that no institute wishes to produce poor data or worthless information. A fundamental action of an international holistic QM program must be to have a supportive rather than a policing role. Assessments made on laboratory performance should be structured toward improvement rather than on creating the "league table". This support is given through positive feedback in reports, workshops, and above all in training/learning programmes to aid improvement.

Within the QUASIMEME studies extreme values in a data set were not rejected as outliers, but were regarded as useful information which can highlight a particular source of error. This can be reinforced by the style of interrogative data analysis used. Formal identification of outliers and systematic parametric test are not used in QUASIMEME (e.g. ISO5725) [22], firstly, because the data are rarely normally distributed and, secondly, by using a critical visual inspection of the data and a robust statistical approach it is possible to obtain a good estimate of the robust mean of the population of *good* laboratories and calculate a score for the performance of each participant (Z score) without eliminating data [23]. By focusing positively on the extreme values it is likely that a well documented explanation can benefit all participants.

This is not to make a virtue out of errors, but to serve as a guide in preventative action for future work. Every effort should be made to assist participants producing extreme values in identifying the root causes of the problem. Poor performance is possible for all participants. The main difference between a good laboratory and one which performs poorly is that the good laboratory takes immediate action as a result of the mistake and remembers the lesson learnt.

One QUASIMEME laboratory which usually produces good trace organic data discovered that poor precision was a result of sample vials containing volatile solvent being store in the refrigerator. It would be reasonable to conclude that it was sound practice to reduce the sample temperature to minimize losses due to volatility. However, on cooling the vial the tightly held screw cap became loose because of the difference between the coefficients of expansion of the glass vial, the plastic top, and the PTFE seal. A finger-tight seal with the vial remaining at room temperature was actually a more secure mode of storage. These apparently trivial details can make a significant difference to the quality of data and are often only identified by painstakingly detailed investigation.

10.3 The elements of an holistic project

The implementation and maintenance of an holistic project are dependent on four key elements

- its focus - do not try to be all things to all people;
- its framework - work well with the team that you have;
- its function - do well what you know and learnt well what you do not;
- its future - see from the beginning what the end might be.

10.3.1 Focus

The area of interest should be positively and clearly defined. Peripheral subjects may be important in their own right, but it is essential to establish the boundaries or any changes as the project progresses. The initial subjects for the QUASIMEME project included all types of physical and physicochemical measurements e.g. temperature, turbidity, and biological parameters e.g.chlorophyll *a*. Once the project was publicized the list of potential participants also grew with extraordinary rapidity within all countries. The size and scope of the project has direct implications on the resources, management, and organization, and a focused decision was required on the absolute priorities and the boundaries of the program. This focus can be obtained with a mission statement. For the QUASIMEME project the acronym itself served the purpose.

Quality Assurance of Information in Marine Environmental Monitoring in Europe

10.3.1.1 Focus on the determinands

A focus was made on information relating to the quality of ***chemical measurement***. The project team was clearly aware of the needs in other areas. Physical measurements, which perhaps have the potential for being more easy to control, and biological measurements, which are currently less developed in terms of QA management, were excluded, not because they are not important, but because it was essential to have a defined program and manageable project within the resources available. The list of chemical determinands initially included in the program was also restricted (Table 10-1). Stepwise introduction of other determinands allowed the working program to be established and participants to be orientated without excessive demands in the initial stages of the project.

Table 10-1: The determinands included in the QUASIMEME laboratory testing schemes. This list is based upon the requirements of MMPs [16,17].

Nutrients	Metals	Chlorobiphenyls	Organochlorine * pesticides	PAH ##
Nitrate	Aluminium	CB28	HCB	Benzo[a]anthrecene
Nitrite	Arsenic*	CB52	ppDDE	Benzo[a]pyrene
Ammonia	Cadmium	CB101	αHCH	Benzo[b]fluoranthene
Phosphorus	Chromium*	CB105	γHCH	Benzo[e]pyrene
Total-N*	Copper	CB118	ppDDD	Benzo[g,h,i]perylene
Total-P*	Lead	CB138	ppDDT	Chrysene
	Mercury	CB153	Dieldrin	Fluoranthene
	Nickel*	CB156	Trans-nonachlor	Indeno[1,2,3-cd]pyrene
	Zinc	CB180		Phenanthrene
				Pyrene

Determinands to be added to the QUASIMEME Laboratory Testing Scheme.

Nutrients	Metals	Chlorobiphenyls	Organochlorine pesticides	PAH
Silicate	Methyl mercury TBT	Non-ortho CBs CB77, CB 126, CB 169 Polybrominated fire retardants	Polychlorinated camphenes Chlordanes	Naphthalene C$_1$& C$_2$alkyl derivatives Anthracene, Dibenzothiophene

* Determinands added in the Laboratory Testing Scheme Round 2
\## Determinands added in the Laboratory Testing Scheme Round 3

10.3.1.2 Focus on the environment

The QUASIMEME project focused on the *marine environment* and data from all other environments were excluded from this program. The marine environment has its own specific problems. The sampling requirements are unique, the storage requirements are different from those of other materials, and the many aspects of the analytical methods are tailored to marine media and have to be optimized to account for the cofactors, e.g. lipids, and the range of concentration found. Specific marine problems require specific solutions obtained through appropriately trained expertise in specialist laboratories.

The current European marine environmental programs have a list of mandatory determinands, a list of voluntary determinands, and a list of "new chemicals[1]". These distinct groups were used as a basis for organizing and developing the program. A specific combination of determinand and matrix was selected on the basis of the current list of mandatory determinands (Table 10-1).

A gradual introduction of the mandatory determinands enabled the program to be established. These stepwise increments were as important for the participants as well as the central organization. It is quite common, when planning, for participants to become euphoric and propose a long list of essential measurements. Subsequently the same group may excuse themselves for not delivering the data by the deadline due to an overload of work! It is essential to maintain a steady, practicable program that can meet the workload of even the busiest laboratory. A good guide in controlling these activities is to ensure that the planning team includes experienced participants so that there is positive feedback and a realistic workload for the exercises. The determinands included in the program also require the measurement of the most likely cofactors which will affect the validity and future interpretation/assessment of the environmental data.

10.3.1.3 Focus on the participation

Although the project is committed to the quality and improvement of measurement there may be a need for restricted participation. There are clear financial limitations when the funding comes from a central organization. A documented demonstration of improvement is only possible with a constant group of participants. A changing membership makes it difficult to determine any improvement, which is equally of little use to those participating and those who may wish to join. It is essential to demonstrate these achievements (i) to the sponsors who initiate the venture and wish to vindicate their decision in the initial (or future) funding, (ii) to encourage participants and to justify their time and effort, and (iii) to future potential participants who will wish to justify their funding by subscription.

Constant membership also aids the initial organization of the project. Even with an experienced project team, additional staff are required to support and operate the program. An advisory team takes time to build and is more efficient as a small group during the early years. Managerial, operational and financial advantages are gained with a finite number of laboratories for the first three years. Once the project has been established and the benefits of the program have been demonstrated then participation can be broadened.

The selection of the participating laboratories can be both politically sensitive and highly emotive when restriction is necessary. Each potential participant will have a justifiable reason for inclusion. Selection should be made on objective principles [24] which are justifiable not only to the organizers, but to those who wish to participate.

The participants in the QUASIMEME project were selected by the steering group who also served as the national contact person in each country. Laboratories which contributed data directly to the MMPs were the first to be invited. A balance was made between the numbers of participants from each the EU member states in relation to the level of fisheries and marine

---

[1]New Chemicals- recent additions by the Commissions to the voluntary list of contaminants which are considered to have sufficient potential environmental impact and require monitoring

activities. Where it was possible, additional participants were invited from institutes with a direct input into national monitoring plans, and then by national marine research institutes. The specific order and balance of the invitations were confirmed by the national contacts on the steering group.

10.3.2 Framework

An holistic project should solve problems, not reorganize structures. It should, primarily, be developed from *real* needs expressed by the participants in addition to the needs of organizations who will assess and form policies on the basis of the data.

10.3.2.1 The participants

The best people to solve quality management problems are those at the bench who have made mistakes in both field and laboratory work. The advantage of involving all participants from the beginning is that they see the program as a direct support to the solution of their problems. They readily see the need to include this additional QA schedule into their work. Most laboratories are primarily interested in their specific performance and only afterwards consider the wider implications of the overall exercise. The project organizers, however, have additional goals. They have to focus on the collective studies describing interlaboratory comparability and create a more unified database of information which can be used to make a coherent assessment of the environmental studies.

The framework and the direction of the project should carefully reflect the needs of the individual participant. It must be based on real identifiable problems that are common to a number of laboratories and which are not easily overcome by individuals or a small group of specialists. It is important to encourage everyone to take responsibility for their problem(s) and to become stakeholders in the solutions to these problems.

It is important to involve people who actually do the work as contact people and for them to participate in the workshops and seminars. The selection of those people attending meetings within any institute is often made on a hierarchical basis rather than on the requirement of the individual person. In such cases it is essential to stress that it is not the best representative of the institute that is always required, but the person who actually *does* the work and who has the problems.

However, a balance is necessary and it is essential to involve both the analyst and the institute management. Formal contacts between the institutes and the project are vital when establishing the laboratory's participation. Agreement should be made between the directors and responsible scientist regarding the commitment and contribution of each institute to the project. This agreement should include a commitment to:

- contribute for the whole project (3 years);
- comply with the project deadlines; and
- any improvement necessary to obtain the agreed quality of data.

Apart from the need to establish the long-term performance of each laboratory, it is important to demonstrate that the approach taken in such a project works and is cost-effective. In the initial stages this is only possible with the same group of laboratories.

10.3.2.2 The project team

The actual structure and make-up of the team(s) may be different for each project, but the following functions are generally necessary:

- central coordination and management;
- scientific planning, critical evaluation, and dissemination of information; and
- communication and feedback network with participants, customers, and sponsors.

Table 10-2: QUASIMEME project teams.

TEAM	FUNCTION	RESOURCES
QUASIMEME project office	Project Management Scientific & Administrative Co-ordination	Principal Scientist Two full time scientific administrators with secretarial support
Scientific Assessment Group	Project planning with the project office Assessment of laboratories Lead/Organization of workshops Preparation of RMs	Six scientists 3-5d per annum 12-15d per annum 10-15d per annum Under separate contract
Steering Group National Contacts representatives of Commissions	Approval of programme Discemination of information & material Provide feedback Organize National/regional QUASIMEME discussion days	18 participants 1meeting pa 5-10 days
Participants	6 monthly laboratory testing scheme Nutrients in seawater Metals in sediment & biota Organics[1] in sediment & biota Sampling programmes QM improvement programme Workshops Questionnaires	89 Institutes, 180 participants which contribute to each of the different sections of the project

1Organics= chlorobiphenyls, organochlorine pesticides, polynuclear aromatic hydrocarbons

The project *teams* have a vital, complementary role in the success of the development of the project. *Team spirit* [25] may be a popular management cliche, but if such a program is to be successful it is essential for people to work well together with the right skills, especially in a multinational framework. The four QUASIMEME *teams* (Table 10-2) are the project office staff, the scientific assessment group, the national contacts who also act as a steering group, and, last but not least, the participants. Although the participants have been discussed above it is imperative part of the holistic philosophy to include them as part of the project team. It is ***not*** a project organized by one group on behalf of the participants. Full involvement and feedback from the participants is a major part of the program development and any organization which ignores this input from the laboratories does so at its peril. The members of the teams should preferably be well known to each other and within each team there should be:

- minimum (or no) hidden agendas;
- open and full cooperative activities;
- common goals; and
- complementary skills between and within teams.

Almost all communication with the participants is channeled through the project office providing a common center for the dissemination of information. The project office must provide highly flexible support to cope with changing demands. All information and data must be fully traceable and participants must feel from the outset that the project staff are highly dependable and trustworthy. A 1% error in a data set may translate into a 100% error for an individual participant!

Although the central office may be the focus of the project it should not be perceived as the authority of the study. This lies with the collective leadership through the scientific group and ultimately the steering group which comprises national contacts from each country. In this way the participants have direct feedback, and influence on the direction and content of the program.

The scientific group should comprise a small number of well known and experienced scientists (6-8 maximum) each of whom bringing unique skills and information; the group which provides:

- supervision and distribution of materials;
- protocols and guidelines;
- critical assessment of exercises;
- progress reports and scientific papers; and
- organization of workshops.

The steering group provides the third tier in the organizational structure and act both as a liaison group and as an advisory board. In this role its members evaluate, modify, and agree the work program of the project. In a complementary role the members of this group are national contacts to provide information, encouragement, and a clearing house for feedback as a result of national correspondence and discussion.

10.3.2.3 The customers

It is vital that the project is focused on the *customers*. For QUASIMEME, the customers are the international commissions who maintain the MMPs by international agreement. Within the framework of these commissions the monitoring and assessment groups (e.g. ASMO-OSPARCOM) regularly reconsider the direction and the content of the monitoring programs. The chemical determinands, matrices, and sampling strategies may change or be modified as different environmental issues evolve. Different compounds are added to the list and the balance between mandatory and voluntary determinands can alter.

The sole purpose of a QA program like QUASIMEME is to underpin the MMPs and aid the environmental data assessors in adjudicating on the quality of these data and their environmental significance. It is essential that the QA program mirrors the environmental monitoring program. The key requirements are:

- good advance planning;
- inclusion of new determinands in the program;
- establishment of a laboratory testing scheme which matches the needs of the MMPs; and
- provision of reference materials (RMs) to match the developing MMPs.

The introduction of a new determinand in an MMP takes time since each participating laboratory will need to include new methodology in their own monitoring program. This additional work will cause some delay as methods are tested and validated. This effort can be shared within the framework of an holistic program. Good dialogue and cooperative evaluation can reduce the risk of introducing poor methodologies and the time needed to obtain the necessary between-laboratory agreement for new determinands.

10.3.2.4 The sponsors

Finally the sponsors are an essential part of the project framework; in many cases they are also the customers. Under normal circumstances some initial funding is essential to establish the project during the first three to four years of the programme and to demonstrate the success of such an approach, after which it would be possible to fund the development from the support of the users or the customers.

The QUASIMEME project was sponsored by the European Union Directorate General (DG XII) Standards, Measurement and Testing (SM&T) programme which has, as a fundamental objective, improvement of the level of agreement between member states, for example, for those chemical determinands listed in the MMP of the Oslo and Paris Commissions. Ongoing sponsorship of such a project may come from different sources: from government, from a local or national agency, or from the participants themselves through subscription.

10.4 Functions

An international holistic QA program has three main functions to develop within and between the participating laboratories [12]:

- management;
- operation; and
- communication.

10.4.1 Management

The management functions relate specifically to the support for the development of the quality systems in each laboratory as it relates to chemical measurement for the MMPs. In addition, guidelines are also necessary to provide more details in those areas not covered by these systems. When participants are brought together in a collaborative program the management practices in each laboratory can be very different. A small number of institutes will operate at a very high level of quality management, under ISO9000, EN29000, or BS5750 and be fully accredited for all the measurements made in the laboratory, while other institutes may have little in the way of quality systems. Most of the institutes, however, are likely to be somewhere between these two extremes.

To assist the institutes in the management of their monitoring programs the following requirements were considered necessary:

- a relatively detailed assessment of the current QM systems used in each laboratory;
- a program to improve and to encourage critical self-assessment of the laboratory's management systems; and
- each laboratory should set specific QM goals for themselves and be encouraged to reach them.

Within the QUASIMEME program each participant was sent a detailed questionnaire containing approximately 200 questions on general and specific aspects of QM in relation to the measurement of the mandatory determinand in marine matrices [26]. As a result of the information from the questionnaire it was possible to assess the most important areas of need, and each participating laboratory was requested to identify a *specific* aspect of QM which could be improved *in their laboratory*. The improvement program was then framed into achievable goals with specific steps and definitive deadlines set by the participants themselves. A copy of this part of the improvement programme was kept at the QUASIMEME project office to monitor progress. In this way each participant improved according to their own ability, direction and agenda.

An example of the type of improvement proposed might be:

"To implement the routine use of control charts for the determination of nickel in marine sediments"
1. obtain sufficient laboratory reference material (LRM) for 12-18 months;
2. establish the nickel content of the LRM with the aid of appropriate certified reference materials (CRMs);
3. include the new LRM with each batch of analyses
4. establish warning and action limits in relation to the laboratory's long-term precision measurements;
5. plot the data obtained and take corrective action when extreme values occur; and
6. review performance.

Workshops and seminars on the organization and QC of working practices are a necessary part of the laboratory improvement scheme. Most participants do not want a lecture on the principles and the theory. That is readily available for all who wish to read. What most people need are the detailed, pragmatic steps and advice that can be offered to address tangible problems that are commonly experienced. As the QUASIMEME program developed, it was clear that a significant number of mistakes were occurring, not as a result of poor analysis, but as a direct result of:

- incorrect calculation;
- transposition of data; and
- transcription errors.

Factors of 2, 5, 10 and 1000 were common for transposition errors, e.g. lead data entered in the column for copper, and the less obvious "69" for "96". To help overcome this type of error the following checklist was suggested.

> Participants should:
> 1. conduct an audit and ensure that sample information is traceable to the sample result;
> 2. read the protocol *carefully* and check the requirements for reporting data, i.e *units!*;
> 3. make regular random checks by calculating by hand to reduce the risk of the wrong factors and sample weights being used;
> 4. check and confirm that the QC data is within specification;
> 5. double check tables prepared by the typist and remember it is not a typing error but a proof-reading error, i.e. the responsibility remains with the scientist not the typist;
> 6. prepare a printout of electronically mailed information and check as for 1-5

In addition, laboratories which were not accredited or which did not operate any recognized QM system were encouraged to do so. Information on courses or contacts in each member country were made available through the national contacts. Many of the national contacts also arrange national QUASIMEME workshops/seminars where representatives from the National Accreditation Board were invited to contribute to the program. In this way the QA Marine Network can encourage participants to use and build on existing quality facilities and avoid duplication.

10.4.2 Operation

In principle, it is logical to establish the quality targets of the MMPs, followed by a systematic investigation of the required quality of sampling, storage, chemical analysis, assessment, and interpretation (Fig. 10-3). Once this circle of events is complete then the next round of improvements would be planned, thus developing each area with a higher quality specification. Such developments have been described elsewhere as the quality spiral [27,28]. However, such sequential developments are difficult to implement and do not necessarily create a characteristic holistic network. In practice it is possible for a number of steps in the holistic cycle to be developed concurrently (Fig. 10-3). The key issues in each area can be identified and improved in an interactive stepwise manner.

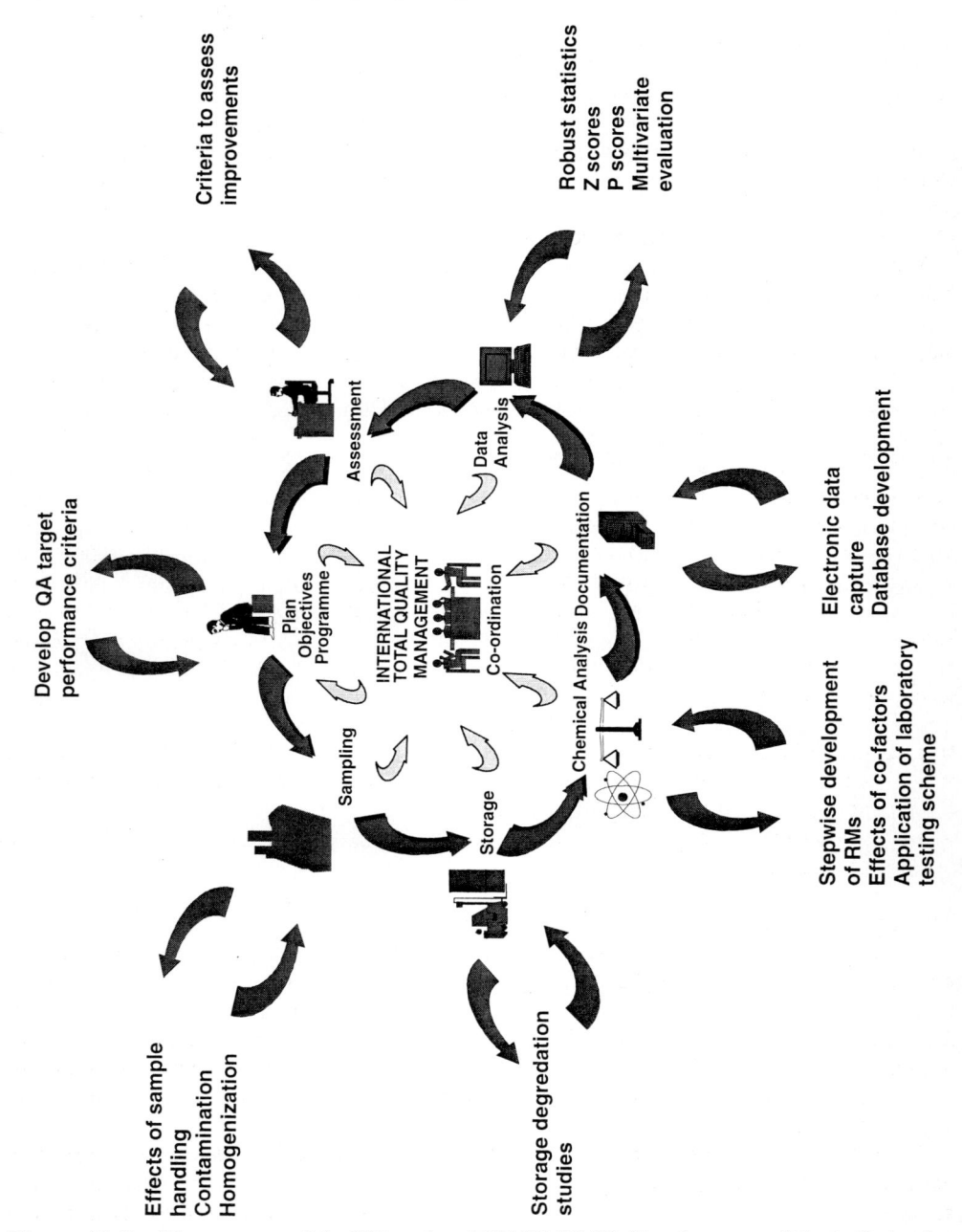

272   *10   An holistic structure for quality management*

Figure 10-3:   The strategy of the EU project QUASIMEME. Development of the holistic plan for the improvement and assessment of the quality of information on chemical measurement for marine environmental monitoring.

10.4.2.1 Target objectives

Before establishing a laboratory testing scheme or improvement program it is necessary to define the accuracy required. It is almost too obvious to state that the minimum data quality requirements must meet the given objectives. Taylor [29,30] proposed that the analytical accuracy of the measurement should exceed the data quality objectives by a factor of three. Considerable resources may be required to reduce the analytical variability by a small, and perhaps insignificant, amount. Overemphasis on the improvement of measurement where it is not required is, at best, a sterile exercise and misdirected effort.

However, for many of the MMPs, the between-laboratory agreement for most mandatory and voluntary determinands has, over the years, been so poor that the collective data was not of sufficient accuracy to enable full environmental assessment. The overall variability of the measurements exceeded the targets of data quality required to fulfil most objectives [18,28,31,32]. Improvement in chemical measurement *and* establishment of data of known quality are both essential [18,28]. Firstly, it is possible to assess international data sets in terms of environmental significance instead of being concerned with artifacts caused by analytical differences. These data can then be used with an acceptably definable level of assurance. Trend or synoptic data can be brought together to give a coherent story. Relationships between the concentration of the chemical and any biological effect can be more reliably established, and multivariate techniques for data interrogation can be used with greater confidence. Models for the selection of targets for analytical accuracy, particularly for trend-monitoring studies, are being developed within the framework of the ICES Working Group on Statistical Aspects of Environmental Assessment and Monitoring Strategies (WGSAEM) and the Working Group on Environmental Assessment and Monitoring Strategies (WGEAMS) [7,9,33]. What is clearly emerging from these studies is the need for:

- a significant improvement in the performance of field and laboratory chemical measurements;
- a program to sustain, monitor, record, and evaluate field and laboratory performance as an integral part of the marine monitoring strategy; and
- a detailed evaluation of the suitability of data from *each* laboratory to provide information for each of the objectives of the MMPs.

10.4.2.2 Laboratory testing and learning schemes

It is essential to establish continuous laboratory testing and learning schemes to obtain the long-term within- and between-laboratory variability of each determination [18, 34-38]. Taylor [34,35], Horwitz [36], Wells et al. [37,38], and de Boer and van der Meer [18] have all shown that when laboratories are brought together in a series of common intercomparison exercises, it takes between two and five steps before the initial rate of improvement begins to plateau for a single group of participants undertaking a specific analysis. In most cases the greatest relative improvement is made during the first and second rounds.

The value of spending time initially on identifying and reducing the main causes of analytical variability will hopefully enable (a) more effective evaluation of the errors associated with storage and sampling, and (b) better target values to be set for the accuracy of the measurement when evaluating the monitoring objectives.

Table 10-3: Stepwise development of the QUASIMEME laboratory testing scheme.

Excercise No	Nutrients	Metals	Chlorobiphenyls	Organochlorine pesticides	PAH
Round 1 February to May 1993	Seawater (low- summer) Seawater (high- winter) Three levels One level 5 replicates Standard solutions 5 replicates	Sandy Sediment Silty sediment Six replicates in duplicate Standard solution Six replicates in duplicate	Cod liver oil Six replicates Standard solution Six replicates in duplicate		
Round 2 January to May 1994	Two seawaters One low and one high levels Standard solutions 5 replicates	Two silty sediment Plaice muscle homogenate Mussel homogenate Standard solution As, Cr & Ni as voluntary determinands	Cod liver oil (five replicate measurements) Two silty sediments Standard solutions	Cod liver oil (five replicate measurements) Two silty sediments Standard solutions	
Round 3 June to November 1994	Two seawaters One low and one high level Standard solution 5 replicates	Two silty sediments (one repeat from round 2 Cod liver homogenate Mussel homogenate (repeat from round 2) As, Cr & Ni included with the other metals	Cod liver homogenate Two silty sediments (one repeated from round 2) Standard solutions	Cod liver homogenate Two silty sediments (one repeated from round 2) Standard solutions	Two standard solutions

QUASIMEME began with a simple series of exercises (Table 10-3) which were extended as the series progressed. The analytical precision and bias of the laboratories were established. The long-term precision was compared with the documented precision given by the participants in the detailed questionnaire [26]. In many intercomparisons the short term precision measured within a day or within batch is usually very small. However, when measured over 5-8 weeks, as in this exercise, a more realistic and meaningful measure of the precision emerged. An alternative strategy for measuring very long term precision of a laboratory would be to repeat replicate blind samples over a period of a few years.

A six-monthly cycle for these exercises enables (i) rapid progress to be made, (ii) the early momentum of the project to be maintained, (iii) the obvious and more extreme sources of error to be reduced at an early stage in the project, and also (iv) gives time for assessment reports to be prepared and distributed. The key information that can come from regular laboratory testing is:

- the long-term performance of laboratories;
- the extent and magnitude of specific errors;
- the improvement made by "poor performers";
- those determinands and matrices which require specific attention.

Participants which have a specific problem can join a learning cycle which has a more focused series of exercises designed to overcome the apparent source of error progressively. Specific advice is given to each laboratory, e.g. to use a specific gas chromatographic column of a particular phase, length and internal diameter and to use it under given optimum conditions for organic trace analysis [18,32]. Should these learning cycles not produce the desired improvement then, depending on the number of participants involved, either a dedicated workshop should be arranged at a well equipped and staffed institute, or an expert should visit the laboratory in question.

This stepwise, focused program tailored to achieve the required level of AQC will produce over 90% of the improvement necessary in all but the few laboratories who elect not to apply the information given.

10.4.2.3 Reference materials

Two vital factors influence the value these materials to marine science. These are:

- the uncertainty of the estimate of the determinand in the RM; and
- the physical nature and type of RM.

With so many RMs now available to support marine science [39] it has become necessary to distinguish between those which are fully certified, e.g. CRMs or standard reference materials (SRMs), and those which have simply become available as a result of intercomparison exercises. In some cases the declared level of uncertainty for some determinands is so high that the RMs is of questionable use for most marine monitoring programs (Table 10-4).

Table 10-4: Recommended contents (ng/g dry mass) for reference materials supplied by IAEA. These materials are available as reference materials, but the assigned value was derived from intercomparison exercises involving a wide range of laboratories of differing capability and quality [39].

Compound	Value	Lower range	Upper range
MA-A-1/OC Copepod homogenate			
aldrin	14	0	33
alpha HCH	10	1.6	18.4
gammma HCH	8.2	1.9	14.5
pp' DDD	5.5	0	11
pp' DDE	6.1	1.5	13.2
pp' DDT	8.3	3.4	13.2
IAEA 351 Tuna homogenate			
aldrin	46	0	110
pp' DDD	25	12	38
pp' DDE	140	77	203
pp' DDT	53	22	84

Certified reference materials aid method development and validation procedures, but are only part of the full validation [3] of laboratory performance. A laboratory's performance is best measured by analyzing the determinand in a matrix matched blind sample. Until recently the main criticism of the CRMs available has been that they do not match the natural matrix. This mismatch has been a particular problem for most marine biota RMs which have been prepared as a dried powder. As a dried powder it is easier to homogenize, stabilize, store, and distribute. However, as an RM it is not possible to check the sample preparation and extraction procedures. Indeed it is sometimes necessary to use different preliminary procedures for the sample and the RM because of this physical difference. These shortcomings have, in part, been addressed by the stabilized emulsions and homogenized suspensions of marine biota developed by the National Institute for Standards and Technology (NIST) and the National Research Council of Canada (NRCC).

However, an holistic laboratory testing and stepwise improvement program requires a succession of RMs which closely matches the specific matrix problems encountered. These include:

- standard solutions of known concentration;
- cleaned-up and raw biota and sediment extracts for organic trace analysis;
- digests from sediments using HF (total) and strong acids (partial) for trace metals;
- digests from biota for trace metals; and
- stabilized tissue homogenates for trace metals and for organic trace analysis.

The preparation of each material requires the same skill and attention given to certified RMs. Homogeneity tests are essential especially for matrices that have a high degree of heterogeneity associated with them e.g. fish liver and some raw biological extracts. Each lot, once prepared and packaged, may revert to a heterogeneous state within the container, and clear instructions on rehomogenization and use are essential if the real value of these materials is to be realized. Misuse of such preparations can lead to the wrong conclusions regarding their preparation, i.e. that they were not homogeneous in the first place, and retard the development and use of these more realistic materials.

10.4.2.4   Documentation - data retrieval

Effective retrieval of data is mandatory. Most complex, interrelated information is usually stored in a database and tailored specifically to the application. It is often difficult to determine at the outset of a project what the content and structure of the stored data should be. Logically, the database should be designed and tested prior to the commencement of the work and developed as the project expands. However, in the absence of a prototype it can be advantageous to delay this development until the direction and the magnitude of the application emerges. It is equally crucial to identify the structure of the data.

One year into the QUASIMEME project, the structure of the database was substantially different from that which would have been produced at the start. The first year's data from the intercomparison exercises were held and interrogated in a spreadsheet, which could be readily exported to statistical packages and graphical outputs. On using these data in this way two overriding needs became apparent. The first was for a fully relational database which enabled rapid selection of information on the basis of cofactors. The second feature resulted from a request from the participants to be supplied with a *pro forma* for data entry on a disk. Conceptually, it was a simple step to extend the development to a complete electronic transfer all of the data either by floppy disk or by email. The QUASIMEME database was developed using Paradox for Windows and the participants' data collector was based on a DOS Fox-Pro system. The database files generated by the participants were read directly into the main system after the mandatory checks for computer viruses and data conformity. The subsequent advantages of such a system outweighed the effort in development particularly for a large project that may continue for some years. The key advantages of the electronic transfer of the laboratory data are that:

- the responsibility for quality remains with the laboratory;
- data transfer is fast and accurate;
- transcription errors are reduced; and
- the operation is cost effective.

The first manual data input took between 2-3 man months to enter the data, fully verify, and confirm with the laboratories that these data were correct. There followed some 20 to 30 changes necessary to incorrect or missing data before data assessment could commence. Subsequently, using the electronic data collector it took 1-2 days to check, load, and verify some 150 disks. No transcription errors were detected. The initial development cost of the database was recouped within 12-18 months.

10.4.2.5 Data assessment
There are two aspects of data assessment relating to quality within the MMPs:

- the assessment of the QA information; and
- the quality control and guidelines for the assessment of the environmental data.

Although these two areas are inextricably linked, it is important to distinguish between them. The assessment of the QA data generated from the laboratory testing schemes should be:

- clear and unambiguous to the participants and to assessors of the MMPs; and
- comparable between exercises so that the quality of trend data can be treated consistently.

Much attention has been paid to the nomenclature, use, design and implementation, and evaluation of laboratory performance studies [36]. Guidelines and protocols for the design and implementation of interlaboratory studies have been issued [40,41]. The classical parametric techniques [42,43] for the statistical treatment of data from such studies assume that the within-laboratory variances are normal and equal and that the between laboratory distribution is normal. The application of ISO5729 or equivalent procedures in studies like QUASIMEME is questionable, both on statistical grounds and on the effect of "rejecting" data when the primary objective is to evaluate the extreme values to establish the cause of the error. Extreme values should be identified, but *remain* within the data set [23]. Cofino and Wells [23] have applied the use of robust statistics [44-46] to the data from the QUASIMEME laboratory testing schemes and the learning schemes.

10.4.2.6 Performance criteria
The primary requirement for most of the determinands is to establish the assigned values. For some determinand groups this is a relatively easy task, e.g. nutrients in sea water where it is possible to spike the RM with accurately known amounts of the determinand [47]. There are several methods for establishing the best estimate of the true value for other matrices [40,48,49]. The QUASIMEME project has used a group of reference laboratories which have a recorded history of valid measurements in certification programs and interlaboratory trials [23].

Z scores were used for an assessment of laboratory performance [40]:
$$Z = (x_i - X_a)/s_b$$
where x_i is the robust mean of the reported value, X_a is the assigned value as determined by the reference laboratories, and s_b is the target standard deviation (Table 10-5).

In addition to the Z score as a normalized indicator of bias, the QUASIMEME project has also introduced a P score as an indicator of precision:

$$P = s_i/s_p$$

where s_i is the standard deviation of the determination for each laboratory and s_p is the target precision [23] (Table 10-5).

Table 10-5: Minimum target performance criteria for the QUASIMEME laboratory testing scheme.

Determinand-matrix	Laboratory target	
	Bias S_b	Precision S_a
Nutrients in seawater (summer)	12.5%	12.5%
Nutrients in seawater (winter)	6%	6%
Nutrients in standard solutions	6%	6%
Trace metals in sediment/biota	12.5%	12.5%
Trace metals in standard solution	6%	6%
Organochlorines in sediment/biota	12.5%	12.5%
PAH in sediment	12.5%	12.5%
Organochlorines in standard solution	6%	6%
PAH in standard solution	6%	6%

The values given above are the minimum allowable criteria.

Within such a program it is necessary to develop techniques for data assessment by refining and adapting the robust models to account for:

- different numbers of extreme values;
- large numbers of missing and/or "less than" values; and
- multivariate treatment of data (e.g. principle components analysis) in the presence of missing data.

In any laboratory testing scheme it is essential to set the performance criteria *prior* to the analysis of the data. In the absence of any clear guidelines from the European MMPs prior to 1994 the QUASIMEME scientific group set *minimum* criteria for the target bias and precision that would be set for any spatial distribution study or trend survey (Table 10-5). It was therefore necessary to compare the actual performance of the laboratories in the testing scheme with these targets. After the 3rd round (Table 10-6) these targets were easily attainable for high concentrations of determinands, e.g. winter levels of nutrients in sea water, but for other determinands that occur close to the detection limit there was still some difficulty in obtaining good agreement between laboratories. Apart from the obvious errors associated with transcription and calculation errors, one of the major contributions to high variance at low concentrations continues to be the contamination of the sample prior to or during analysis. This would appear to particularly true for ammonia in sea water, cadmium and mercury in sediments, and organochlorine pesticides. The improvement in the measurement of chlorobiphenyls in biota has brought the overall level of laboratory agreement of the good laboratories close to the targets set (Table 10-6). However, these targets are the *minimum* performance criteria and, according to Taylor [29] the analytical variance should be at least three times less than the overall allowable variability of the monitoring objectives.

Table 10-6: Comparison of target values for bias and the between laboratory robust coefficient of variation. QUASIMEME laboratory testing scheme, round 3, June-November 1994.

Determinand	Target bias%	CV%	Determinand	Target bias%	CV%	Determinand	Target bias%	CV%
Nutrients in seawater	set	obtained	Metals in sediment	set	obtained	Chlorobiphenyls in cod liver	set	obtained
TOxN low	25	16	Aluminium	25	19	CB28	25	29
TOxN high	12	5.4	Arsenic	25	11	CB52	25	20
Nitrite low	25	19	Cadmium	25	28	CB101	25	28
Nitrite high	12	6.7	Chromium	25	27	CB105	25	29
Ammonia low	25	51	Copper	25	15	CB118	25	26
Ammonia high	12	18	Lead	25	14	CB138	25	23
Phosphorus low	25	79	Mercury	25	23	CB153	25	22
Phosphorus high	12	6.2	Nickel	25	23	CB156	25	26
			Zinc	25	8	CB180	25	25

10.4.2.7 Storage

There is little comparative, comprehensive data on the effects of storage on the matrices and the determinands measured in the MMPs [14,15]. There are no definitive guidelines at present beyond the general range of statements such as:

- contamination of samples should be avoided;
- all biota samples should be frozen as soon as possible after sampling; and
- all samples should be analyzed as soon as possible.

The initial responses from the QUASIMEME questionnaire on QM and laboratory practice [26] indicated that most laboratories selected their storage conditions from literature information alone and that few original studies had been reported. More than a third of the laboratories had no specified storage time prior to analysis and fewer that 40% had any SOP dealing with storage and the quality of samples during storage. Most sample storage conditions are not the ideal, sterile, noninteractive environments that are perhaps, envisaged. Whilst sample modification and determinand and matrix changes occur slowly at temperatures of -20°C or below, such changes, even if only dehydration, *do* occur. Specific studies to determine the fate and effect of different storage conditions are necessary. These are long-term studies and should be planned and should start at the commencement of the overall project. The analysis required to detect changes in composition in the matrix, for example water content, fat residues, and fatty acid content, as well as the determinands normally monitored, must be undertaken in an experienced laboratory which can demonstrate a high level of precision over a sustained period.

10.4.2.8 Sampling and sample handling

Once the sample has been obtained it is particularly easy to allow the local environment or the handling techniques to affect the quality of the sample and the concentration of the determinand in the sample [50]. Again general guidelines are available, but it is the detailed practices of individuals and the interpretation of their own protocols that can exclude those external influences, or contaminate the sample, which are important. Poor sample handling techniques become apparent, e.g., fish livers that are not deep frozen individually and immediately will deteriorate rapidly with cell rupture and liquefaction [51]. The effect is often visually obvious and the samples must be disregarded. However, far more detailed and structured information is required for informative and detailed guidelines.

There are several stepwise developments which are necessary to aid laboratories in preventing contamination and/or deterioration of the sample. The development guidelines, in such a practical and visual area can be augmented by a series of training videos. The cliche *"Every picture is worth a thousand words"* is particularly poignant when communicating fine detail and where there may be language difficulties in describing everything in print. Clear, focused, expert advice is required, as has been given for problems relating to the contamination of sea water by extraneous ammonia [52]. Three key errors may occur with sample handling and sample preparation prior to extraction/digestion:

- contamination/losses;
- poor homogenization; and
- sample degradation.

Participants can prepare samples, for example, from a population of well characterized mussels, and then have the material analyzed in one central laboratory to reduce the analytical variance. Further development in field studies can test the efficacy of sampling equipment and sampling techniques at reference sites for determination of nutrients in sea water, or trace metals and organics in sediments. These aspects of the project can be introduced sequentially as the participants become fully engaged and committed to developing the field aspects of the project.

10.5 Communication

A well developed communications network is an essential feature of any international project both between participants and other interested parties (Table 10-7). Where appropriate, effective communication can be maintained by having a closely related organization represented on the project's steering group (or advisory board) and vice versa. In addition it is advisable to develop a mailing network with interested potential participants in the future. Outside organizations may only hear of the project third hand and receive only part of the project information. Direct communication is always preferable and, at a general level, can be achieved by publishing a project information newsletter, e.g. the QUASIMEME Bulletin. Many organizations now have this style of popular information newsletter for the wider scientific and nonscientific reader. A fuller distillation of scientific progress of the project should be disseminated through the peer review system in the scientific journals, special issues [21,23,37,52], and book chapters. Rapid dissemination of the project's progress is important. Workshops and the results of specific studies should be made available to the wider scientific community within one year to maintain the interest and momentum of the project.

Good communication with participants or with an external organization does not simply depend on the links between the central project office and the institute. One key difficulty in the communication network often lies within the organizations themselves. When establishing key links it is essential to be aware of the internal structures and links of that organization. Confirmatory telephone or personal calls can be a worthwhile investment where important communication must be established and maintained. Tracing the internal links of an organization is one of the most difficult, but necessary, keys to effective communication. Communication between the laboratory manager and the technician and scientists at the bench, which is dependent upon the effectiveness of the internal links, and is vital when transmitting instructions and protocols as well as reports and the recommendations to that laboratory.

Such communication difficulties do not simply rest with the simple transfer of information. In workshop discussions it has emerged that advice given to improve the method and techniques used for a chemical measurement was sometimes difficult to implement back at the laboratory. Technical staff at workshops have admitted privately that it can be almost impossible to have such changes accepted back at the institute. In some cases it is necessary to have a strategy for direct communication from the project at different levels in the same organization, with the same advice to enable a useful improvement to be implemented.

Table 10-7: Communication network for an holistic quality assurance program.

Level of Communication	Type of Communication	Interest Groups	Examples of the Interest Groups for QUASIMEME
General Awareness	Bulletin Executive summaries	Politicians, administrators scientists,	Ministers of the Environment & Fisheries, directors of institutes
Scientific Interest	Bulletin Scientific publications	Analytical chemists & interested agencies Accreditation bodies Other QA projects	Institute libraries, European Environmental Agency, Accreditation organizations, (NAMAS, STERLAB), Western European Laboratory Accreditation Co-operation (WELAC), European Committee for Standardisation (CEN), FAPAS
Specific conclusions and recommendations from the project.	Bulletin Scientific publications Reports of exercises and workshops Videos	Customers, & regulatory authorities Supportive & collaborative organizations, Future prospective participants Associated projects	Oslo & Paris Commissions, Helsinki Commission International Council for the Exploration of the Seas Water Industry, other marine institutes AQUACON, NOAA(USA) NMAQC (UK)
Detailed plans, timetable, and specific information on the project and the programme.	Bulletin Reports of exercises and workshops Guidelines, protocols and exercise details Videos	Participants, interactive groups	Participants and prospective participants, contracted parties for RMs, expert advice and independent assessors.

For explanation of the acronym see text.

In an exact science, such as analytical chemistry, it might initially appear strange that many problems are closely associated with attitude and cultural differences, rather than with methodology or instrumentation. Many key problems relating to errors of measurement are well known, along with the scientific method for their solution and control [34,35,53]. So why, after such a period of analytical enlightenment should so many of our measurements still contain significant errors, i.e. above those routinely expected at the current state-of-the-art. The technology is available, the methods and tools for controlling our measurements are available, so why is it still necessary to have such an intensive international effort to improve the analytical skills of so many institutes? Support programs such as QUASIMEME offer key requirements which:

- define the laboratory's present performance;
- set target performance characteristics;
- provide resources for improvement;
- catalyze the effort necessary to put a quality system in place; and
- encourage personal acceptance of the errors.

A continuous international holistic QA program provides the structure to address these issues and can provide the desire to achieve the quality goals set through effective communication. Most of the main barriers to achieving the necessary standards lie within internal laboratory communication and management.

10.5.1 Within laboratory communication

Effective communication is necessary to create the environment for putting a quality control system in place; it requires:

- creative management;
- adequate time;
- comprehensive and documented training; and
- realistic resources.

"No data is better than bad data" is now a universal slogan, but management must continue to allocate sufficient time for full validation of the methods used. With the increase in contract studies there is an underlying need for work to be externally assessed or "accredited". Contracts can provide the stimulus to provide support for the QA schemes to be put in place.

Management and analytical staff should recognize the time taken a) to establish a sound QA audit-accreditation system; and b) to maintain it. The energy/resources required to establish such a system may appear prohibitive, but once in place it take less effort and resources to follow this program than to revert to "quick and dirty" methodology. Up to 30% of staff time can be required to establish a good QA/QC system over a 12-18 month period, after which 10% of staff time is required to: a) maintain documentation; b) undertake routine in-house QA/QC procedures, e.g. recovery experiments, laboratory or certified reference materials calibration, or blank and replicate analysis; and c) partake in laboratory testing schemes, e.g. FAPAS, QUASIMEME, NMAQC [11,12].

A few laboratories outside the accreditation scheme have a comprehensive, structured, and recorded training scheme. Staff should be encouraged to question and challenge all operations that are not fully understood, both the practice and the intrinsic reason for the operation. This can only come from a well planned and comprehensive training program, which is fully documented and for which a record is kept of each person's knowledge and progress. It sounds simple and logical, but is rarely practised, primarily because of the resources required. However, in economic terms it is sound investment. A 25K ECU instrument can be discarded in three years, but a technician may cost 30-50K ECU *per annum*. Our human resources are the most valuable asset in any institute.

Laboratories need adequate resources. Second-rate materials will not be sufficient even in the most skilled hands. Poor quality calibrants or an unwillingness to purchase certified materials on the grounds of limited resources is not an acceptable approach, if a laboratory is serious about its information base. Adequate laboratory facilities, space and working conditions go a considerable way towards enabling staff to organize and execute the analytical program to the standard required. In any quality assurance program these elements usually rank quite high on the list [12,15,19].

None of these points given above is new, but failure to address the human dimension of problem solving will not be balanced by any effort to address the remaining difficulties. Good interlaboratory communications do much to reduce these intralaboratory difficulties by creating a much wider flow of concepts, ideas, and practical solutions to fulfil the necessary QA requirements.

10.6 Future developments

Project development usually has four phases:

- concept;
- proposal (then funding);
- pilot project; and
- sustained programme.

The pilot project is an essential prerequisite for development of a sustained program which can focus on the specific direction of the established work and concurrently develop proposals for new and related work. At the outset it is important to begin with a focused set of objectives, to operate within an well defined framework, and to develop a limited number of functions that can be performed with a specified number of participants within the time frame of the pilot phase of the project.

There are also disadvantages to this approach which become more evident the longer the first phase takes to complete. For the rest of the scientific community involved in monitoring or research activities such a project can look rather insular, with a small group of people providing, information, cooperation, and travel for themselves. The sharpness of the criticism is perhaps proportional to the desire to participate. Therefore good communication with the wider scientific community during the first phase is vital along with the clear indication that the program will rapidly broaden to encompass all institutes who would like to be part of such a developing program.

10.6.1 Involving wider participation

An open membership to the second phase of the project is highly desirable and brings with it the need for a different strategy. In the first phase the emphasis should be to demonstrate that the stepwise approach to improving the between-laboratory agreement can actually work. It is achieved by building up (i) discrete information on the sources of error and how they can be minimized and (ii) the confidence and ability of the participants to actually solve them. This strategy is more successful with a relatively constant group and a small number of stages (3-4). This has now been very clearly demonstrated in at least two other independent programs [18,37,38] as well as in the QUASIMEME project.

However, it is not necessary to repeat this demonstration in a sustained holistic project. Neither is it imperative to improve the quality of measurement for the *group as a whole*, although this may actually occur given the structure of the program. The primary purpose is to provide *data of known quality* for the MMPs so that the assessors can make a value judgement with some definable assurance. Data produced by laboratories which perform badly in the holistic QA program, whether because of poor sampling techniques, sample handling, storage, or analytical capability, can be identified and the specific data set withdrawn from the environmental monitoring assessment.

In the pilot phase of the QUASIMEME project all data produced were confidential to the participants and no comment was made to outside bodies on the quality of the information produced by individual laboratories. The purpose of this decision was to provide an atmosphere of trust and openness where problems and failings could be discussed without any concern over penalties that might occur as a result of a poor performances. Each *poor performer* was given clear advice on the solution to the problem and was given an opportunity to correct the mistake, since the emphasis was on improvement. However, each laboratory was at liberty to disclose its *own* performance to the MMPs along with its environmental data. Indeed, QUASIMEME positively encouraged this activity since it placed the responsibility of reporting QA data squarely on the laboratory making the environmental data submission. This can be done on a routine basis without compromising the confidentiality of any other institute and negates the need for the managers of programs like QUASIMEME to make formal submissions to the commissions on behalf of the member participants.

Since QUASIMEME participants produce data for different MMPs, disclosure of the quality data to the commissions could remain with the participants. This can be a formal requirement for participants when they submit environmental data, leaving the confidentiality between participant and the QA program intact. This approach maintains the same atmosphere of openness and prevents the QA project from turning into a policing operation. The ultimate evaluation of a laboratory's performance must always remain with that laboratory!

Once the first phase has been accomplished it is important to take advantage of the lessons learnt and build a second phase which takes the emergent project on to maturity. The new framework must provide stable and lasting support of quality assurance to the MMP and to the commissions which initiate these programs.

New participation does not come without cost and the financing of the initial stages of a program, e.g. through the EU for the QUASIMEME project, must be replaced by alternative sources of funding. This may come either from a central organization, which would take overall responsibility for the QA program, or from the participants by subscription.

10.6.2   Developing new strategies

As well as increased participation it is also necessary to include other determinands and support other QA needs since the quality of information not only pertains to the monitoring program but to research and regulatory requirements alike. The first phase of QUASIMEME focused on the quality of data relating to mandatory and some voluntary determinands (Table 10-1). However, since most of these studies can have a relatively long lead-in time it is vital to develop a rolling program to introduce the new determinands into the quality scheme. The arguments with regard to the timing of the introduction of these newer compounds tends to be a circular one. On the one hand there is no need to have an extensive QA program for compounds which are only determined by a small group of laboratories whereas, on the other habd, many laboratories will not undertake these analyses unless there is a real need and a supportive QA program in place!

The holistic approach can be used to cover both aspects. Firstly, with good communications between the institutes and the commissions it is possible to arrange for the development work to be undertaken by a small number of interested and capable laboratories. The resultant developments in methodology can then be disseminated to all participants for wider use in the corporate monitoring plans. This saves a number of years in sequential development and prevents unnecessary mistakes being made with the determination of newer compounds by laboratories that initially have neither the experience nor methodologies.

10.6.3   Developing new links

The holistic style of a project such as QUASIMEME can be used as a role model for future development and to create a wider network between other countries and within member countries of the existing program.

In North America NOAA has established the National Status and Trends Program and is currently developing a QA project which will include the members of the North American Free Trade Association (NAFTA); viz Canada, Mexico, and the United States. International links have been made with QUASIMEME for the exchange of ideas, reference materials, and blind samples to enable better use to be made of the relatively expensive test material. Such independent testing schemes initiated by both programs on a common material enable direct comparison of information and an even more uniform assessment of the whole data set from both groups of participants.

Within the structure of a broadly based international project each country may have, or plan to develop, their own national QA programme, e.g. NMAQC in the UK. It is, therefore, essential to establish cooperative links with each of these programs to :

- prevent duplication;
- share expensive RMs and blind samples;
- prevent an unnecessary tier or organization; and
- maintain a similar level of required performance.

This style of holistic QA is also applicable to research as well as the monitoring program, for example the EU Marine Science and Technology (MAST) projects, the EU Mediterranean Targeted Programme (MTP), and to a network of marine research institutes. It could be equally applied to other disciplines within the marine environment such as hydrology, biology, ecotoxicology, and physical oceanography. Each discipline has a different balance between exact and descriptive science, but each requires the information to be of a known and validated quality. Comparative quality systems are being developed between countries within single disciplines at present, but an holistic approach cast along the lines of the elements given in this paper will also aid development between disciplines and can serve as a generic model for applications beyond the marine environment.

Acknowledgments

The authors would like to acknowledge the members of the QUASIMEME scientific group who have contributed ideas and have taken a active part in developing the solutions to many of the questions of quality management and practice. These are Alain Aminot, Jacob de Boer, Don Kirkwood, and Britta Pedersen. The authors also thank Graham Topping who has contributed much to quality assurance in the marine environment and to Ben Griepink who has acted as a valuable catalyst in the launch of the QUASIMEME project.

10.7 References

[1] ISO9000, International Standards Tech. Comm. ISO/TC 176. International Standards Organization, Geneva (1987)
[2] ISO/REMCO, N229 Fourth International Symposium on the Harmonisation of Quality Assurance Systems in Chemical Analysis. Geneva 2-3 May 1991 (1991)
[3] ISO8402, Quality Management and Quality Assurance Vocabulary 2nd edition (1994)
[4] EN45001, European Standard, Rue Brederode 2, B-1000, Brussel, Belgium (1989)
[5] BS 5750, BSI 2, Park Street, London W1A 2BS (1991)
[6] Anon, Department of Health, *Good Laboratory Practice*, UK Compliance Programme, London, UK (1989)
[7] R.J. Fryer and M.D. Nicholson, *J. of Mar. Sci.*, **50**, 161 (1993)
[8] M.D. Nicholson, *ICES C.M.1992/Poll:1*, ICES Palegade 3, Copenhagen DK (1993)
[9] M.D. Nicholson, in: *Analysis of Contaminants in Edible Aquatic Organisms,* R. Kiceniuk and M. Ray (Eds.) (1994)
[10] Anon, *Protocol for the Food Analysis Performance Assessment Scheme (FAPAS). Organisation and analysis of data*, 3rd edition, FAPAS Secretariat, MAFF Colney, United Kingdom (1993)
[11] M. Thompson and R. Wood, *J. Pure Appl. Chem.*, **65**(9), 2123 (1993)
[12] D.E. Wells, W.P. Cofino, Ph. Quevauviller and B. Griepink, *Mar. Poll. Bull.*, **26**, 368 (1993)
[13] *Monitoring Manual of the Oslo and Paris Commissions*, March 1990, New Court, 48 Carey Street, London WC2A 2JQ, UK (1990)
[14] *The Baltic Sea Joint Comprehensive Action Programme Baltic Sea Environmental Proceedings*, Helsinki Commission, Helsinki, Finland (1993)

[15] G. Topping, *Mar. Poll. Bull.*, **25**, 61 (1992)
[16] *North Sea Task Force*, Report 3, NSTF New Court, 48 Carey Street, London, WC2A 2JE, UK (1990)
[17] *North Sea Task Force*, Report 4, NSTF New Court, 48 Carey Street, London, WC2A 2JE, UK (1991)
[18] J. de Boer and J. van der Meer, *Report on the results of the ICES/IOC/OSPARCOM intercomparison exercise on the analysis of chlorobiphenyl congeners in marine media - step 3b*, ICES Palegade, Copenhagen, Denmark (1993)
[19] G. Topping, D.E Wells and B. Griepink, *Guidelines on Quality Assurance for Marine Monitoring*, DG XII Measurement and Testing Programme, Rue de la Loi 200, B-1049, Brussels (1992)
[20] E.G. Bligh and W.J. Dyer, *Can. J. Biochem. Physiol.*, **37**, 911 (1959)
[21] B. Pedersen and W.P. Cofino, *Mar.Poll. Bull.*, **29**, 166 (1994)
[22] ISO, International Standards ISO 5725-1986E, Tech. Comm. ISO/T 69, *Applications in Statistical Methods* (1986)
[23] W.P. Cofino and D.E. Wells, *Mar.Poll. Bull.*, **29**, 149 (1994)
[24] R. Fisher and W. Ury, *Getting to Yes: How to negotiate to agreement without giving in*, Arrow Books 62-65 Chandos Place London (1981)
[25] D. Cormack, *Team Spirit: People working with people*, Marc Europe (1987)
[26] S.K. Bailey, A.S. Wells and D.E. Wells, *Mar.Poll. Bull.*, **29**, 187 (1994)
[27] B. Broderick, W. Cofino, R. Cornelis, K. Heydorn, W. Horwitz, D. Hunt, R. Hutton, H. Kingston, H. Muntau, R. Baudo, D. Rossi, J.G. van Raaphorst, T. Lub, P. Schramel, F. Smyth, D.E. Wells and A.G. Kelly, *Mikrochim. Acta*, **II**, 523 (1991)
[28] D.E. Wells and A. Kelly, *Mikrochim. Acta*, **III**, 23 (1991)
[29] J.K. Taylor, *Anal. Chem.*, **53**, 1588a (1981)
[30] J.K. Taylor, *Fresenius Z. Anal. Chem.*, **332**, 722 (1988)
[31] A.V. Holden, G. Topping and J.F. Uthe, *Can. J. Fish Aqua. Sci.*, **40**(Suppl 2), 100 (1983)
[32] D.E. Wells and J. de Boer, *Mar.Poll. Bull.*, **29**, 174 (1994)
[33] M.D. Nicholson and R.J. Fryer, *Mar. Poll. Bull.*, **24**, 146 (1992)
[34} J.K. Taylor, *Principles of Quality Assurance of Chemical Measurements*, NBSIR 85-3105 (1985)
[35] J.K. Taylor, *Quality assurance of chemical measurements*, Lewis Publishers Inc, Michigan 48118 (1987)
[36] W. Horwitz, *Anal. Chem.*, **54**, 67A (1992)
[37] D.E. Wells, J. de Boer, L.G.M.Th. Tuinstra and R. Reutergardh, *Fresenius Z. Anal. Chem.*, **322**, 591 (1988)
[38] D.E. Wells, E.A. Maier and B. Griepink, *Int. J. Environ. Anal. Chem.*, **46**, 265 (1992)
[39] A.Y. Cantillo, *Standards and Reference Materials for Marine Science*, IOC Manuals and Guides No.25 (revised edition) (1993)
[40] M. Thompson and R. Wood, *J. Pure Appl. Chem.*, **65**(9), 2123 (1993)
[41] WELAC (Western European Laboratory Accreditation Cooperation), *WELAC Criteria for Proficiency Testing in Accreditation*, (1993)
[42] D.M. Hawkins, *Identification of Outliers*, Chapman and Hall, London (1980)

[43] J.C. Miller and J.N. Miller, *Statistics for Analytical Chemistry*, Ellis Horwood Chichester (3rd Edition) (1993)
[44] Analytical Methods Committee, *Analyst*, **114**, 1693 (1989)
[45] Analytical Methods Committee, *Analyst*, **114**, 1489 (1989)
[46] P. Lischer, *Lebensm. Wiss. Technol.*, **20**, 167 (1987)
[47] A. Aminot and R. Kerouel, *Anal. Chim. Acta*, **248**, 277 (1990)
[48] Anon, *FAPAS - Food Analysis Performance Assessment Scheme - Organisation and Analysis of Data*, MAFF Food Science, Norwich, UK (1980)
[49] Anon, *Harmonised Proficiency Testing Protocol IUPAC/ISP/AOAC Symposium on "Harmonisation of Quality Assurance Systems in Chemical Analysis"*, Geneva, (1991)
[50] K.J.M. Kramer, *Mar.Poll. Bull.*, **29**, 222 (1994)
[51] J.F. Uthe and C.L. Chou, *Sci. Total Environ.*, **71,** 67 (1988)
[52] A. Aminot and D. Kirkwood, *Mar. Poll. Bull.*, **29,** 159 (1994)
[53] F.A.J.M. Vijverberg and W.P. Cofino, *Techniques in Marine Environmental Science*, No. 6, ICES Palægade 2-4, DK 1261 Copenhagen

Subject Index

abundance, relative, aquatic biota   192
accompanying studies, plant sampling   225
accreditation, control systems   20
accumulation
– age/population dependent animals sampling   45
– CRMs   13
– ISO definitions   11 ff
– net mass, sediment sampling   118
– plant sampling   223
– sediment sampling   113
– trace metals, aquatic biota   181
acid-washing, soil sampling   167
acidification, long-term preservation, dissolved nutrients   106
active biomonitoring (ABM)   36
– trace metal analysis   184 f
– – mussels   208
active monitoring, biological specimen   26
additions, aquatic biota   204
adsorption
– storage vessel   81
– surface water sampling quality   69
– systematic errors   7
advices, MMPs   261
affinity, contaminants/supended matter   56
age
– animals sampling strategies   45
– plant sampling   237
– trace metal analysis   193, 208 f
age–size relation, species selection   193
aging, method validation   6
aging condition, RMs/CRMs   15
agricultural run off, surface water sampling   54
Ailanthus altissima, cleaning   250
air
– sulfur dioxide   27
– systematic errors, inorganic analysis   7
air contamination, sediment sampling   121
air-drying, soil samples   169
air pollutants, sampling   18
aliphatic compounds, plastic columns   76

aliquots, plant samples   251
aluminum
– soil sampling equipment   164
– surface contaminations, plant sampling   249
aluminum foil, sediment sampling   122
ammonia
– nitrogen donors, phytoplankton growth   94
– water quality parameter   67
analysis schemes, sampling dissolved nutrients   110
analysis selection, CBs determination   141
analysis validation, organic compounds   136
analyte content, RMs/CRMs preparation   14 ff
analytical procedures   2 ff
– soil sampling   159
animal sample storage   135
animal tissues
– passive biomonitoring   32
– sampling, biological specimen   25
– soil sampling   37
animal uptake rates, plant sampling   249
animals sampling strategies   40 ff
ANOVA test, aquatic biota homogenization   207
anthropogenic influences
– biota, trace metal analysis   185
– plant sampling   228
anti-fouling paints, surface water sampling   69
antimony, plant sampling   249
Aporrectodea longa, indicator animals   41, 44
aquatic biota, trace metal analysis   179 ff
archiving, principles   19
area-related random selection sampling   34
arsenic
– biomonitoring   180
– plants   216, 249
– trace element pollution, soils   159
artificial habitats, sampling, biological specimen   26
artificially enriched samples, assigned values   16
ash-free weight, aquatic biota   205, 207
ashing
– aquatic biota   204
– plant sampling   223

Subject Index

assigning values, RM 16
assimilation, nonliving to living conversion 93
atmophile elements, soil sample depth 164
atmosphere, sediment sample handling 121
auto-correlation, surface water sampling 57
availability, animals sampling strategies 41

background, systematic errors, inorganic analysis 8
background information, sediment sampling 126 f
background sample, soils 167
bacteria, nutrients incorporation 93
Baltic sea, sediment sampling 115
BCR
– certification report 13
– control points 6
– interlaboratory certification studies 16
– soil sampling 158
– trace metals extraction, soils 174
benthic communities, sediment sampling 115, 117
Betula pendula 234
bilberry, site-specific fluctuations 231
bilge water, sampling dissolved nutrients 95
binding manners, RMs/CRMs preparation 14
bioavailability, dustfall cadmium 27
bioconcentration factor, trace metal analysis 183
biodegradation, surface water sampling 81
bioindicators
– plant sampling 238
– sampling, biological specimen 26
– trace metal analysis 182
biological activity 105, 168
biological effects, trace metal analysis 181, 186
biological indicator systems, standardized 28
biological monitoring, passive 30
biological processes
– nutrient concentration 94
– surface water sampling quality 70
biological specimen, sampling 25 ff
Biological System of Elements (BSE) 219 f
biomarkers, trace metal analysis 186
biomass
– aquatic biota, trace metal analysis 192
– earth, plants 215
– phytoplankton, dissolved nutrients 93
biomass synthesis, heterophobic bacteria 94
biomonitoring, trace metal analysis 180
biomonitors, plant sampling 238
biosphere contamination, CRMs 15
biota
– aquatic, trace metal analysis 179 ff
– CBs extraction 137 f
– contamination 29
– organic compounds sampling 133, 135

biota handling, trace metal analysis 180
biota samples, trace metal analysis 189 f
– transport/storage 200 f
bioturbation, sediment sampling 115 f
bivalves
– cleaning, trace metal analysis 198
– organ selection, trace metal analysis 197
– sample preparation, trace metal analysis 203
– trace metal analysis 181 f, 190 f, 196, 207
Black River, PCB sources, passive samplers 76
blank samples, organic compounds 130
blank values, CBs concentration, organic compounds sampling 138
blanks
– quality control samples 84
– sediment sampling 123
– soil 166
blast freezing, biota samples 201
blender, aquatic biota, trace metal analysis 206
boron, plant building elements 216
borosilicate glass, water sampling equipment 70
bottles, sampling 73, 97 ff
box corer, sediment sampling 119
bream, animals sampling strategies 44
brittle fracture technique (BFT), aquatic biota 206
Buchner funnels, sampling dissolved nutrients 102
buckets, sampling 73, 97, 132
bulk elements, soil sampling 40

C_{18}-modified silica
– organic compounds sampling 135
– surface water sampling 75
cadmium
– atmophile elements, soil sample depth 164
– biomonitoring 180
– dustfall 27
– plants 216
– trace metals, surface water sampling 68
caging, ABM, trace metal analysis 185
calcium, plants 216
calculations, methods/errors 3 f
calibration
– CBs analysis 144
– CRMs 13
– GC analysis 141
– plant sampling 223
– systematic errors, inorganic analysis 8
– validation 5
canisters, trace organic compounds 131
capillary columns, organic compounds 76
carbon, plant building elements 216
carry-over contamination 70
case studies, methyl mercury 10

Subject Index

cell death, dissolved nutrients 94
cell lysis, nutrient concentration 94 f
cell rupture, sampling dissolved nutrients 101
cellulose esters fibers, filters 103
centrifugation 80, 104
ceramic knifes, trace metal analysis 188
certification, laboratory approach 16
certification report, method equivalence 13
certified reference materials (CRMs) 1 ff, 12 f
– CBs determination 146 f
– MMPs 270, 275
cesium, sediment storage 123
chain of custody, sediment sample bank 122
chance sampling, biological specimen 26
chemical additions, organism activity inhibition 105
chemical drying, organic compounds sampling 134
chemical methods 3
chemical monitoring, biota 183
chemical multielement analysis, plants 222
chemical properties, soil 29
chemical reactions, dissolved nutrients 93
chemicals, surface water sampling 70 ff
chloride, water quality parameter 67
chlorinated biphenyls, SPM/DOM 80
chlorinated mono aromatic compounds 76
chlorobiphenyls (CBs)
– analysis 137 f
– detection techniques 143
– determinants 264
– marine matrices 129 ff
– testing schemes 274
chloroform, long-term preservation 107
chlorophyll-a, water quality parameters 67
chlorophyll-a levels, Lake Ontario 60
chromatographic columns, aging 6
chrome-nickel oxide layer, canisters 131
chromium
– biomonitoring 180
– plants 216
– *Polytrichium formosum* 232
– trace element pollution, soils 159
circular soil grids 161
clean room, plant sampling 222
clean room/clean bench, trace metal analysis 189
clean-up
– bottles, sampling dissolved nutrients 99
– CBs/biota/sediments 137 f
– equipment, trace metal analysis 189
– organisms, trace metal analysis 198
– PAH analysis 147 f
– plant sampling 249 f
– surface water sampling equipment 70 ff
– systematic errors, inorganic analysis 7
– trace metals, surface water sampling 77

cleaning products, ammonia-rich 96
climatic conditions, plant sampling 228
closures, bottles, sampling dissolved nutrients 98
cluster sampling, soils 162
coastal areas, metal-salinity relationship 86
coastal sediment
– CRMs for PAHs 151
– RCMs for CBs 147
– TBT stability test 15
coastal waters
– dissolved nutrients 92 f, 100
– macrophytes, trace metal analysis 191
cod liver oil, RCMs for CBs 147
cold-vapor atomic absorption spectrometry 10
collection methology, aquatic biota 192 f
column chromatography, fat separation 139
COMAR data bank, CRM producer list 17
communication, MMPs 282 f
comparability
– biota, trace metal analysis 179
– spatial, biological specimen 29
comparative methods 3
compliance, surface water sampling 54
composite soil samples 163
compositions, geno/phenotypical 29
compounds, organometallic 6
computer-aided navigation, sediment sampling 117
concentration estimation, surface water 54
concentration gradient, estuaries 100
concentration levels, dissolved nutrients 93
confidence level
– analyte's identity, surface water sampling 56
– below LOQ, surface water sampling 58
– replicate sediment sampling 118
container materials, trace metal analysis 187 f
containers, soils 165, 172
contaminants, pesticides 56
contaminants affinity, suspended matter 56
contaminants co-elution, CBs determination 139
contamination
– analytical anomalies, surface water sampling 67
– biosphere/food chain, CRMs 15
– biota 29
– blanks, sediment sampling 124
– bottle closures, sampling dissolved nutrients 98
– C_{18}-bonded silica check samples 135
– centrifugation, sampling dissolved nutrients 104
– cleaning, plant sampling 249
– dioxin 38
– drinking water 54
– dust/oil in air, sediment sampling 121
– equipment, sediments subsampling 120
– equipment/procedures-dependent 70
– exhaust fumes 79

Subject Index

- fingerprints, sampling dissolved nutrients 97
- freezing, sampling dissolved nutrients 108
- heavy metals in agricultural areas 37
- helicopter sampling 79
- intrinsic, filters for dissolved nutrients 103
- memory effect 70
- metal, soil solutions storage 173
- nutrient losses, filters 103
- plant sampling 222, 227, 240
- plume around the boat, surface water 69
- poisoning, dissolved nutrients 107
- polyethene equipment 79, 171
- salinity gradients 56
- sampling, biological specimen 28
- – MMPs/QUASIMEME 281
- seasonal effects, surface water sampling 54
- sediment sampling 114
- ship-related, sampling dissolved nutrients 95
- soil particles, plant samples 250
- soil sampling 37, 157, 162, 165 f
- stainless steel buckets, organic compounds 132
- steel hydrowire, organic compounds 131
- stopcock orifice, subsampling nutrients 100
- surface water sampling 68, 81
- systematic errors, inorganic analysis 7
- trace analysis 18
- water 51
- year to year concentration, sediments 118

contamination control, biota 180, 187 ff
contamination path, biological specimen 25
contamination reduction, clean room 189
contamination-controlled sampling facilities 81
control charts 8 ff
- CB105 in cod liver 136
- CRMs 13

control points, method validation 6
control samples, organic compounds 130
control sites, surface water sampling 85
control soil samples 166
controlled atmosphere, sediment sampling 122
conversion, systematic errors, inorganic analysis 7
cooling
- biota samples, trace metal analysis 201
- organism activity reduction 105
- surface water sample preservation 82

cooling water, sampling dissolved nutrients 95
copper
- trace element pollution, soils 159
- trace metals, surface water sampling 68

corer types, sediment sampling 119
coulometric methods 3
CRMs
- for CBs 147
- method validation 4
- for PAHs 151
- preparation, requirements 14 ff
- producers 17

cross-contamination 70, 165
crustacea, trace metal analysis 182
cryogenic homogenization, aquatic biota 206
^{137}Cs, sediment storage 123
customer orientated marine monitoring 268

D'Agostino test, surface water sampling 57
darkening, organism activity reduction 105 f
data quality objectives (DTO), soil sampling 160
DDT degradation, injection 142
decomposition, plant sampling 223
deep subsurface coring, soil sampling 165
deep-freeze chambers, plant sampling 241
deep-freezing, samples 18
degassing, surface water sampling 68, 81
degradation
- biological, surface water sampling 70
- biological effect monitoring 186
- chemical surface water sampling quality 69
- sampling, MMPs 281

deionized water washing, soil sampling 167
denitrification, nutrient concentration 94 f
depuration, trace metal analysis 198
deseasonalization, statistical design 63
desorption 7, 69
destructive methods, error elimination 7
detecting trends, surface water sampling 54
determinants
- MMPs 263 f
- surface water sampling 56

diagenetic processes, sediment sampling 115
dialysis, bioavailable/dissolved contaminants 76
Dietrich's Fixative, aquatic biota 204
digest solution, matrix influences 5
digestion, systematic errors, inorganic analysis 7
dioxin contamination levels, soil sampling 38
dissolution, sediment sampling 115
dissolved inorganic nitrogen (DIN) 94
dissolved nutrients, in marine samples 91 ff
dissolved organic matter (DOM), surface water 80
dissolved organic nitrogen (DON) 94
dissolved organic phosphorus (DOP) 94
disturbances, mechanical, sediment sampling 117
documentation
- plant sampling 241
- principles 19
- sediment sampling 126 f
- soil sampling 168
- spatial comparability 31
- surface water sampling 65 f

dogwhelk, trace metal analysis 180 f
domestic-type water, dissolved nutrients 95
down-hole cross contaminations, soil sampling 165
dried materials homogenization, aquatic biota 207
dried sludge, RCMs for CBs 147
drilling corer, sandy sea beds, sediments 119
drinking water production 54
dry weather, plant sampling 242
dry-weight, aquatic biota, trace metal analysis 205 f
drying
– organic compounds sampling 134
– plant sampling 223, 249 f
– RMs/CRMs selection/preparation 14
– sediment sample treatment 121
– soil samples 168, 172
duplicate samples, quality control samples 85
dust-free garments, trace metal analysis 189
dustfall, cadmium 27
^{162}Dy uptake, mussels 184
dynamics, intrinsic, plant sampling 226

early warning monitoring, surface water 54
earth biomass, plants 215
earthworms, animals sampling 43
echo sounding, sediment sampling 115
ecological representativeness, samples 28
ecosystem, plant sampling 228
ecotoxicological effects, pollutants 30
edaphic factors, plant sampling 228
electron capture detector (ECD)
– CBs determination 139, 143
– methyl mercury 10
electron capture negative ion (ECNI) mass
    spectrometry, CBs 144
element concentration fluctuations, plants 227 f
element content, total, world plant biosphere 219 f
elements, building plant organism 216
elements concentration, plants/earth's crust 217 f
elution patterns, CBs fractionation 139
emission measurements, biological specimen 25
EN29000, accreditated control systems 20 f, 255 f
EN45000, standard series 20
EN45001
– MMPs 255 ff
– surface water sampling 52
enriched samples, assigned values 16
environmental CRMs, types/producers 15 f
environmental monitoring, QA/QC 1 ff
EPA, MMPs 256
equipment
– plant sampling 241
– surface water sampling 68 ff
equipment blank, soils 167

equipment materials
– plant sampling 222
– soil sampling 162 f
– trace metal analysis 187 f
errors
– analytical, plants 221
– calibration validation 5
– inorganic analysis 7
– interfering substances, plant sampling 223
– oil analysis 165
– surface water sampling 53 f
– year-average concentration, surface water 62
estuaries, macrophytes, trace metal analysis 191
estuaries salinity gradients, surface water 56
estuarine areas, metal-salinity relationship 86
estuarine sediments, CRMs for PAHs 151
estuarine water, dissolved nutrients 92 f, 100
ethical considerations, animals sampling 42
exceptions, systematic sampling 33
exclusion criteria, systematic sampling 33
excretion, nutrients concentration 94 f
exhaust fumes, surface water sampling 79
experimental monitoring, biological specimen 26
experimental plants monitoring, ABM 36 ff
exposure, animals, xenobiotic chemicals 41
exposure period, ABM, trace metal analysis 186
extraction
– CBs/biota/sediments 137 f
– PAH analysis 147 f
extraction efficiency, solid material analysis 6

F-test, aquatic biota homogenization 207
fat separation, column chromatography 139
fawns, animals sampling strategies 45
FDA, MMPs 256
fibers, agglomerated, filters for dissolved nutrients 103
field blanks
– sediment sampling 124
– soils 167
– spiked, quality control samples 85
field duplicates, soils 167
field sampling, soil solutions 162
field studies, MMPs 259
filter types, sampling dissolved nutrients 103
filtration
– organism activity reduction 105
– sampling dissolved nutrients 100 ff
– surface water sampling 56, 79
final detection, validation 5
fingerprint pattern, interferences, RMs/CRMs
    preparation 14
fingerprints, nutrient contamination 97

Subject Index

fish
- cleaning 199
- homogenization 204
- human consumption 182
- organ selection 197
- surface water sampling 75
- trace metal analysis 180 f, 191 f, 196 f, 204, 209

fish mortality, passive biomonitoring 31
Floristil columns, CBs fractionation 139
flow chart, chemical multielement analysis 222
fluorescence spectrometry, PAH analysis 148
flux changes, systematic errors, inorganic analysis 8
fluxes, surface water sampling 54
food chain, plants 216
forest soil, organic layers sampling 39
formaldehyde
- aquatic biota, trace metal analysis 204
- biological activity inhibition 83

formalin, aquatic biota, trace metal analysis 204
fortuious sampling, passive biomonitoring 31 f
fossil fuel combustion, PAH release 148
fractionation
- clean-up, CBs/biota/sediments 138
- soil samples 170 f

freeze drying
- aquatic biota, trace metal analysis 204 f
- plants, trace element analysis 251
- RMs/CRMs selection/preparation 14
- sediment sample treatment 121
- soil samples 169

freezing
- animals, organic compounds sampling 135
- biota samples, trace metal analysis 201
- organic compounds samples 134
- sampling dissolved nutrients 108 f
- soil storage 172
- surface water samples 82

frequency, aquatic biota, trace metal analysis 190
fresh tissue, aquatic biota, trace metal analysis 206
funnels, sampling dissolved nutrients 102

gamma irradiation
- plant samples 251
- soil storage 172

garpike tissue, RCMs for CBs 147
Garrett screen, organic compounds 132
gas chromatography (GC), CBs 131, 137 f
gas pressure filtration, surface water sampling 80
Gaussian distribution, soil sample numbers 162
GC analysis, calibration 141
GC/ECD chromatogram, sulfur-containing sediment sample 140
GC/MS, PAH determination 148

gel permeation chromatography, fat separation 139
Gelman GA4, filters for dissolved nutrients 103
genetic variabilities, plant sampling 228
genotype, animals sampling strategies 45
genotype representativeness, samples 28
geographical information systems (GIS) 32
geographical positioning systems (GPS) 32, 117
geographical representativeness, samples 27
geometrical sampling grids 161
geometry, systematic errors, inorganic analysis 8
geostatistics, soil sampling 161
German environmental specimen banking program 38
glass
- equipment material, organic compounds 131
- - trace metal analysis 187
- sediment sampling equipment 120 f
- soil sampling equipment 162, 165
- surface water sampling equipment 70
- trace elements, laboratory ware materials 71

glass bottles, sampling dissolved nutrients 98
glass fiber prefilters, PAH sorption 80
glass fibers, filters, dissolved nutrients 103
glass samplers, silicate leaching 96
glass wool, filters 103
gley, plant sampling 229
glove boxes, athmosphere-controlled sediment sampling 122
Go-Flo sampler, surface water sampling 68, 73
Good Laboratory Practice (GLP)
- MMPs 255 ff
- regulatories 20

grab samplers
- organic compounds sampling 132
- sediment sampling 119

grab samples, surface water sampling 78
gradient sampling, biological specimen 26
grain size, sediment storage 123
graphical plots, surface water sampling 86
grass, active biomonitoring 36
gravimetric methods 3
gravity corer, sediment sampling 119
Green Bay (Lake Michigan), segmented surface water sampling 59
grids
- passive biomonitoring 33
- soil 161

grinding
- plant samples 251
- sediment sample treatment 121
- soil samples 158, 170 f

guidelines, MMPs 261

Subject Index 297

habitat
- animals sampling strategies 45
- aquatic biota, trace metal analysis 181
- sampling, biological specimen 26
- wild population collection 185

harbour sediment
- CRMs for PAHs 151
- RCMs for CBs 147

heat-drying, metal determination, soil samples 169
heating, RMs/CRMs selection/preparation 14
heating, soil storage 172
heavy metal biomonitoring 238
- mosses 242 f
heavy metal samples, shock freezing 82
heavy metals, agricultural areas contamination 37
Helly's Fixative, aquatic biota 204
HEPA filters, clean room, trace metal analysis 189
heterogeneity, soils 158
heterogeneous element distributions, plants 234
heterophobic bacteria, mineralization, dissolved nutrients 94
hexachlorobenzene, soil sampling 38
High Sierra lakes, helicopter contaminations 79
high temperature ashing, aquatic biota 204 f
homogeneity
- aquatic biota, trace metal analysis 207
- bivalves, trace metal analysis 190
- surface water sampling 55
homogenization
- biota, trace metal analysis 180, 203 f, 206 f
- plant sampling 223, 251
- RMs/CRMs selection/preparation 14
- sampling, MMPs 281
- soil samples 170

hot spot
- PBM, trace metal analysis 185
- trace element pollution, soils 159
human consumption
- aquatic biota, trace metal analysis 181
- plants 216
human factor, sample handling 66
human pollution storage, animals sampling 41
humus content, soil sampling 38
hunting seasons, animals sampling strategies 45
hydrochloric acid soaking, trace metal analysis 189
hydrofluoric acid, MMPs 262
hydrogen, plant building elements 216
hydrography, sediment sampling 115
hydrological characteristics, Kentucky River 78
hydrophobic organic compounds, dialysis 76
hydrophobic organic contaminants accumulation, SPM 79
hydrophobic pollutants, surface water 58
hydrosphere control, sediment/water analysis 15

hydrowire
- steel, organic compounds sampling 131
- water sampler 96
Hylocomium splendens 242
Hypogymnie physodes, ABM 36

IAEA, reference materials, MMPs 276
ice crystallization, dissolved nutrients 108 f
ICES guidelines, organic compounds 132
incomplete conversion/digestion, errors 7
Infiltrex sampler, surface water sampling 77
infrared heating, soil samples 169
inhomogeneous element distributions, plants 234
injection techniques, CBs 142
inorganic analysis, errors 7
inorganic nutrients, mineralization 94
inorganic trace analysis, contaminations 18
Instra Analyzed, trace metal analysis 189
integrated (stream) depth sampling 78
interferences, RMs/CRMs preparation 14
interfering substances
- aquatic biota, trace metal analysis 190
- filtration, sampling dissolved nutrients 101
interlaboratory certification studies 16
interlaboratory studies
- CBs analysis 145
- RMs 13
internal quality control procedures, sediments 123
international soil-analytical exchange (ISE) 173
international standards organization *see:* ISO
interpolation technique, soil sampling 161
interval, sediment sampling 114
intrinsic contamination, filtration 103
intrinsic radiation, systematic errors 8
ion-selective electrodes, sampling fishes 75
iron, contaminations, plants 249
irradiation, systematic errors, inorganic analysis 8
irradiation sterilizing, aquatic biota 205
irreversible precipitation, systematic errors 7
ISO, RM/CRMs 12 f
ISO 25, surface water sampling 52 f, 70
ISO 9000
- accreditated control systems 20 f
- CRMs producer list 17
- MMPs 255 ff
ISO 9001, surface water sampling 53
ISO Guide, soil quality/soil samples 169 f
ISO Guide 25
- accreditated control systems 20
- CRMs producer list 17
ISO Guide 33, CRMs 13
ISO/REMCO, CRMs producer list 17

Subject Index

jet miling, soil samples 171
Joint Monitoring Group-Oslo/Paris commissions 132
judgmental soil sampling 160

Kajak corer, sediment sampling 119
kale, passive accumulation, ABM 36
Kendall test, surface water sampling 63
Kentucky River, hydrologic characteristics 78
Kevlar rope, organic compounds sampling 131
kidneys, organ selection, trace metal analysis 197
kriging, soil sampling plan 161
Kuderna-Danish evaporator, CBs 139

labeling
– biota samples, trace metal analysis 200 f
– containers, sediment sampling 122
– soil sampling 165
laboratories comparision, accuracy control 12
laboratory accreditation systems, MMPs 255 ff
laboratory approach, certification 16
laboratory information system, sample handling 19
laboratory performance studies 12
laboratory reference materials (LRMs)
– MMPs 270
– organic compounds sampling 136
laboratory ware materials, trace elements 71
Lake Michigan, surface water sampling 59
Lake Ontario, offshore zone segmentation 60
lanthanides 180, 216
law-applicable animals sampling strategies 42
leaching, phthalate esters 131
lead
– atmophile elements, soil sample depth 164
– biomonitoring 180
– fuel contamination, surface water sampling 69
– plants 216
– sediment storage 118, 123
– trace element pollution, soils 159
– trace metals, surface water sampling 68
lead distribution, mosses 242 f
leakproofness, bottle closure 98
length, animals sampling strategies 45
lichens exposure, active biomonitoring 36
light exposure
– plant sampling 237
– surface water sampling 68, 81
limit of quantification (LOQ), surface water 58
linear sedimentation rate, sediment sampling 118
lipid content
– bivalves, trace metal analysis 208
– fish, trace metal analysis 209

lipids
– CBs/biota/sediments 138
– organic compounds sampling 135
liquid chromatography, PAH analysis 148
liquid-liquid extraction, CBs/biota/sediments 137
literature search 5
LN_2-Dewars, plant sampling 241
location
– aquatic biota, trace metal analysis 190
– plant sampling 238
Lolium multiflorum, active biomonitoring 36
Lombardy poplars, biomonitoring 238
long-term pollution, biological specimen 28
long-term preservation, dissolved nutrients 106
long-term stability, CBs analysis 145
long-term storage, aquatic biota 203
losses, sampling, MMPs 281
low-frequency scanning, sediment sampling 115
low-temperature ashing, aquatic biota 204 f
low-temperature short-term soil storage 172
low-temperature storage, biological materials 188
Lumbricus terrestris 41, 44
lyophilization, aquatic biota 205

mackerel oil, RCMs for CBs 147
macroalgae, trace metal analysis 181 f
macrophytes, trace metal analysis 180 f, 191 f, 196 f, 204, 209
magnesium, plants 216
mammals, trace metal analysis 183
manganese, trace element pollution, soils 159
manganese content
– plants 216
– red whortleberry leaves 223 ff
Mann-Whitney test, surface water sampling 63
manual shallow subsurface coring, soils 165
manual surface grab, soil sampling 164
mapping techniques, sediment sampling 117
marine mammals blubber, organic compounds 133
marine matrices, organic compounds 129 ff
marine monitoring programmes (MMPs) 255 ff
marine samples, dissolved nutrients 91
marine sediments, CRMs for PAHs 151
Markovian persistence order, surface water 63
mass spectrometry (MS)
– CBs determination 144
– methyl mercury 10
materials
– equipment for surface water sampling 70 ff
– sample bottles, chemically inert 97
matrix blanks
– CRMs for PAHs 151
– sediment sampling 124

matrix composition, RMs/CRMs 14 ff
matrix CRMs, assigned values 16
matrix influence, detection validation 5
mature individuals, random distribution 42
membrane filters, dissolved nutrients 103
memory effect, contaminations 70
mercuric chloride, dissolved nutrients 107
mercury, biomonitoring 180
metal biomonitoring 238
metal contaminations, soil solutions storage 173
metal-salinity relationship, estuarine/coastal 86
metal uptake, plants 158
metallic containers, soil storage 165, 172
metallothioneins analysis 187
metals
– determinants 264
– target performance 279
– testing schemes 274
method comparision, accuracy 11
method equivalence, certification report 13
method validation 4 f
methods types/selection 3
methyl mercury, case study 10
Meuse, seasonal effects, surface water 58
micelles formation, surfactants addition 83
microalgae growth, photosynthesis 94
microbial activity, plant sampling 227
microbiological activity, wet soil storage 172
microclimatic conditions, plant sampling 228
microorganisms
– activity reduction 105
– nutrients incorporation 93
Milli-Q system, surface water sampling 71
milling, soil sampling 158
Millipore AA, filters for dissolved nutrients 103
mineralization, heterophobic bacteria 94
minimum number of observations, surface water 65
mixing, soil samples 171
mixing different species, trace metal analysis 193
MMPs 255 ff
mobility
– animals sampling strategies 40, 44
– aquatic biota, trace metal analysis 181
moisture, systematic errors, inorganic analysis 7
molluscs 133, 181
monotonous trends, non-parametric tests 63
mosses, large-area biomonitoring 242 f
multidimensional GC (MDGC), BCs 142
multielement analysis, instrumental, plants 221
municipal waste water 54
mussel samples, methyl mercury 11
Mussel Watch
– PBM 185
– trace metal analysis 190 f, 196

mussels
– ABM exposure, trace metal analysis 208
– freeze drying, organic compounds 135
– organic compounds sampling 133
– trace metal analysis 180, 182
– zebra, animals sampling strategies 46
mussels tissue, CRMs for PAHs 151

NAMAS, MMPs 255 ff
nanoplankton, filtration 105
natural element concentration fluctuations 227 f
nautical means, dissolved nutrients 92 f
NEN3417, surface water sampling 53
net mass accumulation, sediment sampling 118
nickel
– biomonitoring 180
– in lichen RM, bar graphs 17
– plants 216
– trace element pollution, soils 159
Niskin sampler, surface water 67 f, 74
nitrate
– nitrogen donors, phytoplankton growth 94
– water quality parameters 67
nitric acid soaking, trace metal analysis 189
nitrification, nutrients concentration 94 f
nitrite
– nitrogen donors, phytoplankton growth 94
– water quality parameter 67
nitrogen
– dissolved nutrients 91 f
– plant building elements 216
nitrogen donors, dissolved nutrients 94
non-Gaussian distributions, soil sample numbers 163
noncyclic series, statistical design 63
nondestructive methods, error elimination 8
nonhomogeneity, surface water sampling 57
nonprobability approach, soil sampling 160
nutrient concentrations
– process-altered 93 f
– summer–winter cycles 58
nutrient-free detergent cleaning 99
nutrients
– determinants 264
– dissolved, in marine samples 91 ff
– sediment storage 123
– target performance 279
– testing schemes 274
nutrients analysis, biological specimen 25
Nylon plankton net, prefiltration, sampling dissolved nutrients 104

objects, plant sampling   225
oceanic surface water
 – dissolved nutrients   93
 – trace metal levels   52
octadecyl, SPE, surface water sampling   75
OECD
 – GLP guidelines   20
 – MMPs   255 ff
oil, PAH release   148
one-piece closures   98
open sea studies, dissolved nutrients   92
open-sea water sampling   117
organ selection, trace metal analysis   197
organic carbon, sediment storage   123
organic compounds
 – marine matrices   129 ff
 – stability, sodium bisulfate addition   83
 – uncoated capillary columns   76
organic contamination, glass   70
organic layers, forest soil sampling   39
organic trace analysis
 – contaminations   18
 – matrix influences   6
organism, total, trace metal analysis   197
organochlorine pesticides
 – determinants   264
 – ECD   139
 – testing schemes   274
organochlorines, target performance   279
organometallic compounds   6
organotin compounds, trace metal analysis   180
organotropy, passive biomonitoring   32
oven drying
 – aquatic biota, trace metal analysis   204 f
 – organic compounds samples   134
 – sediment samples   121
 – soil samples   169
oxidation, systematic errors, inorganic analysis   7
oxygen-free conditions, sediment sampling   121
oysters, trace metal concentrations   194

PAH
 – determinants   264
 – target performance   279
 – testing schemes   274
PAH analysis   147 f
PAH chromatograms, GC/HPLC   149 f
PAH losses, polyethylene containers   82
PAH sorption, glas fiber prefilters   80
paper bags, plant sampling   241
paper containers, soil sampling   165
paper filters, folded, dissolved nutrients   103
parametric tests, trend detection   63

particle size, soil sampling   158
passive biomonitoring (PBM)
 – plants sampling   31 ff
 – trace metal analysis   184 f
^{210}Pb isotope distribution, sedimentation rate   118, 123
PCB sources, Black River, passive samplers   76
PCB/dioxin-like compound   186
PCBs determination   145
peak overlap, systematic errors   8
peat, plant sampling   229
period frequency, sediment sampling   113, 118
periodicity, statistical design   63
personel, well-trained, sediment sampling   120
pesticides
 – degradation   82
 – surface water sampling   56
Petri dishes, sediment sampling   122
phosphorus
 – biomass synthesis   94
 – dissolved nutrients   91
 – plant building elements   216
 – water quality parameter   67
photophysical degradation, surface water   68
photophysical reactions, surface water   81
photosynthesis
 – nutrients/phytoplankton   93 f
 – plants   215
phthalate ester leaching   131
physical factors, trace metal analysis   199
physical measurements, precision, plants   223
physical status, RMs/CRMs preparation   14
physical treatments, microorganism activity reduction   105
physico-chemical properties, surface water   56
phytoplankton growth, nutrients incorporation   91 f
phytoplanktonic assimilation reduction   106
Picea abies, sampling procedure   35
picoplankton, filtration   105
Pisces, passive time integrated surface water sampling   76
plankton   130
plankton-rich water, cell rupture, filtration   101
planning
 – plant sampling   221, 225
 – surface water sampling   65 f
plant building elements   216
plant sampling   215 ff, 235
 – passive biomonitoring   31 ff
plant storage   215 ff
plant tissues, soil sampling   37
plastic bags, biota samples   201
plastic bottles, sampling dissolved nutrients   98
plastic columns, compounds in water   76

plausibility checks, surface water sampling 85
Pleurozium schreberi, heavy metals 242
podsol, plant sampling 229
poisoning, long-term preservation 106
Poisson distribution, soil sample number 163
pollutant impacts, biological specimen 25
pollutants
− eco/toxicological effects 30
− food chain 216
− organ-dependent accumulation 197
− sex-dependent uptake, trace metal analysis 194
− water/air 18
pollutants concentration
− sediment sampling 118
− in Zebra mussels 46
pollution, sediment sampling 114
pollution storage, in animals 41
poly(vinyl chloride), laboratory ware materials 71
polyamid containers, plant samples 251
polycarbonate
− equipment material, trace elements 71, 187
− filter fibers, sampling dissolved nutrients 103
− filter membranes, surface water sampling 80
polycyclic aromatic hydrocarbons (PAHs)
− analysis 147 f
− marine matrices 129 ff
− soil sampling 40
polyethylene, sampling equipment
− dissolved nutrients 98
− PAH losses 82
− plants 241
− sediments 120 f
− surface water 70
− trace elements 71
polyethylene bags, high-density, soils 165
polyethylene collector washing, soils 167
polyethylene gloves, sample contamination 79
polymer fibers, filters, dissolved nutrients 103
polypropylene, sampling equipment
− dissolved nutrients 98
− surface water 70
− trace metal analysis 71, 187
polystyrene, sediment sampling/storage 120 f
polystyrene cool boxes, biota samples 201
polytetrafluoroethylene, organic compounds 131
polythene, trace metal analysis 187
Polytrichium formosum 220, 232
Polytron blender, aquatic biota 206
poplar leaves, biomonitoring heavy metals 238
population
− animals sampling strategy criteria 45
− geno/phenotypical 29
− heterogeneous, plant sampling 226, 228
− wild, collection 185

population median, symmetric distribution 59
population-dependent tests, surface water 57
Populus nigra italica, heavy metals 238
pore size, filters 100, 103
porous ceramics, soil solutions 162
position, exposed, biomonitoring 238
potassium distribution, *Betula pendula* 234
prawns, trace metal analysis 182, 188
precipitation
− plant sampling 251
− sediment sampling 115
− surface water sampling 68, 81
− systematic errors, inorganic analysis 7
precision
− biological specimen sampling 28
− ISO definitions 11
− method selection 3
− plant sampling 223
− sediment sampling 113
preconcentration, systematic errors 7
precrushing, soil samples 170
predetermination, dissolved nutrients 91 ff
prefiltration, sampling dissolved nutrients 104
preliminary sampling, biological specimen 31
preparation
− RMs/CRMs 14 ff
− systematic errors, inorganic analysis 7
preparation degree, laboratory performance 12
preservation
− aquatic biota, trace metal analysis 204
− dissolved nutrients 105 ff
− surface water sampling 71, 82
pressure filtration, dissolved nutrients 101
pretreatment
− organic compounds samples 133 f
− surface water samples 65
pretreatment procedures, RMs, soils 173
probability approach, soil sampling 160
procedure errors, surface water sampling 55
procedures, sediment sampling 120
producers, environmental CRMs 17
proficiency testing, accuracy control 12
project development, MMPs 285
project-related sampling plan 31
protocols
− aquatic biota, trace metal analysis 209
− dissolved nutrients 91
− sediments 126 f
− soil sampling 158
− specific procedures, surface water 65
  see also: documentation
proton induced X-ray emission (PIXE)
− MMPs 262
− sediment sample treatment 121

Subject Index

PTFE, sampling equipment
- sediments 122
- surface water 70

quality assurance, environmental monitoring 1 ff
quality control, environmental monitoring 1 ff
quality control samples 84 f
quality control soil samples 166
quality management, marine environmental monitoring 255 ff
quality manual, surface water sampling 52
quartz, redistillation, surface water sampling 71
quartz glass, equipment material 222, 187
QUASIMEME 135, 259 ff
- PAH analysis 148
- project teams 267
quick freezing, sampling dissolved nutrients 109

R-chart 9
radiation, systematic errors, inorganic analysis 8
random errors 3, 165
random sampling
- biological specimen 26
- passive biomonitoring 32
- plants 226 f, 235
- soil 160
- trace metal analysis 192
random selection sampling, area-related 35
rare earth elements, biomonitoring 180
reagent blank, sediment sampling 124
reagent contamination, systematic errors 7
red whothleberry 223 f, 230
redox potential
- sediment storage 123
- soils 158
reference locations, surface water sampling 85
reference materials (RM) 3, 12 ff
- CBs determination 146
- MMPs 269, 275
- soils 173
  see also: CRMs
reference methods, assigning values 16
regeneration, heterophobic bacteria 94
regionalized variables, soil sampling 161
registration, samples 19
regulatories, QA/QC 20 ff
relative methods 3
reliability, sampling, biological specimen 30
repeatability
- method validation 4
- sampling, biological specimen 28
replicate samples, confidence level 113, 118

reporting, principles 19
representativeness
- aquatic biota, trace metal analysis 189
- environmental samples 27
- plant sampling 221, 226 f
- population-related 42
- sediment sampling 113
- soil sampling 37, 157
- soil subsampling 171
- spatial, animals sampling strategies 40
- surface water sampling 55
reproducibility
- method validation 4
- plant sampling 223
- sediment sampling 118
resources, interlinked, MMPs 261
Retsch mixer, soil samples 170
reverse osmosis distilled water 77
reverse osmosis distilling, surface water 71
Rhine, seasonal effect, surface water 58
river sediment, RCMs for CBs 147
robustness, method validation 6
roe deer population, animals sampling 44 f
room facilities, sediment sampling 121
root samples, soil separation 250
routine measurements, method validation 4
ruggedness, method validation 6
Ruttner samplers, surface water sampling 68, 73

salinity
- aquatic biota, trace metal analysis 199
- freezing points 108
- soils 158
salinity gradients, surface water sampling 55
sample dividers, polyethylene, soil samples 171
sample equipment *see:* equipment
sample handling
- MMPs 281
- principles 18 ff
- sediments 113
  see also: sampling
sample number
- aquatic biota, trace metal analysis 190
- composite samples 163, 195
- plant sampling 226, 237, 241
- soils 162
- surface water sampling 60
sample preparation
- aquatic biota, trace metal analysis 203 f
- chemical, plants 223 f
- plants 221
- systematic errors, inorganic analysis 7
sample preparation blank, sediments 124

Subject Index   303

sample procedures, sediment sampling   113
sample pumps, surface water sampling   74
sample quality, process-affected   68
sample size
– aquatic biota, trace metal analysis   195
– soils   158
sample storage, sediment sampling   122 ff
sample transportation/storage, surface water
    81 ff
sample treatment, biota, trace metal analysis   179
samplers
– seawater   131
– surface water sampling   68
samples
– enriched, assigned values   16
– hydrosphere control   15
sampling
– biological specimen   25 ff
– biological specimen, QY methods   30 ff
– biota, trace metal analysis   179, 189
– dissolved nutrients   91, 95 f
– MMPs   259, 281
– organic compounds in marine matrices   129 ff
– principles   18 ff
– sediments   113 f, 119 ff
sampling area, soil morphology   38
sampling bottles   96 ff
sampling devices
– properties/selection   72 f
– sediment sampling   119
sampling equipment *see:* equipment
sampling errors   55, 130
sampling period
– aquatic biota, trace metal analysis   191
– passive biomonitoring   36
– plant sampling   242
sampling site
– sediment sampling   114
– surface water sampling   55
sampling strategies
– soils   160
– surface water sampling   53 ff
sampling strategy/techniques   18 f
saponification   139, 137, 147
scanning, low-frequency   115
screening
– *Lumbricus terrestris/Aporrectodea longa*   44
– sampling, biological specimen   31
– soil morphology   38
sea bed structure, sediment sampling   115
seasonal effects
– aquatic biota   181, 191, 199
– plant sampling   226 f
– surface water sampling   54 f, 57 f

Seastar, pumping sampler
– organic compounds   132
– surface water   77
seawater
– analysis   91
– organic compounds   130 f, 134
– target performance   279
– trace elements   67
sediment, CBs extraction   137 f
sediment sample treatment   121
sediment samples, hydrosphere control   15
sediment sampling   113 ff
sediment storage, recommendations   123
sediment waves, surface water sampling   58
sediment whirling, surface water sampling   68
sedimentation holes   115
sedimentation rate   115, 118
sediments, organic compounds sampling   132 f
segmentation, surface water sampling   59
selected sampling   26, 32
selection
– population, animals sampling strategies   45
– RMs/CRMs   14
selenium
– biomonitoring   180
– plants   216
– trace element pollution, soils   159
self-shielding, systematic errors   8
semi-artificial conditions, Zebra mussels   46
sensitivity   4, 113
sewage sludge
– CRMs for PAHs   151
– soil sample depth   164
sex
– animals sampling strategies   45
– aquatic biota, trace metal analysis   193
– fish, trace metal analysis   209
shelf life, biota samples, trace metal analysis   203
Shewart-chart   9
ship positioning, sediment sampling   117 f
ship-related contamination, dissolved nutrients   95 f
shock freezing, heavy metal samples   82
short-term preservation, dissolved nutrients   106
side-scan sonar, sediment sampling   117
sieving, soil sampling   158
silica
– trace elements, laboratory ware materials   71
– water quality parameter   67
silicate, phytoplankton growth   9
silicate determination, glass/dissolved nutrients   98
silicate leaching, water samplers   96
silicate samples, freezing   110
silicon, dissolved nutrients   91
silicone membrane, waste stream monitoring   77

Subject Index

silver, biomonitoring 180
similarity, CRMs 13
sintered glass, filters, dissolved nutrients 103
site blank, soils 167
site selection, sediment sampling 113 f
site-specific element compositions, plants 229
site-specific sampling plan 38
sites, plant sampling 237
size, aquatic biota, trace metal analysis 193
size range definition, trace metal analysis 194
soaking, trace metal analysis 189
sodium bisulfate addition, organic compounds 83
soil, chemical properties 29
soil composition changes, drying 168
soil pollution 157 ff
soil pretreatment 168 f
soil profile, depth selection 164
soil reference material SO-1 173
soil sample depth 164
soil samples, exceptions/exclusion criteria 33
soil sampling 157 ff, 164 f
soil sampling strategies, biological specimen 37 ff
soil separation, root samples 250
soil solution samples 158, 162
soil storage 157, 172 f
soil subsampling 170 f
soil trace metals 157 ff
soil types, plant sampling 229, 237
solid material analysis 6
solid phase extraction (SPE), surface water 75
solubility, dustfall cadmium 27
solubilization, selective, trace metals 168
solutions, matrix influences 5
solvent extraction, PAH analysis 147
solvent extracts, long-period freezing 134
sorption processes, surface water sampling 68
source-related passive biomonitoring 33
Soxhlet extraction
– CBs/biota/sediments 137
– PAH/biota/sediment 147
sparger, surface water sampling 77
spatial burdening, xenobiotic chemicals 40
spatial error, surface water sampling 55
spatial representativeness, samples 27
Spearman test, surface water sampling 63
species selection, aquatic biota 181, 193 f
specimen banking, soil sampling 40
spectrometric methods 3
spiked blanks, soil sampling 167
spiked field blanks, quality control samples 85
spiking
– CRMs for PAHs 151
– matrix influences 6
– methyl mercury 10

– organic compounds 138
– quality control samples 84
split injection, CBs 142
split sediment samples 123
split-tube sampler, top-soil sampling 40
splitting, quality control samples 85
stability test, TBT in coastal sediment 15
stabilization
– aquatic biota, trace metal analysis 204
– RMs/CRMs selection/preparation 14
– samples 18
stainless steel, sampling equipment
– organic compounds 131
– plants 241
– sediments 120
– soil 164 f
standard EPA vials, organic compounds 134
standard operating procedures (SOPs)
– MMPs 262
– soil 157
– spatial comparability 31
– surface water sampling 65
– trace element pollution/soil 160
standard reference materials (SRMs), MMPs 275
standardization, environmental monitoring 20
standardized biological indicator systems 28
standardized grass culture method, ABM 36
standardized methods, validation 4
statistical reference basis, animal population 44
statistics, surface water sampling 57
steam sterilization, soil storage 172
steels, soil sampling equipment 164
sterilization
– aquatic biota, trace metal analysis 204
– plant samples 251
STERLAP, MMPs 255 ff
storage
– biota samples, trace metal analysis 200 f
– dissolved nutrients 91
– MMPs 281
– organic compounds samples 133 f
– plant samples 251
– samples 19
– sampling dissolved nutrients 105 ff
– sediments 113, 122 ff
– soil 157, 168 f, 172 f
– surface water samples 81 ff
storage temperatures/times, sediments 123
stratified random sampling
– biological specimen 26
– less mobile animals 42 f
– mobile animals 44
– passive biomonitoring 36
– soil 161

structure type, filters, dissolved nutrients   103
styrene-divinylbenzene, SPE, surface water   75
subsampling   19
– aquatic biota, trace metal analysis   195
– dissolved nutrients   91, 97 ff
– sediments   113, 118, 120 f
– soils   170 f
  see also: sampling
sulfur, plant building elements   216
sulfur dioxide, ambient air   27
summer-winter cycles, surface water sampling   58
supercritical fluid extraction (SFE)   137 f, 147
Supra Pur, trace metal analysis   189
surface acidity changes, soil sample drying   168
surface contaminations, plant sampling   249
surface mixing depth, sediment sampling   118
surface sediments, organic compounds   132
surface water perturbation, dissolved nutrients   95
surface water sampling   51 ff
surfactants addition, micelles formation   83
suspended matter (SPM)
– contaminant accumulation   79
– surface water sampling   68
SWEDAC, MMPs   255 ff
systematic errors
– inorganic analysis   7
– trueness   3
systematic sampling
– passive biomonitoring   33
– plants   226 f, 235
– soil   160

target performance   279
target positions, ABM, trace metal analysis   185
targets, MMPs   271
taxonomic groups, aquatic biota   181
TBT in coastal sediment, stability test   15
Teflon
– aquatic biota homogenization   206
– sampling equipment
– – soils   164
– – surface water   70
– – trace elements   71
– – trace metal analysis   187
– vibration mill, plant samples   251
temperature
– aquatic biota, trace metal analysis   199 f
– sediment storage   123
– soil storage   158, 168 f
– surface water storage   81
temporal errors, surface water sampling   55
tenability, surface water sampling   66
testing schemes, QUASIMEME   274

thawing, frozen samples dissolved nutrients   109
tidal effects, aquatic biota   199
time-integrated samplers, surface water   76
tin, biomonitoring   180
tissue preparation, aquatic biota   203
tissuemizer blender, aquatic biota   206
titanium concentration   250
titanium knives, aquatic biota   203
titrimetric methods   3
tools materials, trace metal analysis   188
tools/vials, systematic errors, inorganic analysis   7
topography, sediment sampling   115
topsoil sampling, area/volume relation   39
toxic compounds, CRMs   15
toxic metals, soil sampling   40
toxicants, ABM, trace metal analysis   186
toxicokinetics, animals sampling strategies   41
toxicological effects, pollutants   30
toxicological studies, trace metal analysis   180
trace contaminants, sediment sampling   114
trace element pollution, soils   159
trace elements
– biological specimen   25
– laboratory ware materials   71
– plants   216
– seawater   67
– sediment storage   123
– soil sampling   40, 159
trace metal accumulation, SPM   79
trace metal analysis, aquatic biota   179 ff
trace metal biomonitoring, plants   238
trace metal determination, soil/solution   157 ff
trace metal leaching, glass, water samples   70
trace metal level, oceans   52
trace metal uptake, aquatic biota   199
trace metals
– cleaning procedures, surface water   77
– polyethylene, surface water sampling   70
– surface water sampling   68
– target performance   279
trace organic analysis, sediment sampling   120 f
trace organics, sediment storage   123
trace pollutants, ABM, trace metal analysis   186
traceability, CRMs   13
transect sampling, biological specimen   26
transects, passive biomonitoring   33
translocation, ABM, trace metal analysis   185
transparencies, documentation   241
transport
– biota samples, trace metal analysis   179, 200 f
– sediment samples   115
– surface water samples   81 ff
transport blank, soils   167
trends, parametric tests   63

trimmed mean, surface water sampling 59
trip blank, soils 167
tropical rain forest 28
true value, method selection 4
trueness, ISO definitions 11
trueness method selection 3
trunk diameters, plant sampling 240
tuna fish
– methyl mercury 11
– trace metal analysis 182
Turrax blender, aquatic biota 206
two-piece closures 98
type I/II error, surface water sampling 53

ultra clean water, trace metal analysis 198
ultrasonic extraction, CBs/biota/sediments 137
ultrasonication, PAH/biota/sediment 147
Ultrex, trace metal analysis 189
USEPA, control points 6

Vaccinium myrtillus 229, 231
Vaccinium vitis-idaea 230, 233
vacuum drying, plants 251
vacuum filtration
– sampling dissolved nutrients 101
– surface water sampling 80
validation
– analytical methods
– – collection methods 192
– – organic compounds 135
– – soils 173
– – trace metal analysis 209
– fish sampling, trace metal analysis 196
– method selection 4
– surface water sampling 66 ff
van Dorn sampler, surface water sampling 68, 73
van Slyki reaction, acidification 106
vanadium, plant sampling 249
variances, soil analysis 166
vibration mill, Teflon, plant samples 251
volatile organic compounds stability 83
volatilization
– plant sampling 227
– systematic errors 7
volumetric manipulation, systematic errors 7

W test, surface water sampling 57
washing
– plant sampling 223
– polyethylene collectors, soils 167
waste stream on-line monitoring 77
waste water, municipal 54
water characteristics, dissolved nutrients 92
water depth, bivalves, trace metal analysis 190
water quality parameters 67
water samplers 77, 96, 132
water samples
– hydrosphere control 15
– pollutants 18
water sampling 51 ff
weather conditions
– plant sampling 241
– surface water sampling 55
weight, animals sampling strategies 45
weighting
– assigned values 16
– systematic errors, inorganic analysis 7
wet soil storage 172
wet weight, aquatic biota 207
Winchester solvent bottles 131
windsorization, LOQ, surface water sampling 59
worms, trace metal analysis 182

X-chart 9
X-ray fluorescence (XRF)
– MMPs 262
– sediment sample treatment 121
xenobiotic chemicals burdening 40
xenobiotics, in human beings 41

year-average concentration 62
Youden plot, methyl mercury 10

zebra mussels, animals sampling 46
zinc
– atmophile elements, soil sample depth 164
– biomonitoring 180
– trace element pollution, soils 159